Bacteriophages: Biology, Technology and Therapy

Bacteriophages: Biology, Technology and Therapy

Edited by Emma Richardson

www.statesacademicpress.com

States Academic Press,
109 South 5th Street,
Brooklyn, NY 11249, USA

Visit us on the World Wide Web at:
www.statesacademicpress.com

ISBN: 978-1-63989-776-6

Cataloging-in-Publication Data

Bacteriophages : biology, technology and therapy / edited by Emma Richardson.
    p. cm.
Includes bibliographical references and index.
ISBN 978-1-63989-776-6
1. Bacteriophages. 2. Bacteriophages--Health aspects. 3. Bacteriophages--Therapeutic use.
I. Richardson, Emma.
QR342 .B33 2023
579.26--dc23

Table of Contents

Preface.. VII

Chapter 1 **A Hypothesis for Bacteriophage DNA Packaging Motors**...1
Philip Serwer

Chapter 2 **Specific Colon Cancer Cell Cytotoxicity Induced by Bacteriophage**
E Gene Expression under Transcriptional Control of
Carcinoembryonic Antigen Promoter 24
Ana R. Rama, Rosa Hernandez, Gloria Perazzoli, Miguel Burgos,
Consolación Melguizo, Celia Vélez and Jose Prados

Chapter 3 **Structural and Biochemical Investigation of Bacteriophage N4-Encoded**
RNA Polymerases...38
Bryan R. Lenneman and Lucia B. Rothman-Denes

Chapter 4 **Bacteriophages and Biofilms**...59
David R. Harper, Helena M. R. T. Parracho, James Walker,
Richard Sharp, Gavin Hughes, Maria Werthén, Susan Lehman and Sandra Morales

Chapter 5 **Bacteriophage 434 Hex Protein Prevents RecA-Mediated Repressor Autocleavage** 74
Paul Shkilnyj, Michael P. Colon and Gerald B. Koudelka

Chapter 6 **Bacteriophages with the Ability to Degrade Uropathogenic**
Escherichia Coli **Biofilms**..91
Andrew Chibeu, Erika J. Lingohr, Luke Masson, Amee Manges,
Josée Harel, Hans-W. Ackermann, Andrew M. Kropinski and Patrick Boerlin

Chapter 7 **A Genetic Approach to the Development of New Therapeutic Phages to**
Fight *Pseudomonas Aeruginosa* **in Wound Infections**...110
Victor Krylov, Olga Shaburova, Sergey Krylov and Elena Pleteneva

Chapter 8 **Phage Lambda P Protein: Trans-Activation, Inhibition Phenotypes and**
their Suppression 152
Sidney Hayes, Craig Erker, Monique A. Horbay, Kristen Marciniuk,
Wen Wang and Connie Hayes

Chapter 9 **Spatial Vulnerability: Bacterial Arrangements, Microcolonies, and**
Biofilms as Responses to Low Rather than High Phage Densities...........................186
Stephen T. Abedon

Chapter 10   **Utility of the Bacteriophage RB69 Polymerase gp43 as a Surrogate Enzyme for Herpesvirus Orthologs** 212
Nicholas Bennett and Matthias Götte

Permissions

List of Contributors

Index

Preface

I am honored to present to you this unique book which encompasses the most up-to-date data in the field. I was extremely pleased to get this opportunity of editing the work of experts from across the globe. I have also written papers in this field and researched the various aspects revolving around the progress of the discipline. I have tried to unify my knowledge along with that of stalwarts from every corner of the world, to produce a text which not only benefits the readers but also facilitates the growth of the field.

A bacteriophage is a virus that can infect and replicate only inside a bacteria. They can kill bacteria without any negative effect on human or animal cells. They are referred to as phages and are the most common biological entities found on Earth. Bacteriophages vary in size, morphology, and genomic organization. Many phage-based technologies and therapies have been developed, especially in antibacterial treatments. There are several potential advantages of bacteriophages over antibiotics in terms of treating bacterial infections. Some of the benefits of bacteriophages are specificity of action (i.e., they kill only the pathogen they recognize), a considerable spectrum of activity, higher tolerability, and ease of administration. The results of various in-vitro studies have shown that bacteriophages could also be effective in the treatment of cystic fibrosis (CF) disorder. This book is a compilation of chapters that discuss the most vital concepts and emerging trends in the study of bacteriophages. It traces the recent developments in bacteriophage biology, technology and therapy. With state-of-the-art inputs by acclaimed experts of this field, this book targets medical students and professionals.

Finally, I would like to thank all the contributing authors for their valuable time and contributions. This book would not have been possible without their efforts. I would also like to thank my friends and family for their constant support.

Editor

A Hypothesis for Bacteriophage DNA Packaging Motors

Philip Serwer

Department of Biochemistry, The University of Texas Health Science Center, San Antonio, Texas 78229-3900, USA; E-Mail: serwer@uthscsa.edu

Abstract: The hypothesis is presented that bacteriophage DNA packaging motors have a cycle comprised of bind/release thermal ratcheting with release-associated DNA pushing via ATP-dependent protein folding. The proposed protein folding occurs in crystallographically observed peptide segments that project into an axial channel of a protein 12-mer (connector) that serves, together with a coaxial ATPase multimer, as the entry portal. The proposed cycle begins when reverse thermal motion causes the connector's peptide segments to signal the ATPase multimer to bind both ATP and the DNA molecule, thereby producing a dwell phase recently demonstrated by single-molecule procedures. The connector-associated peptide segments activate by transfer of energy from ATP during the dwell. The proposed function of connector/ATPase symmetry mismatches is to reduce thermal noise-induced signaling errors. After a dwell, ATP is cleaved and the DNA molecule released. The activated peptide segments push the released DNA molecule, thereby producing a burst phase recently shown to consist of four mini-bursts. The constraint of four mini-bursts is met by proposing that each mini-burst occurs via pushing by three of the 12 subunits of the connector. If all four mini-bursts occur, the cycle repeats. If the mini-bursts are not completed, a second cycle is superimposed on the first cycle. The existence of the second cycle is based on data recently obtained with bacteriophage T3. When both cycles stall, energy is diverted to expose the DNA molecule to maturation cleavage.

Keywords: bacteriophage structure; biological energy transduction; biological signal noise; cryo-electron microscopy; single-molecule analysis

1. Introduction

Analysis of bacteriophage DNA packaging motors is performed to understand basic principles of energy transduction and associated signaling in multimolecular assemblies. A DNA packaging motor begins as a procapsid assembled from subunits in the absence of DNA (capsid I in the case of the related bacteriophages, T3 and T7; Figure 1). The motor binds a double-stranded DNA molecule and causes the DNA molecule to enter and package within the cavity of a symmetrical protein shell. This DNA packaging is an event that is accompanied by a change in the structure of the shell; this change usually includes increase in size (capsid II in Figure 1). The DNA molecule enters the cavity through an axial hole in two coaxial protein multimers that form a signaling center for the motor. The inner multimer is called the connector and is embedded in the capsid's icosahedral shell at a five-fold rotational symmetry axis of the shell. The connector is made of 12 subunits, each a copy of a single protein, gp8 for T3/T7; the T3 and T7 proteins are named by gp, followed by the gene number from [1]. However, connector-like assemblies with other numbers of subunits are sometimes found when the connector protein is assembled outside of a bacteriophage particle. The monomer of the outer multimer is an ATPase [2-6].

The DNA packaging of some bacteriophages is accompanied by cleavage of the mature DNA genome from a multi-genome concatemer (maturation cleavage). A C-terminal endonuclease domain of the ATPase catalyzes the cleavage [2,3,5,7,8]. In the case of a concatemeric substrate for packaging, the DNA packaging ATPase (with its C-terminal endonuclease domain) is called either terminase (the nomenclature to be used here) or the large terminase protein. Terminase is accompanied by a smaller protein, sometimes called the small terminase protein, that binds DNA, is needed for initiation of packaging and sometimes has other activities (illustrated for T3 and T7 in Figure 1; gp18 is the small terminase protein; gp19 is terminase).

Bacteriophage DNA packaging motors are also models for understanding of both eukaryotic virus assembly and virus evolution, because DNA packaging motors evolved before the prokaryote/eukaryote splits. Specifically, herpes viruses also have a terminase and a connector [9,10]. The herpes virus terminase has sequence similarity to the phage terminases [11-13]. Thus, the connector and terminase proteins had evolved by about 1.6 billion years ago [14]. The high speed and low cost of bacteriophage propagation have been a foundation for relatively thorough genetic/biochemical/biophysical analysis of bacteriophage DNA packaging motors. *In vitro* systems have, for example, shown that DNA with a previously introduced single-stranded break (nick) in the phosphodiester backbone is packaged in the case of both ϕ29 [15,16] and T4 [17]. A nick in a comparatively short packaging substrate (<200 base pairs) did, however, partially inhibit T4 packaging, primarily by slowing it (Figure 5d in [17]). Thus, if transmission of torque is a required aspect of packaging, the protein component of the motor must resist DNA rotation.

The analysis of bacteriophage DNA packaging motors has included the testing of proposed motor mechanisms by single-motor, real time, visible light-based nanometry and fluorescence microscopy of *in vitro* DNA packaging in systems of purified components. Several proposed motor mechanisms require rotation of the connector [18-22]. However, no rotation of the bacteriophage ϕ29 connector was found (with probability >99%), when nanometry was used to measure packaging progression and single-motor fluorescence anisotropy was used to measure packaging-associated connector rotation

[23]. The conclusion of no motor-associated connector rotation had previously been drawn in the case of bacteriophage T4 when packaging continued after the potential for connector rotation was removed by genetically modifying the connector subunits [24].

Figure 1. The DNA packaging pathway of the related bacteriophages T3 and T7 (adapted from [30]). The solid arrows indicate the proposed productive pathway in an infected cell. The dashed arrows indicate the pathways for generating the motor-related particles that have been observed by fractionation and characterization. Duplication of the early stages represents cooperativity detected by single-molecule fluorescence microscopy [93]. The legend at the top indicates the color-coding of both the DNA molecule and the various proteins.

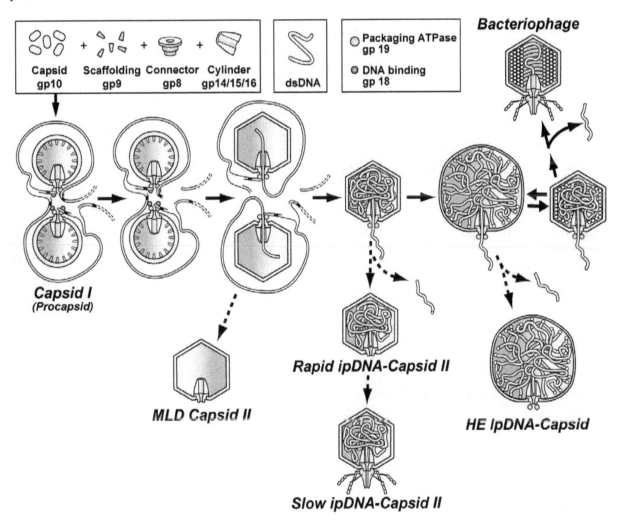

Among the various previously proposed motor mechanisms, some aspects of a cycle based on thermal ratcheting [25,26] did survive the above tests. The definition used here for thermal ratcheting is external force-dependent rectification of thermal motion (see also [27,28]). At the earliest stage of packaging, one proposed rectifier was intracellular osmotic pressure; a second was motor/DNA binding initiated by reverse DNA motion [26]. Other investigators have subsequently adopted motor/DNA binding-based thermal ratcheting, at least in its most general form [23,29]. However, details of a hypothesized thermal ratchet have not been proposed with cognizance of data obtained after 2003.

In past studies of these details, most discussion has included the implicit assumption that each DNA packaging motor has a cycle of one type. However, isolation of motor-related particles from bacteriophage T3-infected cells revealed capsid hyper-expansion that suggested adding to the cycle previously investigated (type 1 cycle) of a second cycle (type 2 cycle) at the later stages of packaging, at least *in vivo*. The proposed type 2 cycle includes ATP-driven hyper-expansion and contraction of the shell [30].

Although shell hyper-expansion has not been detected for other bacteriophages, some data suggest this possibility. First, a sudden decrease in resistance to *in vitro* bacteriophage λ packaging has been observed at 90% packaging by single-molecule nanometry [31]. This decrease is possibly caused by shell hyper-expansion, although capsid rupture is the explanation proposed by the authors. Second, stabilization of a hyper-expanded λ capsid is a possible explanation for the recent observation that a capsid-stabilizing, decoration protein of the λ capsid (D protein) is required not only for DNA retention after packaging, but also for packaging of the last 15-20% of the λ genome [32].

A possibly related phenomenon is that the shell subunits of some bacteriophages, such as HK97, are covalently joined to each other by chain-like cross-links established during assembly [33,34]. Assuming that the HK97 covalent cross-links are produced before packaging is completed, these cross-links could have evolved to stabilize a hyper-expanded capsid at the end of DNA packaging. In this case, any change in cross-linked shell size would occur by refolding of each of the subunits, possibly via a rubber-like (and, by analogy, exothermic) stretching. The cross-links prevent hyper-expansion via inter-subunit translation and rotation. In confirmation, a recent study reveals that the procapsid-to-mature capsid transition for HK97 (equivalent to the capsid I-to-capsid II transition for T3; Figure 1) occurs via change in protein folding [34]. This change is exothermic for bacteriophage P22 [35]. Given the structural analogy of the λ D protein and the HK97 cross-links [36], at least some functions are likely to be the same for these two aspects of bacteriophage capsids.

The objective of the current study is to derive a type 1 cycle that explains all current data and that provides means for initiating a type 2 cycle. To help compare studies of the various bacteriophages, the extent of packaging will be quantified by the ratio (F) of the length of DNA packaged to the mature genome length.

In deriving the type 1 cycle proposed here, the assumption is made that the basics of the type 1 cycle are the same for all bacteriophage DNA packaging motors. This assumption is confirmed by the structure-based similarity of the ϕ29, P22, SPP1 and T7 connectors, without detected amino acid sequence similarity [37,38], and the sequence-based similarity of all terminases [7,11,12]. The N-terminal region of terminases has even been aligned with the DNA packaging ATPase of bacteriophage ϕ29 [39]. Bacteriophage ϕ29 has a monomeric DNA packaging substrate and has a packaging ATPase with no known endonuclease activity [6,41]. The proposed cycle has aspects that should be applicable to all double-stranded DNA viruses, including those of significance for both medicine and ecology. In addition, some aspects should be applicable to energy transducing systems of other types, including systems in which protein folding is assisted by chaperonins.

2. Examination of some past assumptions

As reviewed in [5,6], most of the (type 1) cycles previously proposed for DNA packaging motors were based on the assumption that the ATP usage per base pair packaged is constrained to be uniform throughout packaging (constant ATP assumption) and is equal to 0.5 ATP cleavages per base pair packaged. The sources of the constant ATP assumption are chromatographic measurements of the number of ATP molecules used per base pair packaged during unsynchronized *in vitro* packaging in the case of ϕ29 [41] and T3 [42]. But, the data of the latter two studies revealed ATP usage averaged over all values of F, because the packaging was not synchronized. The ATP usage per base pair packaged has never been measured as a function of F, to the author's knowledge. Performing this measurement is a goal for the future, possibly achievable by single-molecule analysis because single ATP cleavages have been detected by use of single-molecule fluorescence microscopy ([43,44], for example). The constant ATP assumption cannot be rationalized by analogy to non-viral eukaryotic motors because DNA packaging motors evolved before non-viral eukaryotic motors and are not under the same constraints, as further discussed in Section 7.

In addition, the constant ATP assumption implies usage of ATP that is less efficient than it would be if the ATP used per base pair packaged increased as F increased, based on two lines of evidence. First, energetics-based computer simulations of packaging dynamics for bacteriophage ε15 revealed "negligible" (<2 pN) requirement for force when F was less than 0.4 [45]. These simulations have been accurate in the past; they correctly predicted the randomness of newly packaged DNA conformation, for example [46]. Second, single-motor nanometry-based determination of packaging forces for bacteriophages ϕ29 [47,48], λ [31] and T4 [49] had previously yielded basically the same conclusion. Given that the osmotic pressure of the cytoplasm of *Escherichia coli* is about 5 atmospheres [50], an ATP-independent, osmotically derived packaging force of about 6 pN would exist *in vivo*, independent of ATP consumption, if the capsid interior were empty. That is to say, when packaging occurs *in vivo*, ATP-derived energy is possibly not needed to package until $F = \sim 0.3$ [48].

Both the simulations and the nanometry also agree that, as F increases, the force required for packaging is a steeply increasing function of F and can be as high as 125 pN at the end of packaging in the case of the ε15 simulations, and 50-70 pN in the case of the nanometry; the capsid is assumed constant in size. The power delivered by the ϕ29 motor decreases progressively to zero *in vitro* at the lower F values [48]. Thus, the constant ATP assumption implies that roughly half of the ATP to be cleaved is wasted before it is even needed. Viewed in the context of evolution, the constant ATP assumption produces an outcome likely to be subject to negative selection. The result of negative selection can be either F-dependence of the ATP utilization in the type 1 cycle or introduction of the type 2 cycle (or both). The type 1 cycle might not even begin *in vivo* until roughly one third of the DNA was packaged. To the author's knowledge, nanometry-based tests have not yet been made of the effect of osmotic pressure gradients on the *in vitro* packaging of a bacteriophage DNA packaging motor. The type 1 cycle proposed here does not depend on the constant ATP assumption.

Although a previous manuscript [26] does propose two cycles and does propose ATP usage (per base pair packaged) that increases as F increases, the detailed mechanisms proposed have other aspects that are in conflict with more recently obtained data. First, the small terminase subunit was attached to terminase and was the component of the motor that bound the DNA molecule during cycling.

However, the small terminase is not part of the φ29 motor because φ29 does not have this protein. Yet, the φ29 connector has structure-based similarity to the connectors of terminase-dependent bacteriophages. The φ29 packaging ATPase has sequence-based similarity to the terminases. Furthermore, the T4 motor does not need the small terminase protein after initiation of DNA packaging [51]. Thus, the DNA packaging ATPase (gp19 for T3; Figure 1) is assumed here to act without its smaller companion. In addition, several recent studies (described below) have produced data that indicate the need for further modification of the type 1 cycle proposed in reference 26. These data will form the basis for the proposal of a type 1 cycle revised in some, but not all, aspects. The type 2 cycle will be subjected to only limited discussion here.

3. Key data from previous studies of the type 1 cycle

The key data discussed in this section will be used as the basis for the type 1 cycle to be proposed. Of course, other data exist. The author is not aware of any other data with which the type 1 cycle presented below is in conflict.

3.1. Thermal ratcheting and its limitations

An aspect of the type 1 cycle to be proposed is that this cycle includes a time period in which the DNA molecule does not move relative to the connector/ATPase multimer because the DNA molecule is bound to the ATPase multimer. This period of no DNA motion will be called a dwell, as proposed in [52]. The DNA molecule is packaged between dwells in a period called a burst. The evidence for dwells and bursts was the F vs. time relationship observed by high-resolution single-molecule nanometry of in vitro φ29 DNA packaging [52]. The DNA molecule was under constant optical trap-maintained force of 8 pN, equal to the DNA packaging force needed when $F\sim = 0.38$ [48]. In the type 1 cycle to be proposed, both a dwell and a burst will occur.

Details of the nanometry provided additional information. The dwell time decreased with increase in [ATP], although the burst time was independent of [ATP]. This observation was interpreted to mean that binding of ATP caused the dwell (i.e., caused DNA binding) and that the burst occurred in-between ATP binding-associated dwells [52]. However, this observation also implies that the bound ATP caused a change that occurred during the dwell and that increased in magnitude as the number of bound ATP molecules increased. Among the changes observed after ATP binding is decrease in the Gibbs free energy (tighter binding) of the bound ATP [52,53]. In the type 1 motor mechanism to be proposed, this decrease will be essential to activation of a component of the connector. Nonetheless, the ATP binding site is assumed to be on the ATPase (terminase) because, in the case of bacteriophage T3, a strong ATP binding site has been found on the T3 terminase, gp19, although no ATP binding site was observed on the connector, gp8 [42]. DNA binding-associated dwell-periods were also an aspect of the proposed thermal ratchet-based type 1 cycle of [26]. The thermal ratchet was biased by force from an osmotic pressure differential across the capsid's shell (higher outside) and possibly a motor-derived, oscillating electrical field that generated net force on the DNA molecule.

A purely thermal ratchet-based motor mechanism is now in conflict with the following subsequent observation also made by high-resolution nanometry of the φ29 DNA packaging motor. As the optical trap-derived force opposing packaging increased at a fixed F for packaging slowed by the methylation of a patch of 10 DNA phosphates, the time taken to package this patch was reduced by an amount too low by over four orders of magnitude to be explained by diffusion only [16]. That is to say, the DNA movement of the burst was caused by forward-directed force (*i.e.*, a force in the direction of packaging) and was not generated by passive diffusion. This force could not have had a significant component from osmotic pressure, based on the composition of the *in vitro* packaging mixture, which had no compound added to mimic the intracellular water activity. Thus, the motor has to be more than either a purely thermal ratchet or a purely osmotic pressure-biased thermal ratchet. Nonetheless, an osmotic pressure gradient is likely to provide some of the DNA-driving force *in vivo* and also in the case of *in vitro* systems that depend on the presence of polymer, such as those for P22 [54], T7 [55], T3 [56], SPP1 [57] and T4 in a recently developed system [17].

A related aspect of the type 1 cycle is coordination among ATPase multimer subunits in producing ATP binding-induced DNA binding. This coordination was deduced in the case of the ϕ29 DNA packaging motor from the sharpness of a plot of the frequency of any given dwell time *vs.* dwell time. The sharpness of this plot implied a minimum of 2 ATP binding events to generate a dwell [52]. The ATPase multimer was five-membered during packaging in this system [21,58]. One possible mechanism for coordination is ATP binding-induced contraction of a ring of ATPase molecules. The result would be DNA binding via steric clamping by the ATPase ring of the DNA molecule being packaged. Steric clamping will be a non-essential part of the mechanism proposed here (clamping was originally suggested to the author by S. C. Hardies). Other modes of binding are also consistent with the data, although no consensus DNA binding site has yet been reported in the N-terminal ATPase domain (motor domain) of terminases.

3.2. Connector dynamics in the type 1 cycle

Another aspect of the proposed type 1 cycle is that the connector has two essential functions, transmission of information and transduction of force. This aspect is based on the following observations of the bacteriophage SPP1 connector. (1) The DNA channel of the SPP1 connector is narrow enough to contact a DNA molecule in the channel. The channel is ~18.1 Å in diameter at its narrowest point, as found by obtaining the x-ray crystallographic structure of a connector protein 13-mer and extrapolating the structure of the 13-mer to the 12-mer present in bacteriophage particles [38]. This diameter is, in fact, about 20% smaller than the diameter of the narrowest channel that could contain a DNA molecule without contact with the connector. (2) Blockage of packaging occurred when both intra- and inter-connector subunit motion was inhibited by disulfide cross-linking of connector subunits to each other; reversing the cross-links restored the packaging activity [59]. (3) Some point mutations of the SPP1 connector protein either abolished or decreased both DNA packaging and the ATPase activity of the packaging ATPase without disrupting assembly of the connector in the SPP1 procapsid [38,60].

The conclusion drawn was that DNA packaging-associated "cross-talk" existed between the connector and the packaging ATPase and that the connector was part of the DNA packaging motor [38,60]. Assumption of this cross-talk is supported (although not proven) by the observation that the rate of ATP binding to the ϕ29 motor is independent of the nanometry-determined force on the motor [53,61], as though the force is sensed by one motor component and the ATP binding is to another. In the type 1 cycle to be proposed, the force sensor will be the connector; the ATP binding element will be the ATPase multimer. An explanation will be provided for evolution of the partitioning of these functions to two proteins. Both functions are in a single protein in the case of non-viral eukaryotic motors.

The proposed details for the above activities of the connector include change in conformation of peptide segments of the connector subunits. This aspect is based, first, on the analytical calculation that peptide segments have the capacity for achieving up to 4.7 different conformations for which neither stability nor activation energy is comparable to the energy of thermal motion [62]. This number is approximately the number of conformations that can be an independent part of a type 1 cycle of a DNA packaging motor. The value, 4.7, depends primarily on the number of different amino acids (twenty) and is independent of peptide chain length. The conformational mobility aspect is also based on the finding of regions of assembled connector proteins that are channel-proximal and that are likely to be conformationally mobile (to be called mobile peptide segments), in that they are without regular secondary structure.

The connector has subunits each of which has two mobile peptide segments by these criteria. To illustrate the mobile peptide segments in relation to the rest of the motor, Figure 2a shows the DNA molecule (tan), connector (yellow), packaging ATPase (green) and outer shell (blue). The first of these two peptide segments is a C-terminal, 40 residue peptide segment without a unique structure (disordered), seen next to the widest end (crown) of the SPP1 connector, inside of the outer shell (Figure 2b) [38]. In the smaller φ29 connector, this disordered peptide segment is 24 residues long and encompasses the region that corresponds to the crown of SPP1 [22,38]. The second of these two peptide segments forms a loop in the SPP1 connector (tunnel loop) that is attached at both ends to an α-helix and projects into the axial channel at the smallest diameter of the channel (~18.1 Å) (Figure 2b) [38].

For the tunnel loop, genetic analysis supports a function in SPP1 DNA packaging. Among SPP1 connector-associated packaging mutations, five were in the tunnel loop region; four of the 15 amino acids in the SPP1 tunnel loop were mutated. One of these mutations reduced both the efficiency of packaging and the activity of the SPP1 packaging ATPase [38]. The tunnel loop is disordered in the φ29 [21,22], but not the SPP1 [38], connector. The tunnel loops of the 12 SPP1 connector subunits can simultaneously all engage the major groove of the DNA double helix, but they must be translated relative to each other along the connector axis because of the tightness of fit [38]. In the proposed type 1 cycle, the conformational changes of the mobile peptide segments have two functions, motion sensing/signal initiation and force transduction. Both C-terminal and tunnel loop segments are assumed to be participants, although the evidence for the tunnel loop is more complete in the case of SPP1. In the case of φ29, the tunnel loop may be redundant at the earlier stages of packaging because a preliminary study [63] has revealed that deletion of the tunnel loop inhibits packaging only near the end.

Lebedev et al. [38] have already proposed that engagement of tunnel loops with the major groove of the DNA double helix is either signaling or force-delivering in character. In the latter case, the proposed type 1 cycle is a high-detail version of the rotating connector-driven forward motion originally proposed [18], but with the DNA molecule rotating relative to an immobile connector, not the connector rotating relative to the outer shell and the DNA molecule. Rotating the DNA molecule relative to an immobile connector has the disadvantage that supercoils will be introduced in the DNA as it is packaged. In addition, this mechanism is unlikely, based on the observed packaging of mismatched regions in the case of φ29 [16] and T4 [17] and unpaired single-stranded regions in the case of λ [64], as further discussed below. The type 1 cycle proposed here does not have these conflicts with the data.

Figure 2. The components of the proposed type 1 cycle, side view. To simplify the drawing, the correct symmetry of the ATPase multimer, assuming it to be five-fold, is not represented. **(a)** The connector (yellow) is shown as represented in Figure 1 (lower left, in isolation) and also at higher resolution while embedded in the shell (light blue) with terminase (green) attached to it. **(b)** The components of the signaling center of the motor are shown at higher magnification with the labeling used in the text. **(c)** The motor is shown at the beginning of a cycle with DNA molecule undergoing forward thermal motion, *i.e.*, thermal motion into the cavity of the shell. **(d)** The motor is shown with DNA molecule undergoing reverse thermal motion. **(e)** The motor is shown at the beginning of a dwell, with ATP (yellow oval with orange border) bound to terminase and terminase bound to DNA molecule. **(f)** The motor is shown at the end of a dwell with mobile peptide segments activated; coiling illustrates activation, but the true activated conformation is not known. **(g)** The motor is shown at the beginning of a burst, with terminase no longer binding the DNA molecule, ATP cleaved and mobile peptide segments pushing on the DNA molecule as they begin to deactivate. **(h)** The motor is shown in the middle of a burst. As the burst proceeds, the motor returns to its state in (c, d).

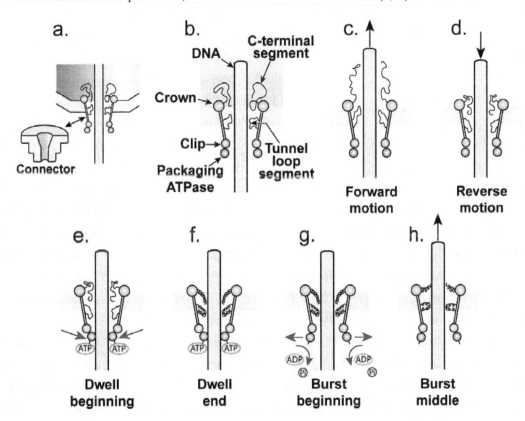

3.3. Fine structure of the single-molecule nanometry

Finally, the type 1 cycle proposed here was developed with the constraint that it must explain the recently observed fine structure of the *F vs.* time relationship obtained by high-resolution, single-molecule φ29 nanometry. When observed at high resolution and force also relatively high, 40 pN, dwells were separated not by a single burst, but by a series of four "mini-bursts". The DNA molecule moved 2.5 base pairs per mini-burst [52]. The non-integral DNA progression during mini-bursts has been a mystery. The type 1 cycle proposed here explains the number of these mini-bursts and their production of non-integral DNA progression during packaging.

4. The proposed type 1 cycle

The type 1 cycle proposed here begins with the connector's mobile peptide segments in contact with the DNA molecule being packaged. The axial channel is so small at its narrowest point (~18 Å in diameter at the tunnel loops, as derived from packaging non-active particles in [38]) that the mobile peptide segments (C-terminal and tunnel loop) are not in a unique relationship to the DNA molecule, although they contact the DNA molecule. The data are not yet sufficient to propose the details for the initiation of packaging, *i.e.*, the events that occur before the state of Figure 2a, b is achieved.

4.1. The ratcheting component

As previously proposed [26], the DNA molecule now moves by diffusion, either into (forward) or out of (reverse) the capsid. An osmotic pressure gradient biases diffusion in the forward direction, as supported by the promotion of DNA packaging in some *in vitro* systems by neutral polymers (discussed above). This polymer-enhancement of DNA packaging is an osmotic pressure-generated, not an excluded volume-generated, effect [65]. Forward motion causes stretching of the mobile peptide segments in the forward direction (Figure 2c). As discussed above, the osmotic pressure present *in vivo* is sufficient so that the type 1 cycle need not proceed further until $F = \sim 0.3$.

Eventually, reverse diffusion occurs (Figure 2d). The probability and magnitude of reverse diffusion increase as obstacles to packaging increase. These obstacles include steric and charge-charge repulsive interactions of both packaged DNA segments (reviewed in [4]) and comparatively small molecules accidentally packaged. One of the proposed purposes of the mobile peptide segments is to sense this reverse motion by changing conformation (Figure 2d). This changed conformation initiates a signal that is part of the connector/ATPase cross-talk previously observed for bacteriophage SPP1 [38]. This signal is transmitted through the connector to the packaging ATPase (terminase) multimer.

The type 1 cycle continues with a dwell-response to the signal sent to the ATPase multimer from the connector. The dwell-response begins with increase in ATP binding by the ATPase subunits. As shown in Figure 2e, ATP binding to the ATPase eventually results in ATPase/DNA binding that stops the reverse motion and produces a dwell. The DNA binding possibly, but not necessarily, occurs by clamping. During the period of the dwell, the mobile peptide segments are held in a conformation (Figure 2e) previously generated by strain during reverse motion of the DNA molecule (Figure 2d). The conformation of mobile peptide segments cannot return to the original because forward DNA motion is prevented and the unoccupied space in the channel is insufficient. The DNA molecule may also be under compressive strain from ATPase-DNA binding during reverse DNA motion and trapping at the other end by the mobile peptide segments (Figure 2e). Analysis of connector-proximal DNA (by Förster resonance energy transfer between dyes at two places on the DNA molecule) has revealed DNA compression during *in vitro* T4 DNA packaging stalled from outside of the capsid [51]. However, the interpretation of this latter study was compression via a power stroke of the ATPase. A power stroke of the ATPase does not occur in the type 1 cycle proposed here. Thus far, the type 1 cycle proposed here is not distinguishable from the cycle of a (power stroke-free) thermal ratchet.

4.2. The non-ratcheting component: ATP-driven mobile peptide segment activation

After the start of the dwell, the proposed type 1 cycle differs from the cycle of a thermal ratchet in that the connector's mobile peptide segments are activated as the dwell continues (Figure 2f; activation is illustrated by coiling of the mobile peptide segments). After the dwell starts and before activation, the activation is made energetically possible by lowering of the Gibbs free energy of the bound ATP molecules. The dwell provides time for both this lowering and the linked, seesaw-like raising of the Gibbs free energy of the mobile peptide segments. The activation occurs while the ATP is bound, not while (or after) the ATP is cleaved. The activated, connector-associated mobile peptide segments have conformations not accessible in the absence of the pathway for Figure 2,c-f. In support of this sequence of events, non-sigmoidal (non-cooperative) [ATP]-dependence of both φ29 packaging velocity [53] and mean φ29 dwell time [52] has led to the conclusion that an initial ATP binding is followed by tightening of ATP binding, thereby short-circuiting the cooperativity otherwise expected [52]. The mobile peptide segments, as drawn in Figure 2, are schematic illustrations and are not accurate representations of conformation.

The apparent paradox of energetic linkage via DNA "stillness" is resolved by the following. Although the DNA molecule does not move relative to the motor, movements within the ATPase and connector subunits still occur. ATPase/connector cross-talk via these movements is physically realistic because movements of this type have already been shown to occur in the case of ABC transporters. In the case of the ABC transporters, movements within dimeric ATP binding domains occur after ATP binding and before ATP cleavage, as proposed above. The movements of the ATP-binding domains cause movements in transport-generating, α-helix-rich transmembrane domains that sometimes, like the connector subunits of bacteriophage DNA packaging motors, are in a separate protein [66,67]. Thermal oscillations potentially assist transfer of energy from ATPase (terminase)-bound ATP to the connector-associated mobile peptide segments. Sufficient information does not exist to propose either complete detail for ABC transporters [66,67] or any details for bacteriophage DNA packaging motors. The process involved for DNA packaging motors may be analogous to protein folding promoted by immobilization in the internal cavity of a chaperonin complex, a process that sometimes occurs at the expense of ATP binding energy [68-70].

The lowering of the Gibbs free energy of ATPase-bound ATP signals the ATPase multimer to cleave the ATP molecules and, thereby, to release (perhaps unclamp) the DNA molecule and start a burst (Figure 2g). During the burst, the connector's activated mobile peptide segments deactivate while pushing the DNA molecule. To illustrate additional details of the deactivation/pushing, Figure 3 begins by schematically illustrating the end of the dwell in projection along the DNA axis (Figure 3a). Each coiled fiber represents both activated C-terminal and activated tunnel loop segments of one connector subunit; each dark yellow ball represents (schematically) the rest of a connector subunit. For clarity, the drawings of Figure 3 are not true projections (see legend to Figure 3).

Figure 3. A transverse section-based representation of the firing of the connector subunits. A coiled peptide segment and a dark sphere represent an activated connector subunit. An uncoiled peptide segment and a light sphere represent a completely deactivated connector subunit. The DNA molecule is in the middle. This representation of the connector is a section with some details missing and modified to avoid confusion. Specifically, the clip and crown are, together, represented by one circle and only one peptide segment is shown; the subunits are not in contact. **(a)** The motor is shown at the end of a dwell with all 12 subunits activated. The motor is shown at the end of the **(b)** first, **(c)** second, **(d)** third and **(e)** fourth mini-bursts. **(f)** The motor is shown during a stalled mini-burst. **(g)** The motor is shown after an activation of the mobile peptide segments that began after a stalled mini-burst.

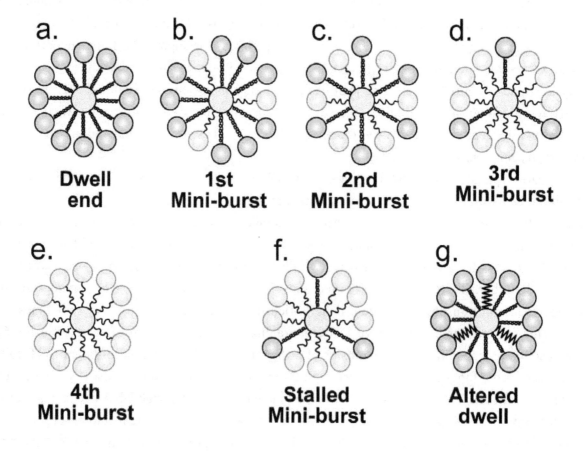

The following aspect of connector-based energy transduction directly satisfies the empirically derived [16] constraint of four mini-bursts per burst. The connector's 12 subunits act sequentially in four groups of three subunits. That is to say, one group of three subunits initially pushes the DNA molecule (fires, in the terminology of an internal combustion engine) and is followed by the other three groups, one firing after the other (Figure 3,b-e). Firing is represented in Figure 3 by lightening of a yellow ball and uncoiling of the attached fiber. To reduce the tightness of fit in the channel [38], (1)

the unfired subunits of Figure 3 rotate at the clip so that their crowns are further from the channel's axis (and away from each other) than crowns of fired subunits and (2) after firing and uncoiling, mobile peptide segments move away from the channel. This latter movement is not represented in Figure 3. The physical basis for firing in groups of three is, therefore, proposed to be steric constraint on the number of mobile peptide segments that fit in the channel together with the DNA molecule.

In addition to explaining the results of nanometry and making room for the mobile peptide segments, firing in four groups of three keeps the DNA molecule centered. Unlike firing in four groups of three, firing in six groups of two would not avoid thermal motion-generated de-centering. The remaining alternative is firing in three groups of four, which would keep the DNA molecule centered, but might cause problems with tightness of fit.

In Figure 3, the conjecture is made that the three firing subunits of a mini-burst are symmetrically located, even though not linked to each other and restricted to thermal motion before firing begins. This conjecture has some basis (but, is not proven) from analysis of electrical systems in that promotion of symmetrical response via combination of more than one independent noise patterns has been found in theory [71].

A type 1 cycle must account for the observations that (1) perturbations of DNA structure do not necessarily inhibit the pushing of the DNA molecule and (2) the number of base pairs (2.5) packaged per mini-burst is non-integral [16]. The perturbations include mismatched, double-stranded regions (φ29: 10 bases [16]; T4: 10 bases [17]) and unpaired single-stranded regions (λ: 19 bases [64]). These observations conclusively demonstrate that recognition of a unique DNA secondary structure is not necessary. (They also indicate that firing can occur less often than one would calculate from the average ATP usage per base pair.) Thus, the firing of the connector subunits is proposed to push the DNA molecule via non-specific forward sweeping of the DNA molecule by the mobile peptide segments. The sweeping is accomplished by steric interaction (i.e., contact with the DNA molecule is involved) and possibly also hydrodynamic interaction of the activated mobile peptide segments with the DNA molecule. No structure-specific engaging of a DNA component, such as the major or minor groove, occurs in the proposed type 1 cycle.

Sweeping drives the DNA molecule forward until either (1) the activation energy for all 12 connector subunits is dissipated and sweeping stops (Figure 3e) or (2) the motor has to work against a force that is high enough so that the sweeping is not completed and the DNA molecule undergoes reverse thermal motion before the final sweep (mini-burst) is completed (stall; Figure 3f). A stall arises from resistance to packaging produced both by the DNA segments already packaged and by small molecules (RNAs and proteins, for example) that have been accidentally packaged in an infected cell. Thus, the proposed type 1 cycle will not stall as often (or maybe not at all) in vitro, unless the macromolecular composition of the cellular cytoplasm is mimicked.

The reverse thermal motion of the stall of Figure 3f initiates a type 2 cycle via the following proposed mechanism. This reverse DNA motion causes a dwell and, then, refolding of the mobile peptide segments, as it did at the beginning of the type 1 cycle. But, the mobile peptide segments from only nine subunits have completed firing. Thus, the dwell is altered in that only nine mobile peptide segments refold to the conformation of Figures 2f and 3a. The remaining three mobile peptide segments have not completed firing and are in a space larger than experienced in an unaltered type 1 cycle. Therefore, these three mobile peptide segments refold to an altered, although activated conformation (Figure 3g) while the DNA molecule is bound to the ATPase multimer and the Gibbs free energy of the bound ATP is lowered. As in a type 1 cycle with an unaltered dwell, the next event is ATP cleavage, followed by a burst. However, the altered activated conformation causes an altered burst. The altered burst includes movements of the connector that initiate capsid hyper-expansion and, then, the rest of the type 2 cycle. The author reserves proposing further details until more information is obtained about the type 2 cycle, while noting that an empirical precedent exists in that the bacteriophage T4 connector initiates a capsid expansion roughly equivalent to the expansion that occurs during the capsid I to capsid II transition in Figure 1 [72].

5. The type 2 cycle and terminase cleavage of a concatemer

The details of the type 2 cycle are also not proposed here beyond what has previously been proposed in [26]. The proposed interaction of the two cycles is the following. (1) The type 2 cycle restarts the type 1 cycle by hyper-expansion-generated reduction of the concentration of packaged DNA segments and pumping-generated removal of the accidentally packaged small molecules from the DNA-containing cavity of the capsid. (2) The pumping works by expansion/contraction, coupled with changes in permeability, of the shell. (3) The type 2 cycle stops when not triggered by the type 1 cycle. (4) The type 1 cycle continues running and re-triggers the type 2 cycle when the type 1 cycle again stalls. (5) Eventually, both cycles undergo a stall (called a co-stall). In the case of T3 DNA packaging *in vivo*, the first type 2 cycle occurs at $F \sim 0.28$ [30]; in the case of λ DNA packaging *in vitro*, the F value at packaging force reduction suggests that the only type 2 cycle occurs at $F \sim 0.9$ [31].

The hypothesis presented here is extended to propose that a co-stall initiates the maturation cleavage of a concatemer (Figure 1). The proposed mechanism is alternative channeling of the energy of bound ATP, since this energy can no longer be channeled to DNA packaging. Specifically, as the time of co-stalling increases, the probability increases that the energy of bound ATP is alternatively channeled to expose the DNA molecule to the endonuclease domain of terminase. When that happens, the maturation DNA cleavage occurs. Sufficient information does not exist to propose further details. By this proposal, the probability of maturation cleavage is never zero at any stage of packaging. Thus, a background of erroneous maturation cleavages occurs and erroneous cleavages are made more frequent by a dwell or a stall during the normal cycling of the motor.

In support of these proposals, observation has been made of prematurely cleaved, incompletely packaged T3 genomes in capsids obtained from bacteriophage T3-infected cells [30,46] (see Figure 1). Based on the lengths of the cleavage products, cleavage positions are sometimes quantized, as though occurring during the type 1 cycle stalls that trigger the type 2 cycle [30]. In further support of the proposed cleavage-promoting alternative pathway, artificial slowing of *in vitro* T3 packaging accelerates the maturation cleavage of concatemers [73]. Involvement of the connector is suggested by the observation that some bacteriophage P22 connector mutations cause delay of maturation cleavage. The observed result is that P22 packages an oversized genome [74]. Unlike the T3/T7 maturation cleavage, the P22 maturation cleavage is not nucleotide sequence-specific. The packaging of an oversized genome by the P22 mutants is explained by assuming that the mutants have more rapid transmission of energy from ATPase ring to connector-associated peptide segments. That is to say, packaging continues further than it normally does, at the expense of cleavage.

In the case of bacteriophage T4, the above pathway for the terminase-catalyzed maturation cleavage explains the following otherwise puzzling observation. The small terminase protein, although not needed for the packaging motor, stimulates the ATPase activity of terminase and, thereby, inhibits premature (random) cleavage of a concatemer during packaging [8]. The blockage of premature cleavage is explained by the small terminase protein's blocking of the alternative (DNA-cleaving) pathway for energy usage by removing the source of energy, the terminase-bound ATP. The small terminase protein also forms a multi-subunit ring that has a symmetry mismatch with the terminase, in the case of both T4 [8] and P22 [75].

Unlike the observation for bacteriophage λ [31], a drop in nanometry-observed, packaging-resisting force has not been observed at the later stages of bacteriophage φ29 *in vitro* packaging, in spite of the extensive nanometry-based studies of φ29, as referenced above. Bacteriophage φ29 also has a procapsid (equivalent to T3 capsid I) with a shell that is not smaller than the mature φ29 shell [58,76]. These two observations suggest that, in fact, φ29 does not have an operative type 2 cycle. In support, φ29 packages a monomeric DNA molecule, not a concatemer, *in vivo* and has a relatively short genome [6,40], thereby reducing the selection for mechanisms to accelerate packaging in the later stages. The crown region of the φ29 connector may have evolved to be relatively small because the primary function of the crown is either initiating or conducting the type 2 cycle. An outstanding question is whether or not φ29 is the product of reductive evolution from ancestors that had a complete terminase and, by inference, a type 2 cycle.

6. Comparison with thermal ratcheting in other systems

The feedback-controlled, ATP-dependent, DNA binding/releasing aspect of the mechanism proposed here is in the category of an ATP-dependent thermal ratchet. A thermal ratchet rectifies thermal motion and, as a general concept, includes miniaturized mechanical ratchets, as articulated in [77]. The sweeping-of-DNA aspect of the type 1 cycle proposed here is outside of the concept of a thermal ratchet, as are reach/bind/pull mechanisms for DNA packaging, one version of which has been called a "ratchet model" [78]. That is to say, some terminology is inconsistent.

Thermal ratcheting, as opposed to bind/pull, is not a new idea. Huxley [79] proposed an ATP-driven, bind/release thermal ratchet-based mechanism for muscle contraction a long time ago. Even earlier, Donnan ([80]; page 320) approached this concept in his discussion of the interaction of statistical mechanics and biology, but did not fully articulate it, because of the absence of knowledge of molecular biology. Currently, thermal ratchet-based cycles are in active consideration for actin/myosin and kinesin/microtubule motors (reviewed in [28,29,43,81,82]). Recent data indicate that thermal ratcheting of these eukaryotic motors is, as proposed here, part of a cycle that also has a non-ratcheting component [28].

In contrast to what occurs in non-viral eukaryotic motors, the bacteriophage DNA packaging ATPase does not perform the non-ratcheting component of the type 1 cycle proposed here. If this type 1 cycle is correct, an explanation is needed for separation of function to not only two different proteins, but also two different multimers. This explanation will ideally include an explanation for the symmetry mismatches among the components of the motor. Bacteriophages have had 1.6 billion years of evolution to match non-viral eukaryotic motors in incorporating these functions in one protein.

7. Symmetry mismatches, evolution and non-viral eukaryotic motor proteins

The type 1 cycle proposed here provides a new explanation for previously unexplained symmetry mismatches, including the connector/ATPase, 12/5 symmetry mismatch of the bacteriophage φ29 motor [58] and the symmetry mismatches of the T4 motor [8,83]. This explanation is that symmetry mismatching reduces thermal motion (noise)-derived signaling errors between the connector and ATPase multimers. Thus, evolutionary selection for symmetry mismatching occurs, even though the motor would work with a symmetry match that would simplify assembly of the motor.

This explanation has support from calculations that reveal the following in the case of communication networks. Although non-random events (signaling errors, in the case of the connector/ATPase multimer) can be generated from adding the thermal noise from two symmetry related sources, this occurrence of non-random events is suppressed by breaking the symmetry [84] and potentially yields greater signal amplification via stochastic resonance [84,85]. In the case of the proposed type 1 cycle, breaking the symmetry means having a symmetry mismatch between two multimers. By this reasoning, separating signal source (connector) and target (terminase/packaging ATPase) to separate multimers evolved to reduce thermal motion-induced signaling errors that occur in a single protein.

Symmetry mismatching to reduce thermal motion-derived signaling errors is a concept that extends to the long-unexplained symmetry mismatch between connector ring and outer shell of the capsid [5,6,76] and also to the symmetry mismatches in some chaperonin-protease complexes [86]. As shown directly for the connector during DNA packaging [23], facilitation of rotation [18] is not likely to be the reason for evolution of the mismatch in the case of the chaperonin-protease complexes either [87]. In the case of eukaryotic cellular (in contrast to viral) motors, the larger number of cooperating, non-symmetrically placed motors would, by the ideas presented here, reduce thermal motion-induced signaling errors enough to allow evolution of motor proteins with signal source and target in the same protein molecule.

Eukaryotic non-viral motors, such as kinesin/tubulin and myosin/actin motors, also (1) do not work against a predictably increasing load and (2) evolved long after bacteriophage DNA packaging motors. Thus, in the type 1 cycle proposed here, the relatively long distances for transmission of both information and energy are not in conflict with the finding of shorter transmission distances in eukaryotic cellular motors.

In support of the above interpretation of the symmetry mismatches, evidence exists that the φ29 connector/ATPase symmetry mismatch is not biophysically necessary for DNA packaging. An artificially generated, non-mismatched hexameric ATPase (and associated hexameric packaging RNA molecule) can be effective for *in vitro* DNA packaging [19,88]. Even assuming that the *in vitro* system used does not precisely mimic packaging *in vivo*, this observation means that φ29 could have evolved to make use of the assembly advantages of symmetry matching [89], unless something else promoted evolution toward symmetry mismatching. The motor mechanism proposed here predicts that this "something else" is suppression of signaling errors generated by thermal noise. A test of the proposed role of errors in the evolution of symmetry mismatching is the determining of how the frequency of errors (other than premature concatemer cleavage) depends on whether or not the φ29 motor is symmetry mismatched.

A pure thermal ratchet-based cycle is more primitive than the type 1 cycle proposed here, especially when ratchet-associated binding is chemically non-specific, as it is for clamping. Thus, the working assumption is that, whatever the most advanced bacteriophage DNA packaging motors are today, they started as thermal ratchets, perhaps without the packaging ATPase and with ATP-derived energy coming from only a type 2 or related cycle. In this case, the DNA would be packaged less tightly than has been described [4,90,91] for packaging ATPase-dependent bacteriophages such as φ29, λ, P22, T3, T4, T5 and T7. Perhaps, such low DNA density, packaging ATPase-less bacteriophages still exist in environmental niches that favor them.

8. Other, recently proposed type 1 cycles: Predictions of the cycle of Figures 2 and 3

At this point in time, the type 1 cycle proposed here appears to be alone in meeting all of the data-based constraints described in Section 3. Other proposed type 1 cycles include bind/pull cycles with all pulling dynamics occurring within the packaging ATPase multimer. One such proposed cycle is based on details of structure for the φ29 and T4 terminases [83], but (1) does not account for the subsequently obtained data of [52] in that it uses uncoordinated (though regulated) cleavage of ATP and (2) does not account for the data of [38] and [59] in that it does not incorporate the connector in the energetics. Moffitt *et al.* [52] have proposed a bind/pull mechanism with the pull generated by lock washer-like distortions of the relationships of at least four subunits within the ATPase ring (thereby explaining coordination), but again without considering the connector and introducing the concept that one of the 5 ATPase molecules in the multimer is different from the others. Yu *et al.* have more recently proposed a sterically driven push and roll mechanism whereby eccentric DNA motion assists the movement between ATPase molecules and ATP binding is delayed after four ATPase molecules have fired [92]. However, again, the connector is not incorporated in the motor.

Among the previous proposed type 1 cycles, one does have the connector and DNA packaging ATPase integrated in the motor. In this type 1 cycle, the connector initially blocks DNA motion from an ATPase-delivered power stroke; the connector later releases the DNA molecule in a burst [17,51]. This cycle basically has the roles of the connector and packaging ATPase inverted in relation to the type 1 cycle presented here. The proposed cycle of [17,51] presumably will be updated to explain the four minibursts subsequently revealed.

The type 1 cycle proposed here makes at least two predictions that are not made by bind/pull-based cycles and that can be tested. The first prediction is that the previously demonstrated dwells of the type 1 cycle are preceded by reverse motion of the DNA molecule. In fact, high-resolution nanometry of *in vitro* φ29 packaging does have an approximation of this pattern when packaging is dramatically slowed (to create a pause) by methylating the phosphates of a 10 base pair patch of DNA. In this case, forward movement interrupts the pause and is followed by several backward movements, each followed closely by forward movement (Figure 2 of [16]). One interpretation of the repeated backward/forward motion is the operation of the type 1 cycle with time scale stretched and forward motion inhibited by DNA methylation. That is to say, movements normally too rapid and small to resolve are made resolvable by the methylation. However, this potential interpretation of the data was not a focus of [16]. More probing of the interpretation of the repeated backward/forward motion is needed before the nanometry can be considered a test.

The second prediction is that the motions of the connector-associated, mobile peptide segments are both the signaling and DNA-driving aspects of the motor. This prediction can be tested by real time single motor fluorescence microscopy/nanometry of the DNA packaging process ([23], for example) with the various regions of the connector labeled with fluorescent probes and use of Förster resonance energy transfer, for example, to monitor packaging associated changes in peptide conformation.

In addition, the structural details of the type 1 cycle proposed here (or any proposed type 1 cycle) can tested by cryo-EM with 3-D reconstruction of DNA packaging motors with DNA still in the connector/ATPase channel. The cryo-EM will be dependent on the isolation and preliminary characterization of these various "motor intermediates" in a state as native as possible.

Acknowledgements

The author thanks David F. Baker for computer drawings. This work was supported by grants from the National Institutes of Health (GM24365 and GM069757) and the Welch Foundation (AQ-764).

References and Notes

1. Pajunen, M.I.; Elizondo, M.R.; Skurnik, M.; Kieleczawa, J.; Molineux, I.J. Complete nucleotide sequence and likely recombinatorial origin of bacteriophage T3. *J. Mol. Biol.* **2002**, *319*, 1115-1132.

2. Fujisawa, H.; Kimura, M.; Hashimoto, C. *In vitro* cleavage of the concatemer joint of bacteriophage T3 DNA. *Virology* **1990**, *174*, 26-34.

3. Catalano, C.E. The terminase enzyme from bacteriophage lambda: a DNA-packaging machine. *Cell Mol. Life Sci.* **2000**, *57*, 128-148.

4. Petrov, A.S.; Harvey, S.C. Packaging double-helical DNA into viral capsids: structures, forces, and energetics. *Biophys. J.* **2008**, *95*, 497-502.

5. Rao, V.B.; Feiss, M. The bacteriophage DNA packaging motor. *Annu. Rev. Genet.* **2008**, *42*, 647-681.

6. Lee, T.J.; Schwartz, C.; Guo, P. Construction of bacteriophage phi29 DNA packaging motor and its applications in nanotechnology and therapy. *Ann. Biomed. Engineering* **2009**, *37*, 2064-2081.

7. Ponchon, L.; Boulanger, P.; Labesse, G.; Letellier, L. The endonuclease domain of bacteriophage terminases belongs to the resolvase/integrase/ribonuclease H superfamily: a bioinformatics analysis validated by a functional study on bacteriophage T5. *J. Biol. Chem.* **2006**, *281*, 5829-5836.

8. Al-Zahrani, A.S.; Kondabagil, K.; Gao, S.; Kelly, N.; Ghosh-Kumar, M.; Rao, V.B. The small terminase, gp16, of bacteriophage T4 is a regulator of the DNA packaging motor. *J. Biol. Chem.* **2009**, *284*, 24490-24500.

9. Bogner, E. Human cytomegalovirus terminase as a target for antiviral chemotherapy. *Rev. Med. Virol.* **2002**, *12*, 115-127.

10. Cardone, G.; Winkler, D.C.; Trus, B.L.; Cheng, N.; Heuser, J.E.; Newcomb, W.W.; Brown, J.C.; Steven, A.C. Visualization of the herpes simplex virus portal in situ by cryo-electron tomography. *Virology* **2007**, *361*, 426-434.

11. Sheaffer, A.K.; Newcomb, W.W.; Gao, M.; Yu, D.; Weller, S.K.; Brown, J.C.; Tenney, D.J. Herpes simplex virus DNA cleavage and packaging proteins associate with the procapsid prior to its maturation. *J. Virol.* **2001**, *75*, 687-698.

12. Serwer, P.; Hayes, S.J.; Zaman, S.; Lieman, K.; Rolando, M.; Hardies, S.C. Improved isolation of undersampled bacteriophages: finding of distant terminase genes. *Virology* **2004**, *329*, 412-424.

13. Baker, M.L.; Jiang, W.; Rixon, F.J.; Chiu, W. Common ancestry of herpesviruses and tailed DNA bacteriophages. *J. Virol.* **2005**, *79*, 14967-14970.

14. Kutschera, U.; Niklas, K.J. The modern theory of biological evolution: an expanded synthesis. *Naturwissenschaften* **2004**, *91*, 255-276.

15. Moll, W.D.; Guo, P. Translocation of nicked but not gapped DNA by the packaging motor of bacteriophage phi29. *J. Mol. Biol.* **2005**, *351*, 100-107.

16. Aathavan, K.; Politzer, A.T.; Kaplan, A.; Moffitt, J.R.; Chemla, Y.R.; Grimes, S.; Jardine, P.J.; Anderson, D.L.; Bustamante, C. Substrate interactions and promiscuity in a viral DNA packaging motor. *Nature* **2009**, *461*, 669-673.

17. Oram, M.; Sabanayagam, C.; Black, L.W. Modulation of the packaging reaction of bacteriophage t4 terminase by DNA structure. *J. Mol. Biol.* **2008**, *381*, 61-72.

18. Hendrix, R.W. Symmetry mismatch and DNA packaging in large bacteriophages. *Proc. Natl. Acad. Sci. U. S. A.* **1978**, *75*, 4779-4783.

19. Chen, C.; Guo, P. Sequential action of six virus-encoded DNA-packaging RNAs during phage phi29 genomic DNA translocation. *J. Virol.* **1997**, *71*, 3864-3871.

20. Grimes, S.; D. Anderson, D. The bacteriophage phi29 packaging proteins supercoil the DNA ends. *J. Mol. Biol.* **1997**, *266*, 901-914.

21. Simpson, A.A.; Tao, Y.; Leiman, P.G.; Badasso, M.O.; He, Y.; Jardine, P.J.; Olson, N.H.; Morais, M.C.; Grimes, S.; Anderson, D.L.; Baker, T.S.; Rossmann, M.G. Structure of the bacteriophage phi29 DNA packaging motor. *Nature* **2000**, *408*, 745-750.

22. Guasch, A.; Pous, J.; Ibarra, B.; Gomis-Ruth, F.X.; Valpuesta, J.M.; Sousa, N.; Carrascosa, J.L.; Coll, M. Detailed architecture of a DNA translocating machine: the high-resolution structure of the bacteriophage phi29 connector particle. *J. Mol. Biol.* **2002**, *315*, 663-676.

23. Hügel, T.; Michaelis, J.; Hetherington, C.L.; Jardine, P.J.; Grimes, S.; Walter, J.M.; Falk, W.; Anderson, D.L.; Bustamante, C. Experimental test of connector rotation during DNA packaging into bacteriophage phi29 capsids. *PLoS Biol.* **2007**, *5*, e59.

24. Baumann, R.G.; Mullaney, J.; Black, L.W. Portal fusion protein constraints on function in DNA packaging of bacteriophage T4. *Mol. Microbiol.* **2006**, *61*, 16-32.

25. Serwer, P. The source of energy for bacteriophage DNA packaging: an osmotic pump explains the data. *Biopolymers* **1988**, *27*, 165-169.

26. Serwer, P. Models of bacteriophage DNA packaging motors. J. Struct. Biol. **2003**, *141*, 179-188.

27. Alberts, B.; Miake-Lye, R. Unscrambling the puzzle of biological machines: the importance of the details, *Cell* **1992**, *68*, 415-420.

28. Howard, J. Motor proteins as nanomachines: The roles of thermal fluctuations in generating force and motion. *Séminaire Poincaré* **2009**, *12*, 33-44.

29. Mickler, M.; Schleiff, E.; Hügel, T. From biological towards artificial molecular motors. *Chemphyschem.* **2008**, *9*, 1503-1509.

30. Serwer, P.; Wright, E.T.; Hakala, K.; Weintraub, S.T.; Su, M.; Jiang, W. DNA packaging-associated hyper-capsid expansion of bacteriophage T3. *J. Mol. Biol.* **2010**, *397*, 361-374.

31. Fuller, D.N.; Raymer, D.M.; Rickgauer, J.P.; Robertson, R.M.; Catalano, C.E.; Anderson, D.L.; Grimes, S.; Smith, D.E. Measurements of single DNA molecule packaging dynamics in bacteriophage lambda reveal high forces, high motor processivity, and capsid transformations. *J. Mol. Biol.* **2007**, *373*, 1113-1122.

32. Yang, Q.; Maluf, N.K.; Catalano, C.E. Packaging of a unit-length viral genome: the role of nucleotides and the gpD decoration protein in stable nucleocapsid assembly in bacteriophage lambda. *J. Mol. Biol.* **2008**, *383*, 1037-1048.

33. JConway, J.F.; Cheng, N.; Ross, P.D.; Hendrix, R.W.; Duda, R.L.; Steven, A.C. A thermally induced phase transition in a viral capsid transforms the hexamers, leaving the pentamers unchanged. *J. Struct. Biol.* **2007**, *158,* 224-232.

34. Gertsman, I.; Komives, E.A.; Johnson, J.E. HK97 maturation studied by crystallography and H/2H exchange reveals the structural basis for exothermic particle transitions *J. Mol. Biol.* **2010**, *397*, 560-574.

35. Tuma, R.; Prevelige, P.E., Jr.; Thomas, G.J., Jr. Mechanism of capsid maturation in a double-stranded DNA virus. *Proc. Natl. Acad. Sci. U. S. A.* **1998**, *95*, 9885–9890.

36. Lander, G.C.; Evilevitch, A.; Jeembaeva, M.; Potter, C.S.; Carragher, B.; Johnson, J.E. Bacteriophage lambda stabilization by auxiliary protein gpD: timing, location, and mechanism of attachment determined by cryo-EM. *Structure* **2008**, *16*, 1399-1406.

37. Agirrezabala, X.; Martin-Benito, J.; Valle, M.; Gonzalez, J.M.; Valencia, A.; J. Valpuesta, J.M.; Carrascosa, J.L. Structure of the connector of bacteriophage T7 at 8A resolution: structural homologies of a basic component of a DNA translocating machinery. *J. Mol. Biol.* **2005**, *347*, 895-902.

38. Lebedev, A.A.; Krause, M.H.; Isidro, A.L.; Vagin, A.A.; Orlova, E.V.; Turner, J.; Dodson, E.J.; Tavares, P.; Antson, A.A. Structural framework for DNA translocation via the viral portal protein. *EMBO J.* **2007**, *26*, 1984-1994.

39. Hardies, S.C.; Serwer, P. Alignment and structural modeling of DNA packaging ATPases: Utilization of a steric clamp to hold the DNA. *XXth Biennial Conference on Phage/Virus Assembly* 2007, Toronto, Canada.

40. Grimes, S.; Jardine, P.J.; Anderson, D. Bacteriophage phi 29 DNA packaging. *Adv. Virus Res.* **2002**, *58*, 255-294.

41. Guo, P.; Peterson, C.; Anderson, D. Prohead and DNA-gp3-dependent ATPase activity of the DNA packaging protein gp16 of bacteriophage phi 29. *J. Mol. Biol.* **1987**, *197*, 229-236.

42. Hamada, K.; Fujisawa, H.; Minagawa, T. Characterization of ATPase activity of a defined *in vitro* system for packaging of bacteriophage T3 DNA. *Virology* **1987**, *159*, 244-249.

43. Yanagida, T. Fluctuation as a tool of biological molecular machines. *Biosystems* **2008**, *93*, 3-7.

44. Verbrugge, S.; Lechner, B.; Woehlke, G.; Peterman, E.J. Alternating-site mechanism of kinesin-1 characterized by single-molecule FRET using fluorescent ATP analogues. *Biophys. J.* **2009**, *97*, 173-182.

45. Petrov, A.S.; Lim-Hing, K.; Harvey, S.C. Packaging of DNA by bacteriophage epsilon15: structure, forces, and thermodynamics. *Structure* **2007**, *15*, 807-812.

46. Fang, P.A.; Wright, E.T.; Weintraub, S.T.; Hakala, K.; Wu, W.; Serwer, P.; Jiang, W. Visualization of bacteriophage T3 capsids with DNA incompletely packaged *in vivo*. *J. Mol. Biol.* **2008**, *384*, 1384-1399.

47. Smith, D.E.; Tans, S.J.; Smith, S.B.; Grimes, S.; Anderson, D.L.; Bustamante, C. The bacteriophage straight phi29 portal motor can package DNA against a large internal force. *Nature* **2001**, *413*, 748-752.

48. Rickgauer, J.P.; Fuller, D.N.; Grimes, S.; Jardine, P.J.; Anderson, D.L.; Smith, D.E. Portal motor velocity and internal force resisting viral DNA packaging in bacteriophage phi29. *Biophys. J.* **2008**, *94*, 159-167.

49. Fuller, D.N.; Raymer, D,M,; Kottadiel, V.I.; Rao, V.B.; Smith, D.E. Single phage T4 DNA packaging motors exhibit large force generation, high velocity, and dynamic variability. *Proc. Natl. Acad. Sci. U. S. A.* **2007**, *104*, 16868-16873.

50. Koch, A.L. Shrinkage of growing *Escherichia coli* cells by osmotic challenge. *J. Bact.* **1984**, *159*, 919-924.

51. Ray, K.; Sabanayagam, C.R.; Lakowicz, J.R.; Black, L.W. DNA crunching by a viral packaging motor: Compression of a procapsid-portal stalled Y-DNA substrate. *Virology* **2010**, *398*, 224-232.

52. Moffitt, J.R.; Chemla, Y.R.; Aathavan, K.; Grimes, S.; Jardine, P.J.; Anderson, D.L.; Bustamante, C. Intersubunit coordination in a homomeric ring ATPase. *Nature* **2009**, *457*, 446-450.

53. Chemla, Y.R.; Aathavan, K.; Michaelis, J.; Grimes, S.; Jardine, P.J.; Anderson, D.L.; Bustamante, C. Mechanism of force generation of a viral DNA packaging motor. *Cell* **2005**, *122*, 683-692.

54. Gope, R.; Serwer, P. Bacteriophage P22 *in vitro* DNA packaging monitored by agarose gel electrophoresis: rate of DNA entry into capsids. *J. Virol.* **1983**, *47*, 96-105.

55. Son, M.; Hayes, S.J.; Serwer, P. Optimization of the *in vitro* packaging efficiency of bacteriophage T7 DNA: effects of neutral polymers. *Gene* **1989**, *82*, 321-325.

56. Shibata, H.; Fujisawa, H.; Minagawa, T. Characterization of the bacteriophage T3 DNA packaging reaction *in vitro* in a defined system. *J. Mol. Biol.* **1987**, *196*, 845-851.

57. Oliveira, L.; Alonso, J.C.; Tavares, P. A defined *in vitro* system for DNA packaging by the bacteriophage SPP1: insights into the headful packaging mechanism. *J. Mol. Biol.* **2005**, *353*, 529-539.

58. Morais, M.C.; Koti, J.S.; Bowman, V.D.; Reyes-Aldrete, E.; Anderson, D.L.; Rossmann, M.G. Defining molecular and domain boundaries in the bacteriophage phi29 DNA packaging motor. *Structure* **2008**, *16*, 1267-1274.

59. Cuervo, A.; Vaney, M.C.; Antson, A.A.; Tavares, P.; Oliveira, L. Structural rearrangements between portal protein subunits are essential for viral DNA translocation. *J. Biol. Chem.* **2007**, *282*, 18907-18913.

60. Oliveira, L.; Henriques, A.O.; Tavares, P. Modulation of the viral ATPase activity by the portal protein correlates with DNA packaging efficiency. *J. Biol. Chem.* **2006**, *281*, 21914-21923.

61. Michaelis, J.; Muschielok, A.; Adrecka, J.; Kügel, W.; Moffitt, J.R. DNA based molecular motors. *Phys. Life Rev.* **2009**, *6*, 250-266.

62. Fink, T.M.; Ball, R.C. How many conformations can a protein remember?, *Phys. Rev. Lett.* **2001**, *87*, 198103.

63. Atz, R.; Ma, S.; Aathavan, K.; Bustamante, C.; Gao, J.; Anderson, D.L.; Grimes, S. Role of φ29 connector channel domains in late phase DNA-gp3 packaging. *XXth Biennial Conference on Phage/Virus Assembly* 2007, Toronto, Canada.

64. Pearson, R.K.; Fox, M.S. Effects of DNA heterologies on bacteriophage lambda packaging. *Genetics* **1988**, *118*, 5-12.

65. Louie, D.; Serwer, P. Quantification of the effect of excluded volume on double-stranded DNA. *J. Mol. Biol.* **1994**, *242*, 547-558.

66. Higgins, C.F. Multiple molecular mechanisms for multidrug resistance transporters. *Nature* **2007**, *446*, 749-757.

67. Linton, K.J. Structure and function of ABC transporters. Physiology **2007**, *22*, 122-130.

68. Lin, Z.; Madan, D.; Rye, H.S. GroEL stimulates protein folding through forced unfolding. *Nature Struct. Mol. Biol.* **2008**, *15*, 303-311.

69. Horwich, A.L.; Fenton, W.A. Chaperonin-mediated protein folding: using a central cavity to kinetically assist polypeptide chain folding. *Q. Rev. Biophys.* **2009**, *42*, 83-116.

70. Jewett, A.I.; Shea, J.E. Reconciling theories of chaperonin accelerated folding with experimental evidence. *Cell Mol. Life Sci.* **2010**, *67*, 255-276.

71. Zaikin, A.; Garcia-Ojalvo, J.; Bascones, R.; Ullner, E.; Kurths, J. Doubly stochastic coherence via noise-induced symmetry in bistable neural models. *Phys. Rev. Lett.* **2003**, *90*, 030601.

72. Ray, K.; Oram, M.; Ma, J.; Black, L.W. Portal control of viral prohead expansion and DNA packaging. *Virology* **2009**, *391*, 44-50.

73. Fujisawa, H.; Kimura, M.; Hashimoto, C. *In vitro* cleavage of the concatemer joint of bacteriophage T3 DNA. *Virology* **1990**, *174*, 26-34.

74. Casjens, S.; Wyckoff, E.; Hayden, M.; Sampson, L.; Eppler, K.; Randall, S.; Moreno, E.T.; Serwer, P. Bacteriophage P22 portal protein is part of the gauge that regulates packing density of intravirion DNA. *J. Mol. Biol.* **1992**, *224*, 1055-1074.

75. Nemecek, D.; Lander, G.C.; Johnson, J.E.; Casjens, S.R.; Thomas, G.J., Jr. Assembly architecture and DNA binding of the bacteriophage P22 terminase small subunit. *J. Mol. Biol.* **2008**, *383*, 494-501.

76. Morais, M.C.; Tao, Y.; Olson, N.H.; Grimes, S.; Jardine, P.J.; Anderson, D.L.; Baker, T.S.; Rossmann, M.G. Cryoelectron-microscopy image reconstruction of symmetry mismatches in bacteriophage phi29. *J. Struct. Biol.* **2001**, *135*, 38-46.

77. Feynmann, R.P.; Leighton, R.B.; Sands, M.L. *The Feynman Lectures on Physics*, 2nd ed.; Addison-Wesley, Reading, MA, USA, 1963, Volume I, Chapter 46.

78. Fujisawa, H.; Morita, M. Phage DNA packaging. *Genes Cells* **1997**, *2*, 537-545.

79. Huxley, A.F. Muscle structure and theories of contraction. *Prog. Biophys. Biophys. Chem.* **1957**, *7*, 255-318.

80. Donnan, F.G. The mystery of Life, *Smithsonian Report for 1929*. 1929; pp. 309-321.

81. Ishii, Y.; Taniguchi, Y.; Iwaki, M.; Yanagida, T. Thermal fluctuations biased for directional motion in molecular motors. *BioSystems* **2008**, *93*, 34-38.

82. Iwaki, M.; Iwane, A.H.; Shimokawa, T.; Cooke, R.; Yanagida, T. Brownian search-and-catch mechanism for myosin-VI steps. *Nat. Chem. Biol.* **2009**, *5*, 403-405.

83. Sun, S.; Kondabagil, K.; Draper, B.; Alam, T.I.; Bowman, V.D.; Zhang, Z.; Hegde, S.; Fokine, A.; Rossmann, M.G.; Rao, V.B. The structure of the phage T4 DNA packaging motor suggests a mechanism dependent on electrostatic forces. *Cell* **2008**, *135*, 1251-1262.

84. Singh, K.P.; Ropars, G.; Brunel, M.; Le Floch, A. Lever-assisted two-noise stochastic resonance. *Phys. Rev. Lett.* **2003**, *90*, 073901.

85. Ropars, G.; Singh, K.P.; Brunel, M.; Dore, F.; Le Floch, A. The dual stochastic response of nonlinear systems. *Europhys. Lett.* **2004**, *68*, 755-761.

86. Beuron, F.; Maurizi, M.R.; Belnap, D.M.; Kocsis, E.; Booy, F.P.; Kessel, M.; Steven, A.C. At sixes and sevens: characterization of the symmetry mismatch of the ClpAP chaperone-assisted protease. *J. Struct. Biol.* **1998**, *123*. 248-259.

87. Bewley, M.C.; Graziano, V.; Griffin, K.; Flanagan, J.M. The asymmetry in the mature amino-terminus of ClpP facilitates a local symmetry match in ClpAP and ClpXP complexes. *J. Struct. Biol.* **2006**, *153*, 113-128.

88. Chen, C.; Sheng, S.; Shao, Z.; Guo, P. A dimer as a building block in assembling RNA. A hexamer that gears bacterial virus phi29 DNA-translocating machinery. *J. Biol. Chem.* **2000**, *275*, 17510-17516.

89. Klug, A. Architectural design of spherical viruses. *Nature* **1983**, *303*, 378-379.

90. Jiang, W.; Chang, J.; Jakana, P.; Weigele, J.; King, J.A.; Chiu, W. Structure of epsilon15 bacteriophage reveals genome organization and DNA packaging/injection apparatus. *Nature* **2006**, *439*, 612–616.

91. Johnson, J.E.; Chiu, W. DNA packaging and delivery machines in tailed bacteriophages. *Curr. Opin. Struct. Biol.* **2007**, *17*, 237-243.

92. Yu, J.; Moffitt, J.; Hetherington, C.L.; Bustamante, C.; and Oster, G. Mechanochemistry of a viral DNA packging motor. *J. Mol. Biol.* **2010**, *400*, 186-203.

93. Sun, M.; Louie, D.; Serwer, P. Single-event analysis of the packaging of bacteriophage T7 DNA concatemers *in vitro*. *Biophys. J.* **1999**, *77*, 1627-1637.

Specific Colon Cancer Cell Cytotoxicity Induced by Bacteriophage *E* Gene Expression under Transcriptional Control of Carcinoembryonic Antigen Promoter

Ana R. Rama [1,2,*], Rosa Hernandez [2,3,4], Gloria Perazzoli [2,3], Miguel Burgos [5], Consolación Melguizo [2,3,4], Celia Vélez [2,3,4] and Jose Prados [2,3,4]

[1]   Department of Health Science, University of Jaén, Jaén 23071, Spain

[2]   Institute of Biopathology and Regenerative Medicine (IBIMER), Granada 18100, Spain;
      E-Mails: r_faraya@hotmail.com (R.H.); gperazzoli@ugr.es (G.P.); melguizo@ugr.es (C.M.);
      mariaceliavelez@ugr.es (C.V.); jcprados@ugr.es (J.P.)

[3]   Biosanitary Institute of Granada (ibs.GRANADA), SAS-University of Granada, Granada 18071, Spain

[4]   Department of Human Anatomy and Embryology, School of Medicine, University of Granada,
      Granada 18012, Spain

[5]   Institute of Biotechnology and Department of Genetics, University of Granada, Granada 18071, Spain;
      E-Mail: mburgos@ugr.es

*   Author to whom correspondence should be addressed; E-Mail: arama@ujaen.es

Academic Editor: Sabrina Angelini

Abstract: Colorectal cancer is one of the most prevalent cancers in the world. Patients in advanced stages often develop metastases that require chemotherapy and usually show a poor response, have a low survival rate and develop considerable toxicity with adverse symptoms. Gene therapy may act as an adjuvant therapy in attempts to destroy the tumor without affecting normal host tissue. The bacteriophage *E* gene has demonstrated significant antitumor activity in several cancers, but without any tumor-specific activity. The use of tumor-specific promoters may help to direct the expression of therapeutic genes so they act against specific cancer cells. We used the carcinoembryonic antigen promoter (*CEA*) to direct *E* gene expression (*pCEA-E*) towards colon cancer cells. *pCEA-E* induced a high cell growth inhibition of human HTC-116 colon adenocarcinoma and mouse MC-38 colon cancer cells in comparison to normal human CCD18co colon cells, which have practically undetectable levels of CEA. In addition, *in vivo* analyses of mice bearing tumors induced using MC-38 cells showed a significant decrease in tumor volume after *pCEA-E* treatment and a low level of Ki-67 in relation to untreated tumors. These results suggest that the *CEA* promoter is an excellent candidate for directing E gene expression specifically toward colon cancer cells.

Keywords: carcinoma embryonic antigen; colorectal cancer; *E* gene; suicide gene therapy; promoter tissue specific

---

1. Introduction

Colon cancer, along with breast and lung cancer, is one of the most prevalent cancers in the world [1]. While in early stages, colon cancer is characterized by a good prognosis, in more advanced, metastatic stages, the five-year survival rate is only 10%. Approximately 25% of all colon cancer patients reach this stage and are principally treated with 5-fluorouracil (5-FU) alone or a combination of oxaliplatin (FOLFOX, a combo of oxaliplatin, 5-FU and leucovorin), irinotecan (FOLFIRI, a combo of irinotecan, 5-FU and leucovorin), angiogenesis inhibitors and/or epidermal growth factor receptor inhibitors [2]. However, the results from current treatments are poor and may be accompanied by tissue damage. In this context, gene therapy tries to modify or destroy the tumor cell uniquely from within, without causing damage to any other tissues. Recent studies have investigated several aspects of gene therapy related to cancer treatment; one of these approaches is suicide gene therapy [3], which may enhance the potential of the drugs typically used to treat cancer [4], including colon cancer [5,6].

Classic systems of suicide gene therapy rely on the administration of a prodrug. The prodrug is catalyzed by suicide enzymes to produce a toxic substance capable of inducing cancer cell death. The most representative enzyme of this therapeutic strategy, thymidine kinase (TK), has been assayed in clinical trials against gliomas [7], prostate cancer [8] and hepatocellular carcinoma [9], among others. However, the conversion of a non-toxic prodrug into toxic metabolites and the bioavailability of the activated drug severely limit the system's efficacy. These causes of treatment failure are currently overcome by using genes that encode for cytotoxic proteins, which have a direct antitumor action. Some of these genes are taken from non-eukaryotic organisms, such as viruses, bacteria and plants [4,10–12]. We have recently shown how the toxic *E* gene from the bacteriophage ϕX174, which codes for a 91-amino acid membrane protein with lytic function [6,13,14], significantly decreased colon cancer cell proliferation, inducing mitochondrial apoptosis. Analysis of the mechanism suggests the formation of a "transmembrane pore" through which the cell loses cytoplasmic content. Interestingly, this gene did not need a prodrug to induce cell death [15]. The use of tumor-specific promoters that are overexpressed in cancer could drive transcription of these proteins known to be selectively active in tumor cells, thus obtaining a therapeutic system with a more specifically localized activity. Recently, survivin promoter [16], human telomerase reverse transcriptase promoter [17] and epithelial cell adhesion molecule (EpCAM) promoter [18] have been assayed to delivery *TK* or *CD* (*cytosine deaminase*) genes in cancer cells.

Carcinoma embryonic antigen (CEA) is an oncofetal tumor marker, which is overexpressed in over 90% of colorectal cancer cells, but not in normal colon cells [19–21]. In addition, CEA can be detected in 70% of colorectal cancer diagnoses [22,23]. Michl *et al.* [24] discovered significantly elevated CEA serum concentrations in patients in the final stages of the pathology; hence, they used CEA as a prognosis marker. Shibutani *et al.* [22] corroborated the utility of CEA levels for predicting the prognosis and also for monitoring recurrence and metastasis after potentially curative surgery in patients with stage II colorectal cancer. Wang *et al.* [25] concluded that high levels of tissue mRNA expression and CEA serum are associated with the incidence and progression of colorectal cancer, while Patel *et al.* [26] used CEA as a clinical and pathologic prognostic marker of local recurrence and overall survival after resection. Thus, the *CEA* promoter has been used in gene therapy to direct the expression of therapeutic genes toward CEA-positive cancer cells [16]. In fact, Zhang *et al.* [27] demonstrated the selective expression, under the transcriptional control of the *CEA* promoter, of the cytosine deaminase (CD) enzyme in colon cancer cells.

The aim of this study was to investigate the activity of the *E* gene, a toxic gene for colon cancer cells, under *CEA* promoter transcriptional control, which is frequently overexpressed in this type of tumor cell. We analyzed different colon cancer cell lines in order to select those with differential CEA expression. Colon cancer cells were then transfected *in vitro* to assess the anti-proliferative effect of the *E* gene under the influence of the *CEA* promoter. We also analyzed the *in vivo* cytotoxicity of the *pCEA-E* gene to demonstrate its activity with respect to tumor growth and mouse survival rates. Furthermore, we analyzed the toxicity of this new gene therapy and its effect over proliferation rates.

2. Results

2.1. Transcriptional Activity of CEA Promoter

CEA activity was detected in all seven colon cancer cell lines by luciferase assay. Human HTC-116, CACO-2, RKO colon adenocarcinoma cell lines and mouse MC-38 colon cancer cells showed high levels of luciferase expression, whereas human HT-29, T-84 and SW480 adenocarcinoma cell lines showed a low degree of luciferase expression. Specifically, HTC-116 showed the highest level of luciferase among all of the colorectal cancer cells, while T-84 cells presented the least expression of cancer cells. In contrast, normal intestinal epithelial CCD18co cells demonstrated the lowest levels of *CEA* activity (Figure 1).

2.2. In Vitro Inhibition of Cell Growth by pCEA-E

To analyze the *E* gene antiproliferative effect under *CEA* promoter transcriptional control, we selected the colon cancer cell lines with the highest (HTC-116) and lowest (T-84) *CEA* promoter activity, as well as normal CCD18co cells, which presented practically null *CEA* promoter activity. Furthermore, we analyzed MC-38 cells in order to carry out *in vivo* experiments. As shown in Figure 2A, HTC-116 cells transfected with the *E* gene showed a significant and time-dependent decrease in growth. Cell growth inhibition was 15%, 31% and 48% at 24, 48 and 72 h, respectively, after transfection and in relation to control cells. Similar results were observed for the MC-38 colon cancer cells, which were also characterized by strong *CEA* promoter activity (Figure 2C). By contrast, T-84 cells with weak *CEA* promoter activity showed proliferation inhibition levels of only 4%, 8.5% and 13.5% at 24, 48 and 72 h after transfection, respectively. CCD118co colon cells with no *CEA* promoter activity showed an insignificant change in the % of proliferation (see "control").

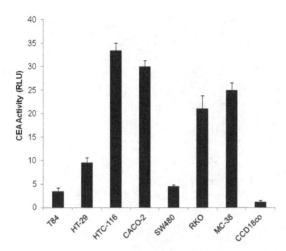

Figure 1. Transcriptional activity of CEA. Human CACO-2, HT29, HCT-116, SW480, RKO and T-84 colon adenocarcinoma cell lines, mice MC-38 colon cancer cell line, and the normal human colon CCD18co cell line Cells were co-transfected with luciferase expression vectors pPGL2/CEA or pGL2 and the Renilla expression vector CMV/Renilla. Luciferase activity of each transfection was normalized by the Renilla reading. Luciferase activity is represented by the ratio of the specific promoter over the activity of pGL2. Data represent the mean value of three replicates ± the standard error of the mean (SEM).

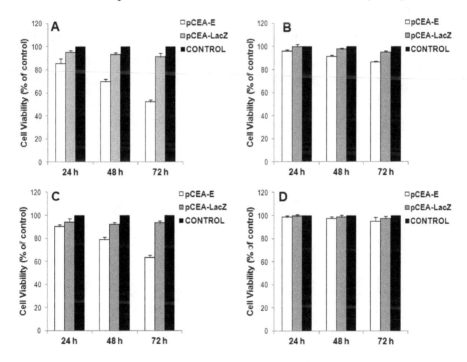

Figure 2. Effect of *pCEA-E* on cell proliferation. Cells from cell lines HTC-116 (**A**), T-84 (**B**), MC-38 (**C**) and CCD18co (**D**) were transfected with *pCEA-E* to determine proliferation rate modulation after 24, 48 and 72 h. Data represent the mean value of three replicates ± the standard error of the mean (SEM).

Expression of E protein in cells transfected with *pCEA-E* was confirmed using the anti-V5-FITC antibody. Positive staining was detected in HTC-116 and MC-38 cells, but not in T-84 and CCD118co cells (Figure 3).

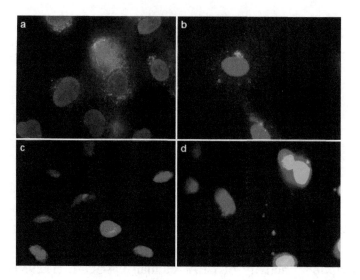

Figure 3. Subcellular localization of E/V5 fusion protein. Recombinant E protein (E-V5) was detected using anti-V5-FITC antibody (green) in HTC-116 (**a**) and MC-38 (**b**). No stain was observed in T-84 (**c**) and CCD18co (**d**) cells following *pCEA-E* treatment. Twenty-four hours after transfection, the fluorescence pattern was dotted and located in the cell cytoplasm. Cell nuclei were stained by DAPI (blue). Magnification: 40×.

2.3. In Vivo Tumor Growth Inhibition and Survival Analysis

As Figure 4 shows, intratumoral treatment with *pCEA-E* produced significant reductions in tumor volumes after 33 days. Specifically, *pCEA-E* was able to induce a 36% tumor volume reduction in comparison to the control ($p < 0.05$). *pCEA-LacZ* treatment, on the other hand, did not bring about any modifications in tumor growth rates, yielding similar results to those observed in the untreated mouse group. Nevertheless, although *pCEA-E* treatment produced reductions in tumor volumes, it failed to increase mouse survival rates compared to the untreated control group (data not shown).

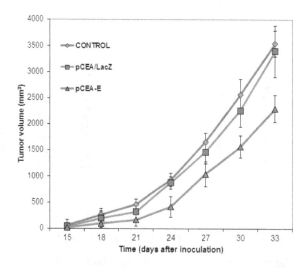

Figure 4. Tumor growth inhibition induced by *pCEA-E*. Treatment with *pCEA-E* induced a significant reduction in tumor volume at the end of the study period (33 days) in comparison with the growth of *pCEA-LacZ*-treated or untreated tumors (control) ($p < 0.05$). Data represent the mean value ± SEM.

2.4. In Vivo Toxicity

Mouse body mass was monitored throughout the treatment in order to determine the *in vivo* toxicity of the *pCEA-E* transfected gene. Both mice treated with *pCEA-E* or *pCEA-LacZ* revealed no significant weight loss in comparison to the control group, suggesting that neither of the plasmid administrations induced any toxicity (Figure 5).

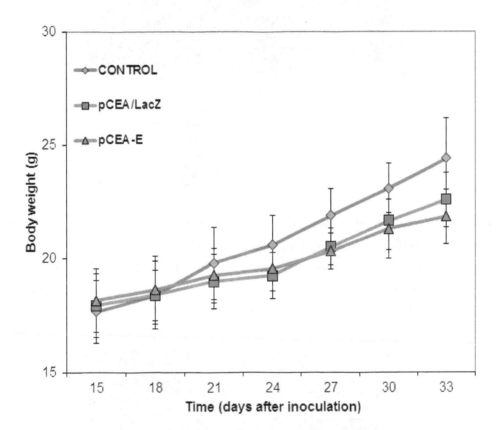

Figure 5. Weight evolution after *pCEA-E* treatment. The weights of mice bearing subcutaneous MC-38 tumors and treated with *pCEA-E* were measured. *pCEA-E*-treated mice experienced a similar weight evolution to that of *pCEA-LacZ*-treated and untreated mice (control). Data represent the mean value ± SEM.

2.5. Effect of pCEA-E on Cell Proliferation and Apoptosis

To determine the effect of *pCEA-E* on cell proliferation and apoptosis, we examined Ki-67 expression using immunohistochemical staining and detected apoptotic cells using the TUNEL assay. Our results showed that Ki-67 expression significantly decreased (55.4% of expression relative to the control tissue) in tumors that received *pCEA-E* treatment, suggesting significant inhibition of cell proliferation ($p < 0.05$) (Figure 6).

By contrast, the *pCEA-LacZ* treatment did not modulate Ki-67 expression, which was similar to the control group. On the other hand, analysis of the apoptosis-linked DNA fragmentation using a TUNEL assay showed similar staining between tumors treated with *pCEA-E* and the control tissue (Figure 7).

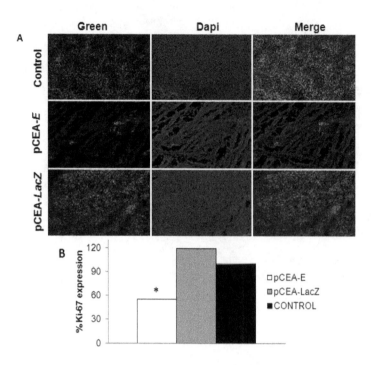

Figure 6. Immunohistochemical staining of Ki-67 expression in mice bearing subcutaneous MC-38 tumors treated with *pCEA-E*. (**A**) Images of Ki-67 expression (green) and cell nuclei stained with DAPI (blue). Original magnification: 10×; and (**B**) Ki-67 histogram protein expression profiles. Ki-67 expression was lower in the *pCEA-E*-treated group than in the *pCEA-LacZ* and control groups (* $p < 0.05$).

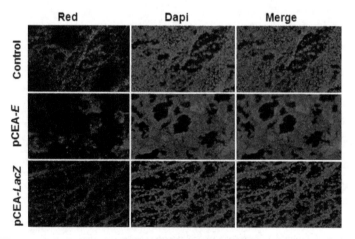

Figure 7. TUNEL assay in mice bearing subcutaneous MC-38 tumors. Images presenting apoptotic cells (red) and nuclei stained by DAPI (blue). *pCEA-E* treatment did not lead to an increased number of apoptotic cells in comparison with *pCEA-LacZ* or untreated tumors (control). Original magnification: 20×.

3. Discussion

Despite significant advances in the development of new therapies and improvements in traditional treatments for patients with advanced colon cancer, the prognosis and survival rate for these patients remains poor. New gene therapy strategies are being developed to treat cancers, but to date have only demonstrated low levels of *in vivo* efficacy in terms of gene delivery and tumor specificity [17,28].

The utility of tumor-specific promoters for gene targeted expression has been described in various cancers, such as melanoma, brain, lung, pancreas and colon [16,21,27,29,30]. Our study described the use of the *CEA* promoter to increase the colon cancer cell specificity of *E* gene expression and the subsequent tumor growth inhibition induced by expressing this cytotoxic gene.

We have previously shown that *E* gene expression can induce apoptosis, dilated mitochondria with disrupted cristae, caspase-3 activation and the release of cytochrome C into the cytoplasm of some cancer cells [5,6,13]. In this study, we analyzed CEA activity through luciferase expression in several cell lines. We found the highest expression among all colon cancer cells in HTC-116 cell lines (33.4%) and the lowest in T-84 cells (3.49%); whereas normal CCD18co cells revealed practically undetectable levels of CEA. These results for *CEA* promoter activity are similar to those described by Gou *et al.* [31] and Dąbrowska *et al.* [32], who also observed a greater *CEA* promoter activity in colon cancer cell lines than in non-tumor cancer cell lines using a luciferase assay.

Analysis of luciferase expression was correlated against the *E* gene's inhibitor effect and revealed a decrease in the proliferation according to HTC-116 > MC-38 > T-84 > CCD18co. Furthermore, *E* gene transfection induced a significant and time-dependent decrease in HTC-116 and MC-38 cell growth, as has occurred in other cancer cells lines [6,13,14]. In fact, the greatest level of growth inhibition was found in both HTC-116 cells and MC-38 cells 72 h after transfection (48% and 36.6%, respectively). By contrast, low levels of growth inhibition were detected in T-84 and CCD18co cells (13.6% and 5.1% at 72 h after transfection, respectively). Other experiments using suicide genes under *CEA* transcriptional expression control corroborated a specific growth inhibition in cells with high levels of CEA expression. In fact, Zhang *et al.* [27] studied the effect of the *CD* suicide gene under *CEA* promoter control in the LoVo human colon cancer cell line (CEA-positive) in comparison to HeLa cells (a CEA-negative human adenocarcinoma cell line). Both cell lines were transfected with CEA-*CD* and treated with 5-FC for five days. HeLa cells transfected with CEA-*CD* did not sensitize the cytotoxicity caused by 5-FC, whereas CEA-*CD* LoVo cells showed a 72.7% growth inhibition. A similar study by Liu *et al.* [33] revealed a greater inhibition of growth (89.8%) in a CEA-positive human gastric cancer cell line (SGC7901 cells) than in a HeLa cell line (2%); in this case, both lines were treated with a double suicide gene system (*TK* and *CD* genes*)* under CEA control.

To demonstrate the *in vivo* efficiency of our therapeutic system, we investigated the activity of the *CEA* promoter-*E* gene in mice bearing colon cancer tumors from MC-38 CEA-positive cells. Statistical evaluation of tumor growth rates obtained from mice treated with *pCEA-E* revealed a significant decrease (36%) of tumor volume in comparison to the control group. These results support the hypothesis that the *E* gene under *CEA* post-transcriptional control was able to reduce the proliferation rate of a tumor generated by CEA-positive cells. Studies by Zhang *et al.* [27] demonstrated a similar effect in LoVo mice xenografts treated with the *CD* gene under the *CEA* promoter. Liu *et al.* [33] studied a human xenograft gastric carcinoma during treatment with a double gene therapy system (*CD* and *TK* gene) for 36 days. A tumor growth inhibition of 54% was reported in treated tumors compared to control tumors. This greater decrease in tumor volume compared to our findings may be explained by the use of calcium phosphate nanoparticles (CPNP) that increase transfection efficiency and the use of a stable cell line to express therapeutic genes. Interestingly, analysis of Ki-67 showed a strong degree of expression in the untreated and *pCEA-LacZ* groups (controls) in relation to tumors intratumorally treated with the *pCEA-E* vector. The decreased level of Ki-67 expression in *pCEA-E* genes is representative of the colon cancer cells' loss of proliferative capacity. In fact, Okabe *et al.* [34] have demonstrated that the *CEA* promoter in an adenovirus expressing the *TK* gene was able to improve its antitumor activity in mice bearing colorectal tumors from RCM-1 CEA-positive cells and to reduce the number of liver metastases (after 42 days of treatment) in relation to an untreated group of mice.

We previously reported the successful use of the *E* gene in cancer therapy. However, a lack of selective antitumor activity is its main limitation. The use of tumor-specific promoters of cytotoxic genes (but which have a low activity in normal cells) may direct therapeutic gene expression toward cancer cells. In this study, we demonstrated that the *CEA* promoter can induce *E* gene expression, which is selective of colon cancer cells, while the gene is practically unexpressed in normal cell lines. *E* gene expression under *CEA* control rapidly and efficiently inhibited cell growth in CEA-positive colon cancer cells, which, in turn, induced a significant decrease in the tumor volume of mice bearing such cancer cells. In addition, a decrease in Ki-67 tumor expression indicated a significant decrease in post-treatment cell proliferation. In conclusion, we propose that the *CEA* promoter provides a means of specifically directing *E* gene expression toward colon cancer cells in order to mediate suicide gene therapy.

4. Experimental Section

4.1. Cell Culture

The human colon adenocarcinoma CACO-2, HT29, HCT-116, SW480, RKO and T-84 cell lines, the mice colon cancer MC-38 cell line (kindly provided by Jeffrey Schlom from Public Health Service, National Institutes of Health, Bethesda, MD, USA) and the normal human colon CCD18co cell line (Instrumentation Service Center, University of Granada, Granada, Spain) were used in this study. All cell types were grown in Dulbecco's Modified Eagle's Medium (DMEM) (Sigma, St. Louis, MO, USA), supplemented with 10% fetal bovine serum (FBS) and 1% streptomycin-penicillin (Sigma), under air containing 5% CO_2 and in an incubator at 37 °C.

4.2. Construction of Luciferase and E Expression Vectors

CEA promoter from human genomic DNA was amplified by PCR using the primers: *CTCGAG*CCA TCCACCTTGCCGAAA and *AAGCTT*GCTGTCTGCTCTGTCCTC. *XhoI* and *HindIII* restriction sites were introduced into the forward and reverse primers, respectively. The *CEA* promoter fragment and pGL2 vector (Promega Biotech Ibérica, Madrid, Spain) were digested with *XhoI* and *HindIII* restriction enzymes, and the *CEA* promoter fragment was finally cloned in the pGL2 vector (pGL2/CEA). We used a previously developed vector, pcDNA3.1/E (Rama *et al.*, 2010), to construct the E expression vector under the control of the *CEA* promoter. Firstly, the *CEA* promoter was amplified by PCR from pGL2/CEA using the primers *CAATTG*CCATCCACCTTGCCGAAA and *GAGCTC*GCTGTCTGCTCTGTCCTC with *MfeI* and *SacI* restriction sites. Secondly, the *pcDNA3.1/E CMV* promoter was removed by digestion with *MfeI* and *SacI*. Finally, CEA was digested with the same restriction enzymes and ligated in place of the *CMV* promoter, thus generating pcDNA3.1/CEA/E (*pCEA-E*). pcDNA3.1/CEA/*LacZ* (*pCEA-LacZ*) was used as a control and generated from pcDNA3.1/*LacZ* using a method similar to the one for *pCEA-E* [13].

Subcloning Efficiency™ DH5α™-competent *E. coli* cells (Qiagen, Barcelona, Spain) were transformed with the generated plasmids and their correct sequences confirmed by DNA sequencing.

4.3. Luciferase Assay

All transfections were performed using FUGENE6 reagent (Roche Diagnostic, Barcelona, Spain) according to the manufacturer's instructions. Cells (7×10^3) were seeded into 96-well culture plates. After 24 h, the cells were co-transfected with 0.2 µg/well of luciferase reporter vectors (pGL2/CEA or pGL2) and 0.05 µg/well of CMV/Renilla vector as internal controls for normalization of the transcriptional activity of the reporter vectors (provided by Miguel Burgos of the Biotechnology Institute, Granada, Spain). Experiments were performed in three groups: cells transfected with pGL2/CEA and CMV/Renilla; cells transfected with pGL2 and CMV/Renilla; and untransfected cells (control). Forty-eight hours after transfection, luciferase activity was determined using the Dual-Glo Luciferase Assay System (Promega) according to the manufacturer's instructions. The luminescence was measured in a luminometer (96 GloMax® microplate). The luciferase activity in each well was normalized to CMV/Renilla using the formula $Ln = L/R$ (*Ln*: Normalized luciferase activity; L: Luciferase activity reading; R: Renilla activity reading). The *Ln* was further standardized according to the transcriptional activity of pGL2 using the formula of $RLU = Ln/pGL3$-basic (RLU: Relative luciferase unit). All transfections were performed in triplicate.

4.4. In Vitro Proliferation Assay

To determine the rate of proliferation, transfection was performed using FUGENE6 reagent (Roche Diagnostic) as described previously. Cells were seeded in 24-well plates (10^4 cells per well) to analyze proliferation. After 24 h, the cells were transfected with 1 µg/well of the respective vector (*pCEA-E* or *pCEA-LacZ*). Untransfected cells were used as a control. After 24, 48 and 72 h, MTT (3-(4,5-dimethylthiazol-2-yl)-2,5-diphenyltetrazolium bromide) solution (5 mg/mL) was added to each well (20 µL) and incubated for 4 h at 37 °C. Two hundred microliters of dimethylsulfoxide (DMSO) were then added to each well after the medium had been removed. Optical density was determined using a Titertek multiscan colorimeter (Flow Laboratories, Irvine, CA, USA) at 570 and 690 nm. Cells transfected with empty vectors were used as controls.

4.5. Microscopic Analysis

pcDNA3.1 provides a V5 epitope tag for efficient detection of recombinant proteins. Therefore, E protein expression was confirmed using the anti-V5-FITC antibody (Invitrogen, Madrid, Spain). The cells were grown on coverslips and transfected with *pCEA-E* and *pCEA-LacZ*. As above, untransfected cells were used as a control. After 24 h of transfection, cells were washed with PBS, fixed in 100% methanol at room temperature for 3 min, blocked with 10% bovine serum albumin/PBS for 20 min and then incubated with anti-V5-FIFC antibody diluted (1:500) in 1% bovine serum albumin/PBS. DAPI solution (100 nM) (Invitrogen, Madrid, Spain) was used for nuclear staining. The cells were then rinsed with PBS, mounted and visualized using fluorescent microscopy analysis (Nikon Eclipse Ti, Nikon Instruments Inc., Melville, NY, USA). V5 was excited at 488 nm and DAPI nuclear stain at 364 nm.

4.6. In Vivo Study

Female C57BL/6 mice (Scientific Instrumentation Centre, University of Granada, Granada, Spain) were used in the *in vivo* studies. All mice (body weight: 25–30 g) were kept in a laminar airflow cabinet located in an ambient-controlled room (37.0 ± 0.5 °C and a relative humidity of 40%–70%) and subjected to a 12-h day/night cycle under specific pathogen-free conditions. All studies were approved by the Ethics Committee of the Medical School at the University of Granada and complied with international standards (European Communities Council Directive 86/609). Tumors were induced by subcutaneous injection of 7×10^5 MC-38 cells in 200 µL of PBS into the left flanks of mice. Tumors were allowed to grow to 75 mm³ (a minimal size for ideal intratumoral injection) before starting intratumoral treatment. Mice were randomly assigned to the following groups: treated with *pCEA-E*, treated with *pCEA-LacZ* and untreated (control). *In vivo* JetPEI (Polyplus-transfection Inc., New York, NY, USA) was used as a transfection enhancer reagent according to the manufacturer's instructions. PEI/DNA complexes with a ratio of 1:6 were formed for 15 µg DNA/50 µL 10% glucose plus 1.8 µL *in vivo*-JetPEI/50 µL 10% glucose. Each mouse received intratumoral injections every three days, up to a maximum total of six administrations. Mice weights and deaths were recorded throughout this period. Tumor volumes (V, mm³) were estimated by using a digital caliper to measure the largest diameter "a" and the second largest diameter "b" perpendicular to "a", then calculating the volume from, $V = ab^2 \times \pi/6$.

4.7. Histological Studies

Tumors were cryopreserved at −80 °C then cut into 15 µm-thick cryostat sections, collected on gelatin-coated slides and stored at −20 °C until used. The presence of apoptotic cells within the tumor sections was evaluated after 20 min of fixation with 4% paraformaldehyde at room temperature using the TUNEL technique with the In Situ Cell Death Detection Kit TMR red (Roche, Mannheim, Germany) according to the manufacturer's recommendations. Cell nuclei were counterstained with DAPI and fluorescence images captured using a Nikon eclipse 50i microscope. Sections were probed with Ki-67(M-19) antibody (Santa Cruz Biotechnology Inc., Heidelberg, Germany) in order to evaluate proliferation. Tissue sections were fixed for 20 min with 4% paraformaldehyde at room temperature. The sections were then washed with PBS and blocked for 1 h with donkey serum before being incubated for 1 h at room temperature with the primary antibody (1:100). The sections were then washed with PBS and incubated at room temperature with an Alexa 488 anti-donkey secondary antibody (1:500) (Invitrogen, Madrid, Spain) for 1 h. Cell nuclei were again counterstained with DAPI and fluorescence images captured with a Nikon eclipse 50i microscope. Representatively, stained areas on all slides were digitally imaged, and TUNEL and Ki-67 protein expressions were quantified using ImageJ software plugins.

4.8. Statistical Analysis

All of the results were represented as the mean ± standard deviation (SEM). Statistical analysis was performed using the Student's *t*-test (SPSS version 15, SPSS, Chicago, IL, USA). The probability of mice survival was determined by the Kaplan–Meier method, and the log-rank test was used to compare the fraction of surviving mice between groups ($\alpha = 0.05$). Data with $p < 0.05$ and $p < 0.001$ were considered as significant and very significant, respectively.

5. Conclusions

New gene therapy strategies are being developed to treat cancers. The use of tumor-specific promoters are being developed may help to direct the expression of therapeutic genes so they act against specific cancer cells. We have proved that carcinoma embryonic antigen is an excellent tumor-specific promoter to direct *E* gene expression towards colon cancer cells but not to normal colon cells, inducing cell growth inhibition, Ki-67 expression reduced and decrease of tumor volume. We propose the system of the *E* gene under *CEA* post-transcriptional as a novel gene therapy strategy for the treatment of colorectal cancer.

Acknowledgments

This research was funded by FEDER, Plan Nacional de Investigación Científica, Desarrollo e Innovación Tecnológica (I+D+I), Instituto de Salud Carlos III- Fondo de Investigaciones Sanitarias (FIS) through Projects PI11/01862 and PI11/0257.

Author Contributions

Ana R. Rama performed the genetic construction and luciferase analysis and wrote the paper; Rosa Hernandez performed proliferation assay and statistical analysis and wrote the paper; Gloria Perazzoli performed histological studies; Miguel Burgos performed the genetic construction; Consolación Melguizo performed *in vivo* study; Celia Vélez revised the paper; Jose Prados supervised the study and revised the paper.

References

1. Labianca, R.; Beretta, G.D.; Kildani, B.; Milesi, L.; Merlin, F.; Mosconi, S.; Pessi, M.A.; Prochilo, T.; Quadri, A.; Gatta, G.; *et al.* Colon cancer. *Crit. Rev. Oncol. Hematol.* **2010**, *74*, 106–133.
2. Van Cutsem, E.; Kohne, C.H.; Hitre, E.; Zaluski, J.; Chang Chien, C.R.; Makhson, A.; D'Haens, G.; Pinter, T.; Lim, R.; Bodoky, G.; *et al.* Cetuximab and chemotherapy as initial treatment for metastatic colorectal cancer. *N. Engl. J. Med.* **2009**, *360*, 1408–1417.
3. Cao, S.; Cripps, A.; Wei, M.Q. New strategies for cancer gene therapy: Progress and opportunities. *Clin. Exp. Pharmacol. Physiol.* **2010**, *37*, 108–114.
4. Prados, J.; Melguizo, C.; Rama, A.R.; Ortiz, R.; Segura, A.; Boulaiz, H.; Velez, C.; Caba, O.; Ramos, J.L.; Aranega, A. *Gef* gene therapy enhances the therapeutic efficacy of doxorubicin to combat growth of MCF-7 breast cancer cells. *Cancer Chemother. Pharmacol.* **2010**, *66*, 69–78.
5. Ortiz, R.; Prados, J.; Melguizo, C.; Arias, J.L.; Ruiz, M.A.; Alvarez, P.J.; Caba, O.; Luque, R.; Segura, A.; Aranega, A. 5-Fluorouracil-loaded poly(ε-caprolactone) nanoparticles combined with phage *E* gene therapy as a new strategy against colon cancer. *Int. J. Nanomed.* **2012**, *7*, 95–107.
6. Rama, A.R.; Prados, J.; Melguizo, C.; Alvarez, P.J.; Ortiz, R.; Madeddu, R.; Aranega, A. *E* phage gene transfection associated to chemotherapeutic agents increases apoptosis in lung and colon cancer cells. *Bioeng. Bugs* **2011**, *2*, 163–167.

7. Zhao, F.; Tian, J.; An, L.; Yang, K. Prognostic utility of gene therapy with herpes simplex virus thymidine kinase for patients with high-grade malignant gliomas: A systematic review and meta analysis. *J. Neuro Oncol.* **2014**, *118*, 239–246.

8. Nasu, Y.; Saika, T.; Ebara, S.; Kusaka, N.; Kaku, H.; Abarzua, F.; Manabe, D.; Thompson, T.C.; Kumon, H. Suicide gene therapy with adenoviral delivery of *HSV-tK* gene for patients with local recurrence of prostate cancer after hormonal therapy. *Mol. Ther.* **2007**, *15*, 834–840.

9. Sangro, B.; Mazzolini, G.; Ruiz, M.; Ruiz, J.; Quiroga, J.; Herrero, I.; Qian, C.; Benito, A.; Olague, C.; Larrache, J.; *et al.* A phase I clinical trial of thymidine kinase-based gene therapy in advanced hepatocellular carcinoma. *Cancer Gene Ther.* **2010**, *17*, 837–843.

10. Amit, D.; Hochberg, A. Development of targeted therapy for a broad spectrum of cancers (pancreatic cancer, ovarian cancer, glioblastoma and HCC) mediated by a double promoter plasmid expressing diphtheria toxin under the control of H19 and IGF2-P4 regulatory sequences. *Int. J. Clin. Exp. Med.* **2012**, *5*, 296–305.

11. Thakur, M.; Mergel, K.; Weng, A.; von Mallinckrodt, B.; Gilabert-Oriol, R.; Durkop, H.; Fuchs, H.; Melzig, M.F. Targeted tumor therapy by epidermal growth factor appended toxin and purified saponin: An evaluation of toxicity and therapeutic potential in syngeneic tumor bearing mice. *Mol. Oncol.* **2013**, *7*, 475–483.

12. Wang, C.; Zhang, Y. *Apoptin* gene transfer via modified wheat histone H4 facilitates apoptosis of human ovarian cancer cells. *Cancer Biother. Radiopharm.* **2011**, *26*, 121–126.

13. Rama, A.R.; Prados, J.; Melguizo, C.; Ortiz, R.; Alvarez, P.J.; Rodriguez-Serrano, F.; Hita, F.; Ramos, J.L.; Burgos, M.; Aranega, A. *E* phage gene transfection enhances sensitivity of lung and colon cancer cells to chemotherapeutic agents. *Int. J. Oncol.* **2010**, *37*, 1503–1514.

14. Ortiz, R.; Prados, J.; Melguizo, C.; Rama, A.R.; Segura, A.; Rodriguez-Serrano, F.; Boulaiz, H.; Hita, F.; Martinez-Amat, A.; Madeddu, R.; *et al.* The cytotoxic activity of the phage E protein suppress the growth of murine B16 melanomas *in vitro* and *in vivo*. *J. Mol. Med. Berl.* **2009**, *87*, 899–911.

15. Witte, A.; Wanner, G.; Blasi, U.; Halfmann, G.; Szostak, M.; Lubitz, W. Endogenous transmembrane tunnel formation mediated by phi X174 lysis protein E. *J. Bacteriol.* **1990**, *172*, 4109–4114.

16. Qiu, Y.; Peng, G.L.; Liu, Q.C.; Li, F.L.; Zou, X.S.; He, J.X. Selective killing of lung cancer cells using carcinoembryonic antigen promoter and double suicide genes, thymidine kinase and cytosine deaminase (*pCEA*-TK/CD). *Cancer Lett.* **2012**, *316*, 31–38.

17. Higashi, K.; Hazama, S.; Araki, A.; Yoshimura, K.; Iizuka, N.; Yoshino, S.; Noma, T.; Oka, M. A novel cancer vaccine strategy with combined IL-18 and HSV-TK gene therapy driven by the hTERT promoter in a murine colorectal cancer model. *Int. J. Oncol.* **2014**, *45*, 1412–1420.

18. Danda, R.; Krishnan, G.; Ganapathy, K.; Krishnan, U.M.; Vikas, K.; Elchuri, S.; Chatterjee, N.; Krishnakumar, S. Targeted expression of suicide gene by tissue-specific promoter and microRNA regulation for cancer gene therapy. *PLoS ONE* **2013**, *8*, e83398.

19. Long, H.; Li, Q.; Wang, Y.; Liu, T.; Peng, J. Effective combination gene therapy using CEACAM6-shRNA and the fusion suicide gene yCDglyTK for pancreatic carcinoma *in vitro*. *Exp. Ther. Med.* **2013**, *5*, 155–161.

20. Zhou, X.; Xie, G.; Wang, S.; Wang, Y.; Zhang, K.; Zheng, S.; Chu, L.; Xiao, L.; Yu, Y.; Zhang, Y.; *et al.* Potent and specific antitumor effect for colorectal cancer by CEA and Rb double regulated oncolytic adenovirus harboring *ST13* gene. *PLoS ONE* **2012**, *7*, e47566.

21. Xu, C.; Sun, Y.; Wang, Y.; Yan, Y.; Shi, Z.; Chen, L.; Lin, H.; Lu, S.; Zhu, M.; Su, C.; *et al.* *CEA* promoter-regulated oncolytic adenovirus-mediated Hsp70 expression in immune gene therapy for pancreatic cancer. *Cancer Lett.* **2012**, *319*, 154–163.

22. Shibutani, M.; Maeda, K.; Nagahara, H.; Ohtani, H.; Sakurai, K.; Toyokawa, T.; Kubo, N.; Tanaka, H.; Muguruma, K.; Ohira, M.; *et al.* Significance of CEA and CA19-9 combination as a prognostic indicator and for recurrence monitoring in patients with stage II colorectal cancer. *Anticancer Res.* **2014**, *34*, 3753–3758.

23. Vukobrat-Bijedic, Z.; Husic-Selimovic, A.; Sofic, A.; Bijedic, N.; Bjelogrlic, I.; Gogov, B.; Mehmedovic, A. Cancer antigens (CEA and CA 19-9) as markers of advanced stage of colorectal carcinoma. *Med. Arch.* **2013**, *67*, 397–401.

24. Michl, M.; Koch, J.; Laubender, R.P.; Modest, D.P.; Giessen, C.; Schulz, C.; Heinemann, V. Tumor markers CEA and CA 19-9 correlate with radiological imaging in metastatic colorectal cancer patients receiving first-line chemotherapy. *Tumour Biol.* **2014**, *35*, 10121–10127.

25. Wang, W.; Li, Y.; Zhang, X.; Jing, J.; Zhao, X.; Wang, Y.; Han, C. Evaluating the significance of expression of CEA mRNA and levels of CEA and its related proteins in colorectal cancer patients. *J. Surg. Oncol.* **2014**, *109*, 440–444.

26. Patel, M.M. Getting into the colon: Approaches to target colorectal cancer. *Expert Opin. Drug Deliv.* **2014**, *11*, 1343–1350.

27. Zhang, G.; Liu, T.; Chen, Y.H.; Chen, Y.; Xu, M.; Peng, J.; Yu, S.; Yuan, J.; Zhang, X. Tissue specific cytotoxicity of colon cancer cells mediated by nanoparticle-delivered suicide gene *in vitro* and *in vivo*. *Clin. Cancer Res.* **2009**, *15*, 201–207.

28. Deng, L.Y.; Wang, J.P.; Gui, Z.F.; Shen, L.Z. Antitumor activity of mutant bacterial cytosine deaminase gene for colon cancer. *World J. Gastroenterol.* **2011**, *17*, 2958–2964.

29. Yawata, T.; Maeda, Y.; Okiku, M.; Ishida, E.; Ikenaka, K.; Shimizu, K. Identification and functional characterization of glioma-specific promoters and their application in suicide gene therapy. *J. Neuro Oncol.* **2011**, *104*, 497–507.

30. Kagiava, A.; Sargiannidou, I.; Bashiardes, S.; Richter, J.; Schiza, N.; Christodoulou, C.; Gritti, A.; Kleopa, K.A. Gene delivery targeted to oligodendrocytes using a lentiviral vector. *J. Gene Med.* **2014**, *16*, 364–373.

31. Guo, X.; Evans, T.R.; Somanath, S.; Armesilla, A.L.; Darling, J.L.; Schatzlein, A.; Cassidy, J.; Wang, W. *In vitro* evaluation of cancer-specific NF-κB-CEA enhancer-promoter system for 5-fluorouracil prodrug gene therapy in colon cancer cell lines. *Br. J. Cancer* **2007**, *97*, 745–754.

32. Dabrowska, A.; Szary, J.; Kowalczuk, M.; Szala, S.; Ugorski, M. CEA-negative glioblastoma and melanoma cells are sensitive to cytosine deaminase/5-fluorocytosine therapy directed by the carcinoembryonic antigen promoter. *Acta Biochim. Pol.* **2004**, *51*, 723–732.

33. Liu, T.; Zhang, G.; Chen, Y.H.; Chen, Y.; Liu, X.; Peng, J.; Xu, M.H.; Yuan, J.W. Tissue specific expression of suicide genes delivered by nanoparticles inhibits gastric carcinoma growth. *Cancer Biol. Ther.* **2006**, *5*, 1683–1690.

Structural and Biochemical Investigation of Bacteriophage N4-Encoded RNA Polymerases

Bryan R. Lenneman [1,*] and Lucia B. Rothman-Denes [1,2,*]

[1] Committee on Genetics, Genomics, and Systems Biology, The University of Chicago, 920 East 58th Street, Chicago, IL 60637, USA

[2] Department of Molecular Genetics and Cell Biology, The University of Chicago, 920 East 58th Street, Chicago, IL 60637, USA

* Authors to whom correspondence should be addressed;
E-Mails: blenneman@uchicago.edu (B.R.L.); lbrd@uchicago.edu (L.B.R.-D.)

Academic Editors: Sivaramesh Wigneshwararaj and Deborah M. Hinton

Abstract: Bacteriophage N4 regulates the temporal expression of its genome through the activity of three distinct RNA polymerases (RNAP). Expression of the early genes is carried out by a phage-encoded, virion-encapsidated RNAP (vRNAP) that is injected into the host at the onset of infection and transcribes the early genes. These encode the components of new transcriptional machinery (N4 RNAPII and cofactors) responsible for the synthesis of middle RNAs. Both N4 RNAPs belong to the T7-like "single-subunit" family of polymerases. Herein, we describe their mechanisms of promoter recognition, regulation, and roles in the phage life cycle.

Keywords: bacteriophage N4; single-subunit polymerases; N4 virion RNA polymerase; N4 RNAPII

1. Introduction

Regulation at the transcriptional level is the primary means used by bacteriophage to progress through distinct developmental stages during infection. Gene products expressed immediately after infection are primarily involved in the takeover of essential host processes. Predominantly, phage utilize the host RNAP to recognize the phage early promoters; subsequently, a product of phage early transcription either modifies the properties of the host RNAP to overcome transcription termination signals (*i.e.*, λN, λQ, HKO22 Put RNA), redirects the host RNAP to middle promoters through sigma factor remodeling (*i.e.*, T4 AsiA-MotA) or replaces the host vegetative sigma factor by phage-encoded sigma factors that direct the transcription of middle and late genes (*i.e.*, SPO1 gp28, gp34) [1–3]. Alternatively, host RNAP transcription of the phage early genes results in the synthesis of a phage early gene product (*i.e.*, ø29 gp4) that activates the host RNAP to utilize the phage weak late promoter while inhibiting the transcription of the phage early genes, eliciting the transition from early to late transcription [4]. In all above-mentioned cases, the host RNAP core enzyme is essential throughout the phage growth cycle. Host RNAP transcription of the early genes of other phage leads to the synthesis of host RNAP inhibitors (*i.e.*, T7 gp2, *Xanthomonas oryzae* phage Xp10 P7) and of a phage-encoded RNAP whereby the phage becomes transcriptionally independent of the host (*i.e.*, T7 RNAP, Xp10 RNAP) [5–8]. Middle transcription then commences, focusing largely on the synthesis of proteins involved in phage genome replication. Genome replication is followed by late gene transcription, which includes the production of both morphogenetic proteins, involved in virion assembly and DNA packaging, and of proteins required for host cell lysis. In contrast, the recently described *Pseudomonas aeruginosa* giant bacteriophage øKZ provides a unique example of a transcriptional strategy that is completely independent of the host. Two sets of phage-encoded polypeptides that have homology to the β and β′ subunits of bacterial RNAPs have been identified, with one set present in virions. Three classes of putative promoter sequences have been identified upstream of genes transcribed at early, middle, and late times during infection. The role of these polypeptides and sequences in transcription of the phage genome has not yet been confirmed [9].

The *Escherichia coli* K-12 strain-specific bacteriophage N4, isolated from the sewers of Genoa, has evolved a unique and "reversed" transcriptional strategy (Figure 1). N4 establishes transcriptional independence of the host immediately upon infection through the injection of a virion-encapsidated RNAP that transcribes the phage early genes. This independence is maintained through middle transcription catalyzed by a second phage-encoded RNAP and its cofactors, which transcribe genes encoding phage replicative functions. In contrast to the previous transcriptional strategies described above, late N4 transcription is carried out by the host σ^{70}-RNAP directed to the late promoters by N4 single-stranded DNA-binding protein (N4SSB), essential for N4 DNA replication.

2. N4 Transcriptional Architecture

2.1. N4 vRNAP Synthesizes N4 Early RNAs

Unlike most phage, where the host RNAP holoenzyme is responsible for the synthesis of their early mRNAs, a burst of RNA synthesis is observed immediately after N4 infection under conditions where the host RNAP is inhibited [10]. This transcription occurs even under conditions where post-infection protein synthesis is inhibited, suggesting the existence of a novel activity and leading us to postulate the presence of a RNAP in N4 virions [10,11]. This hypothesis was confirmed by detecting a RNAP activity in extracts from purified N4 virions disrupted through a denaturing protocol [12,13]. A single polypeptide of molecular weight 320 kDa was purified to homogeneity [13]. *In vitro* analysis showed that the protein requires the four ribonucleotide triphosphates, Mg^{2+}, and denatured N4 DNA as a template for RNA synthesis [12–14]. vRNAP was completely inactive on native N4 DNA, but transcribed denatured N4 DNA efficiently and, most notably, with *in vivo* specificity. In contrast with other RNAPs, N4 vRNAP cannot use other denatured DNA templates (T4, T7, salmon sperm) [12,14–16].

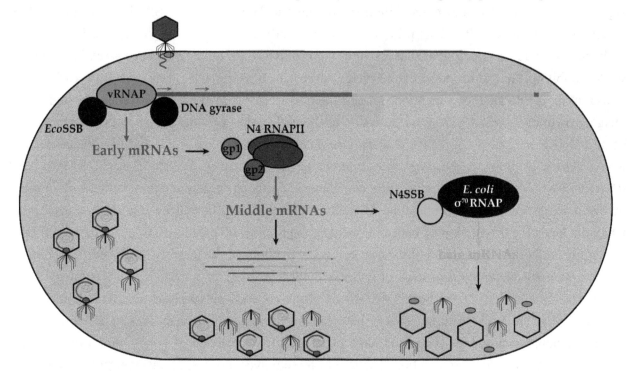

Figure 1. Schematic of transcriptional program controlling the N4 life cycle. Upon interaction with its receptor, NfrA, N4 injects vRNAP and genomic DNA into the host. Host DNA gyrase introduces negative supercoils into the phage genome, driving early promoter hairpin extrusion. *Eco*SSB stabilizes the hairpin structures, allowing vRNAP to initiate transcription from early promoters to transcribe early genes required for middle transcription and transport the genome into the host cytoplasm. The early gene products gp1 and gp2 act as cofactors for the heterodimeric polymerase N4 RNAPII (gp15 and gp16) to carry out middle transcription. These transcripts encode functions required for N4 DNA replication. One such protein, N4SSB, interacts with host σ^{70}-RNAP, redirecting it to late promoters. The late genes encode morphogenetic proteins involved in virion assembly, DNA and vRNAP packaging, and host lysis. Host proteins depicted in black, phage-encoded early transcriptional machinery in blue, phage-encoded middle transcriptional machinery in shades of red, and phage-encoded late transcriptional machinery in green. Arrows indicate polarity of transcription.

Given the inactivity of vRNAP on native N4 DNA and its preference for single-stranded DNA, we proposed that the structure of phage DNA might become modified upon its injection into the host, rendering it competent for transcription by vRNAP [12,14]. Host DNA gyrase, which introduces negative superhelical turns into circular DNA, was tested for its involvement in N4 early RNA synthesis [17]. Treatment of cells prior to N4 infection with coumermycin, an inhibitor of *E. coli* DNA gyrase, significantly reduced N4 early transcription *in vivo*, suggesting that host DNA gyrase, and therefore negative supercoiling, is required for vRNAP cognate promoter recognition [14,18]. Considering that the N4 double-stranded DNA genome is linear, these results are surprising but can be explained if the genome is topologically constrained during injection. In support of this mechanism, the leftmost portion of the genome containing all vRNAP promoters is injected first [19].

To elucidate the template requirements for specific vRNAP transcription, sites of transcription initiation were mapped both *in vivo* and *in vitro* to the nucleotide level, revealing three sites of vRNAP initiation (P1, P2, and P3) within the leftmost 10% of the genome [19–21]. Sequences spanning −17 to +1 (relative to the transcription start site at +1) are conserved across all three promoters and include a GC-rich heptamer centered at −11 flanked by inverted repeats [21]. The conservation of inverted repeats, coupled with the requirement of host DNA gyrase *in vivo*, suggests a model where vRNAP transcription depends on the supercoil-induced extrusion of hairpin sequences mediated by DNA gyrase.

Subsequent studies confirmed key features of this model. The conserved hairpin stem contains both conserved and variable nucleotides. The conserved bases in all three promoters were shown to be required for both vRNAP binding and transcription initiation, while the identity of the non-conserved bases was shown to be extraneous as long as they maintain the promoter hairpin, suggesting that hairpin formation and direct contacts with vRNAP are required for promoter recognition [22]. Hairpin structure was detected by the cleavage patterns of single-stranded DNA templates upon treatment with enzymatic and chemical probes [22]. Although these data clearly prove the presence and function of hairpins in vRNAP promoters using *in vitro* systems, theoretical predictions of hairpin extrusion suggested that short hairpins can only form at superhelical densities significantly greater than those observed under physiological conditions [23,24]. Circles of phage DNA with differing superhelical densities and containing two early promoters were generated and probed for hairpin extrusion by four different chemical and enzymatic methods. DNA cruciform structures were detected at sub-physiological superhelical densities (−0.035) in the presence of Mg^{2+} [25]. The hairpins were shown to form *in vivo* by introducing the promoter P1 hairpin sequence between the −35 and −10 regions of the *E. coli rrnB* gene promoter. In this strain, promoter activity was abolished, while inhibiting DNA gyrase with the drug novobiocin restored activity [25]. These results clearly demonstrate that the N4 promoter hairpin is not only capable of extrusion under physiological conditions, but that extrusion occurs *in vivo* in a DNA gyrase activity-dependent manner.

The unexpected ability of N4 promoter hairpins to extrude at physiological DNA superhelical densities led to a series of experiments to determine the DNA sequence requirements for this process. The results of runoff transcription and hairpin-extrusion probing on promoter mutant templates are summarized in Figure 2a. The observation that the non-template strand hairpin loop was sensitive to single–stranded probes while the template strand loop was resistant was surprising and indicated structural differences determined by the loop sequences between the two DNA strands [25,26]. Analysis of the sequence requirements revealed that the 3'G:C5' loop closing base pair and the 3'A and 5'G bases of the loop are required for hairpin extrusion and hairpin stability [22,25,26]. Results of runoff transcription assays on mutant promoters and of crosslinking experiments with templates containing the photocrosslinking nucleotide analog 5-iododeoxyuracil (5-IdU) at specific positions showed that vRNAP specifically recognizes −11A/G, −10G, −8G, and +1C [26–28]. Contact with the −8 base was shown to occur through the major groove of the hairpin stem, while the site of crosslinking to the −11 purine residue was mapped to W129 in vRNAP [28].

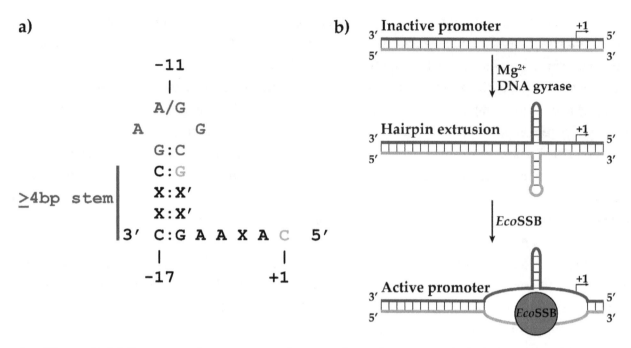

Figure 2. (a) Summary of sequence requirements for early promoter hairpin-extrusion (blue), vRNAP binding (green), and both processes (red) as determined by hairpin extrusion probing and runoff transcription experiments. Template strand sequences are shown relative to the transcription start site (+1) in 3' to 5' orientation. (b) Model of early promoter activation. N4 genomic DNA is injected into the host cytoplasm in an inactive linear double-stranded conformation. Introduction of negative supercoils by DNA gyrase induces the extrusion of the promoter hairpin in a Mg^{2+}- and sequence-dependent manner. Subsequent binding of EcoSSB leads to melting of the non-template strand hairpin and stabilization of the template strand hairpin to provide the active promoter conformation required for vRNAP binding. Template and non-template strands of promoter DNA are represented as purple and pink, respectively.

The initial model of promoter recognition by vRNAP involved the introduction of negative supercoils into N4 DNA concurrent with genome injection. However, supercoiled promoter-containing plasmids were not active templates for vRNAP transcription *in vitro*, suggesting the requirement of another factor for promoter activation. Genetic analysis identified the missing component as *E. coli* single-stranded DNA-binding protein (EcoSSB). No *in vivo* transcription of early promoters contained within plasmids in *ssb-1* (ts) mutant hosts was detected at the restrictive temperature [29]. *In vitro*, EcoSSB activates vRNAP transcription initiation 40-fold from supercoiled templates with *in vivo* specificity [29]. This process is due to invasion and melting of the non-template strand hairpin while the template strand hairpin is not perturbed, an unexpected behavior since single-stranded DNA binding proteins' role is to erase DNA secondary structures [29–31]. In this context, EcoSSB is acting as an architectural transcription factor by providing an active promoter conformation for vRNAP binding (Figure 2b) [27].

Along with its role in promoter activation, EcoSSB plays a role in transcript displacement. vRNAP transcription on single-stranded templates leads to RNA:DNA hybrid formation. At limiting template concentrations, addition of EcoSSB activates transcription [29]. S1 nuclease protection assays in reactions containing EcoSSB showed that the protein binds to the transcript as it exits vRNAP and facilitates template recycling *in vitro* [29,32]. Interaction of EcoSSB with the emerging transcript was confirmed by the results of crosslinking experiments with transcripts containing 5-IdU substitutions at specific positions [32]. Comparison of the T7 RNAP and N4 vRNAP sequences indicated that part of the T7 RNAP N-terminal domain responsible for RNA separation and exit is missing from vRNAP; therefore, we have proposed that *Eco*SSB fulfills this role in vRNAP transcription [32,33]. Our inability to detect *Eco*SSB-N4 vRNAP interactions with purified proteins suggests that these might occur as the polymerase transitions into the elongation complex.

The large size (3500 aa, 320 kDa) of vRNAP suggests that it may have multiple functions in phage development, along with several domains responsible for each activity. In order to define the minimally active transcriptional domain of vRNAP, the protein was subjected to limited trypsin digestion. A 1106 aa (998–2103), 122 kDa domain was identified that possesses the same transcription initiation, elongation, and termination properties as the full-length polypeptide [34]. A BLAST search with the minimally active transcriptional domain (mini-vRNAP) of vRNAP showed that this protein belongs to the family of T7-like "single-subunit" RNAPs, which encompasses the bacteriophage-encoded, nuclear-encoded mitochondrial and chloroplast, and linear plasmid-encoded enzymes [34,35]. Mini-vRNAP represents a highly-diverged member of this family, with little sequence homology outside of the conserved sequence blocks that have been implicated in catalysis [34]. Indeed, motifs A, B, and C, along with the DX_2GR (TX_2GR in mini-vRNAP) motif are all present within mini-vRNAP and mutational analyses show that their catalytic functions are conserved between T7 RNAP and mini-vRNAP [34].

Despite the low level of sequence similarity between mini-vRNAP and T7-like polymerases, they share a common architecture. The "cupped right hand" architecture shared by all related DNA and RNA polymerases is evident in the crystal structure of the apo form of mini-vRNAP, solved at 2.0 Å resolution [36]. A comparison with the crystal structure of T7 RNAP shows that the active sites of T7 RNAP and mini-vRNAP superimpose, reinforcing results of mutational analyses [34,36,37]. This comparison also identified three structural motifs in mini-vRNAP shown to be required for promoter recognition in T7 RNAP: the AT-rich recognition loop, β-intercalating hairpin (β-IH), and specificity loop [38,39]. Surprisingly, the crystal structure of apo mini-vRNAP reveals two structural motifs, the "plug" and the "motif B loop," that block the pathway of the DNA to the active site, suggesting that the structure represents an inactive conformation (Figure 3a) [36].

This model was confirmed by comparing (Figure 3) the apo structure with that of the binary complex (BC) of mini-vRNAP and its promoter P2 solved at 2.4 Å resolution (Figure 3b). The structures show a 25.1° rotation of the plug and β-IH motif, allowing for a 32.6 Å movement of the motif B loop and its rearrangement into the O helix upon promoter binding, which then grants single-stranded DNA access to the active site [36,40]. Crosslinking studies tethering the plug inside the active site resulted in transcriptionally inactive enzymes, confirming the necessity of these rearrangements [40]. The apo and BC mini-vRNAP structures present a model where vRNAP is present in an inactive conformation until it recognizes its promoter hairpin, explaining why vRNAP does not transcribe single-stranded DNA templates devoid of promoter hairpins. This DNA conformation is an allosteric effector that acts as a key to unlock the pathway of the template to the active site and render vRNAP competent for transcription initiation.

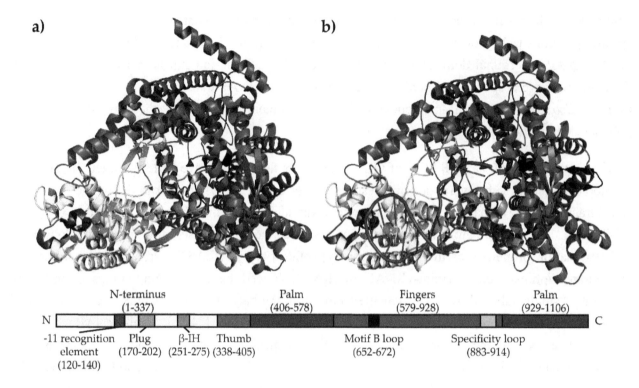

Figure 3. Comparison of the mini-vRNAP apo (Accession number 2PO4) **(a)** and BC (Accession number 3C3L) **(b)** structures highlighting rearrangements near the enzyme active site. Both structures show "cupped right hand" architecture (thumb, palm, fingers) characteristic of this family of RNAPs. Upon promoter binding, large scale structural rearrangements of the plug and β-IH motif occur, allowing the rearrangement of the motif B loop into the O helix to allow single-stranded DNA access to the active site. Bottom bar represents the mini-vRNAP primary sequence with amino acid numbering indicated in parentheses. Domains and structural motifs are labeled and colored as in the crystal structures, with template strand DNA represented as a purple ribbon.

A comparison of the sequences and structures required for mini-vRNAP and T7 RNAP promoter recognition, shown in Figure 4, is striking. T7 RNAP recognizes a bipartite double-stranded DNA sequence that spans from −17 to +6. Bases −17 to −5 constitute the upstream binding sequence, while bases −4 to +6 constitute the initiation region [41–44]. vRNAP, in contrast, recognizes a conserved promoter-hairpin structure along with sequences within it and its 3 bp loop (Figure 4a) [21,22,27].

Despite these drastic differences in promoter architecture, the RNAP motifs responsible for their recognition are remarkably similar (Figure 4b). Both RNAPs have a specificity loop that makes sequence-specific contacts with the promoter through the major groove, a β-IH structure responsible for defining the double-stranded to single-stranded DNA junction, and an upstream recognition element (−11 recognition element in mini-vRNAP and AT-rich recognition loop in T7 RNAP) [36,38–40]. Although the upstream recognition elements are more complex in mini-vRNAP, they share common structures nonetheless. This suggests that T7-like RNAPs developed unique strategies using the same set of basic tools to recognize their cognate promoters and ensure specificity.

a)

b)

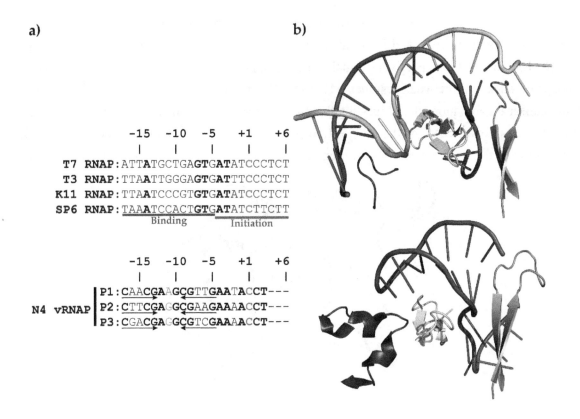

```
              -15  -10   -5   +1   +6
               |    |    |    |    |
   T7 RNAP: ATTATGCTGAGTGATATCCCTCT
   T3 RNAP: TTAATTGGGAGTGATTTCCCTCT
  K11 RNAP: TTAATCCCGTGTGATATCCCTCT
  SP6 RNAP: TAAATCCACTGTGATATCTTCTT
                   Binding     Initiation


              -15  -10   -5   +1   +6
               |    |    |    |    |
           P1:CAACGAAGCGTTGAATACCT---
N4 vRNAP   P2:CTTCGAGGCGAAGAAAACCT---
           P3:CGACGAGGCGTCGAAAACCT---
```

Figure 4. (a) Consensus promoter sequences of T7 and other related phage RNAPs compared with N4 vRNAP promoters. Template strand sequences spanning −17 to +6 are shown relative to the transcription start site (+1) in 3' to 5' orientation. Bold: conserved bases in all promoters both within and across species. Blue: +1 nucleotide. Arrows: inverted repeats. Orange: central base of hairpin loop. Purple underline: −17 to −5 binding region. Green underline: −4 to +6 initiation region. **(b)** Comparison of the three structural motifs responsible for promoter recognition between T7 RNAP (Accession number 1CEZ) (top) and mini-vRNAP (bottom). The AT-rich recognition site (T7 RNAP) and −11 recognition element (mini-vRNAP) (grey) are responsible for recognition of upstream sequences. The specificity loops (cyan) recognize the promoter by sequence-specific contacts through the major groove. The β-IH motif (orange) is responsible for defining the single-stranded to double-stranded DNA boundary. The template strand and non-template strand of promoter DNA are represented as purple and pink ribbons, respectively.

The N-terminal −11 recognition element contacts bases in the conserved loop through hydrogen bonding between K114 and R119 with −10G and −11G, respectively. The structure confirmed W129 base stacking with −11G [28,36,40]. The conformation of the hairpin loop explains its remarkable stability. Bases −9, −10, and −11 stack together and a sharp turn in DNA between bases −11 and −12 enables further base stacking interactions between −12 and −13 bases, while −10G and −12A form a sheared 5'G:A3' base pair. This unusual DNA structure explains both the sequence conservation and the remarkable stability of the promoter hairpin [22,25–27,40]. Three other elements are involved in promoter recognition. The β-IH defines the junction between double-stranded and single-stranded DNA between bases −5 and −4. In addition to defining the junction, this element enforces it by melting the 2 bps at the base of the 7 bp stem (structure P2_7a) with residues K267 and K268. Furthermore, specificity loop residues D901 and R904 make contacts through the major groove of the hairpin stem with −9C/−10G and −8G, respectively. Finally, finger residues K849 and K850 form salt bridges with the phosphate backbone between residues −12 and −13. Interestingly, there are no base contacts with the four As near the initiation site, suggesting that these bases act as a molecular ruler to start transcription eight nucleotides downstream of the −8 position [40]. These contacts, summarized in Figure 5, aid in the recognition of the promoter hairpin and render mini-vRNAP competent for transcription initiation.

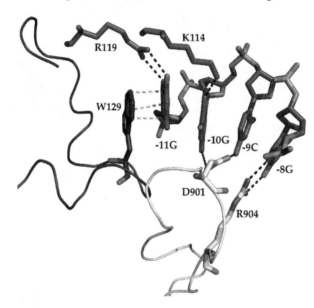

Figure 5. Summary of all sequence specific contacts required for mini-vRNAP recognition of early promoter P2. Promoter DNA represented as purple sticks. Mini-vRNAP structural motifs involved in recognition are represented as ribbons, with individual residues highlighted as sticks. Direct interactions are represented as dark grey (hydrogen bonds) or green (base stacking) dashed lines. W129 of the −11 recognition element (light grey) base-stacks with the −11G residue in the hairpin loop. Fingers domain (dark blue) residues K114 and R119 hydrogen bond with −10G and −11G, respectively. Specificity loop (cyan) residues D901 and R904 contact −9C/−10G and −8G through the major groove of the hairpin stem, respectively.

A unique aspect of DNA-dependent RNAPs is the ability to carry out first dinucleotide bond formation. A previous structural study using T7 RNAP failed to capture this event since the catalytic metal ion was absent and the substrate reactive groups were misaligned due to the use of the substrate analog 3'-deoxyGTP [45]. In an attempt to visualize this event with substrates containing all pertinent functional groups, GTP or the nonhydrolysable analog guanosine-5'-[(α,β)-methyleno] triphosphate (GMPCPP) and Mg^{2+} were soaked into preformed mini-vRNAP binary complex crystals [46]. Four structures were solved: (i) the precatalytic complex I (SCI) with two GTP molecules aligned with DNA bases +1 and +2 along with Mg^{2+} as the nucleotide-binding metal; (ii) the precatalytic complex II (SCII) with two GMPCPP molecules aligned with DNA bases +1 and +2 along with two Mn^{2+} catalytic metals; (iii) a mismatch complex (MC) with a GTP aligned with DNA base +1, but misaligned with the DNA base T at the +2 site along with two Mg^{2+} catalytic metals, and (iv) the product complex (PC) with a 2-mer RNA product and PPi [46]. These structures show that the initiating nucleotide base-stacks with the purine at the −1 position of the template strand, explaining the conservation of purines at the −1 position of many T7-like RNAP promoters. Secondly, the structure revealed for the first time the interactions of the initiating nucleotide with RNAPs: the palm residues K437 and R440 stabilize the binding of the initiating nucleotide [46]. Both catalytic metal ions, brought in along with the nucleotide substrates, cause a conformational shift in the catalytic aspartates and reorganization of the phosphates in the +1 NTP to break contact with R440 and initiate new contacts with E557, positioning the 3'OH 3.1 Å from the +2 NTP αP (Figure 6) [46]. In the product structure, the catalytic Mg^{2+}s leave the active site and the catalytic aspartates shift back to their binary complex states [46]. These structures define the contacts required for coordination of the initiating nucleotide phosphate groups in the process and suggest that binding of the catalytic metal is the last step before catalysis of the dinucleotide bond. This notion has been supported by direct observation through time-resolved X-ray crystallographic studies [47].

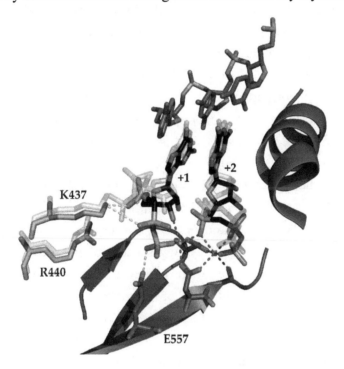

Figure 6. Structural transitions between mini-vRNAP SCI (Accession number 3Q22) and SCII (Accession number 3Q23) complexes required for stabilization of the initiating nucleotide. SCII complex structural motifs required for catalysis, O helix (dark blue) and motifs A and C (dark red), are represented as cartoons. Template strand DNA from −1 to +2 is represented as purple sticks. +1 and +2 GMPCPP molecules are represented as black sticks. Catalytic metals are represented as violet spheres. Palm domain residues required for stabilization of +1 NTP phosphate interactions are represented as green sticks. Catalytic aspartates and motif A residue E557 required for stabilization of +1 NTP phosphates are represented as red sticks. Interactions between residues and the catalytic aspartates are represented as dark grey dashes and interactions required for stabilization of +1 NTP phosphates are represented as yellow dashes. SCI residues K437 and R440, along with +1 and +2 GTP, are represented as transparent light grey sticks. Upon transition to the SCII complex, the phosphate residues of the initiating nucleotide break contact with R440 and are re-stabilized by E557 after a large conformational change. This change facilitates the repositioning of the 3'OH within 3.1 Å of the +2 NTP αP.

2.2. N4 RNAPII Synthesizes N4 Middle RNAs

The existence of a second N4-encoded transcriptional activity was postulated based on the 100-fold decrease in the rate of post-infection RNA synthesis when cells were pretreated with chloramphenicol [10,11,48]. Therefore, most phage transcription requires the synthesis of N4 early gene products. Indeed, infection with N4 phage containing mutations in ORF15 (N4am15) or ORF16 (N4am23) mimics the chloramphenicol-pretreated wild type phage-infected cell phenotype, indicating that the products of these genes (gp15 and gp16) encode a second transcriptional activity responsible for middle transcription [10,11,48,49].

Attempts to purify the phage proteins responsible for this activity proved difficult due to their tight association with the host cytoplasmic membrane [48,50]. Characterization was facilitated by the discovery, through additional mutagenesis screens, of a third N4 gene product (gp2) required for middle transcription. N4am126S mutant phage, which has a mutation in ORF2, was defective in middle transcription but could complement infections by either N4ORF15am or N4ORF16am mutants, indicating the involvement of gp2 in this process [51]. An *in vitro* complementation assay was developed where gp15 and gp16 present in the supernatant of N4am126S-infected cells were combined with gp2 and template DNA present in the membrane fraction of N4ORF15am- or N4ORF16am-infected cells [51,52]. This assay was competent for *in vitro* RNA synthesis and allowed for the purification and characterization of N4 RNAPII (gp15–gp16).

Purified N4 RNAPII, a heterodimer of gp15 and gp16, could utilize denatured DNA or the gp2-DNA bound membrane complex as templates [52]. Transcription originating from denatured DNA displayed no sequence specificity, while *in vivo* and *in vitro* transcription from the gp2-membrane complex displayed increased specificity and activity. This suggests that gp2 may have a role in providing specificity to N4 RNAPII either by direct interaction or providing a specific secondary structure for promoter recognition in a sequence-dependent fashion [52,53]. Further *in vitro* studies showed that N4 RNAPII binds only to single-stranded DNA with no sequence specificity while addition of a high salt-wash from N4-infected cell membranes containing gp2 provided some specificity in runoff transcription assays [53]. Sites of N4 RNAPII transcription initiation were mapped to a series of overlapping transcripts confined to the leftmost 50% of the genome [19,54]. N4 RNAPII recognizes a minimal promoter spanning −7 to +2 that is highly enriched in AT pairs, which may be an important factor in promoter melting [54,55].

Sequence analysis of ORF15 and ORF16 (gp15 and gp16) showed that these proteins align to non-overlapping portions of T7 RNAP (883 aa), confirming that N4 RNAPII is a heterodimer and identifies the enzyme as a member of the T7-like "single-subunit" RNAP family [35,49,52]. Together, gp15 and gp16 contain all four conserved motifs and thirteen blocks of conserved sequence within members of the T7-like RNAP family across both subunits. Gp15 (269 aa) aligns to the N-terminus, thumb, and DX$_2$GR domains, while gp16 (404 aa) aligns to the palm and fingers domains along with motifs A, B, and C [49,56]. N4 RNAPII (673 aa) represents a rather minimal RNAP, as it is one of the smallest members of this family, with a truncated thumb domain and a truncated N-terminal domain shown to be involved in promoter recognition in vRNAP and T7 RNAP [49]. However, N4 RNAPII shows much greater sequence homology to T7 RNAP than vRNAP and likely shares a similar architecture [34,49].

To elucidate the role of gp2 in middle transcription, ORF2 was cloned and sequenced, revealing no homology to proteins of known function [57]. The purified protein was tested for DNA binding and transcription-enhancing properties *in vitro*. Surprisingly, gp2 was shown to be a single-stranded DNA-binding protein exhibiting no sequence specificity [57]. It activates transcription through a recruitment mechanism, as binding of gp2 to single-stranded DNA increased the affinity of N4 RNAPII for the same template. Recruitment occurs through binding to N4 RNAPII; the N4 RNAPII-gp2 complex withstands treatment with high salt [57]. The data presented above has led us to propose a model for middle transcription whereby an unknown protein binds to double-stranded DNA to unwind the AT-rich promoter sequence, allowing gp2 to bind to the single-stranded DNA at middle promoters and recruit N4 RNAPII through direct interactions to the promoter sequences. N4 RNAPII then recognizes specific sequences in the template strand and initiates transcription (Figure 7). Preliminary evidence suggests that the product of ORF1, gp1, is required for middle promoter utilization *in vivo*. Experiments to characterize gp1's role in middle transcription and define the contacts required for N4 RNAPII promoter recognition are currently in progress.

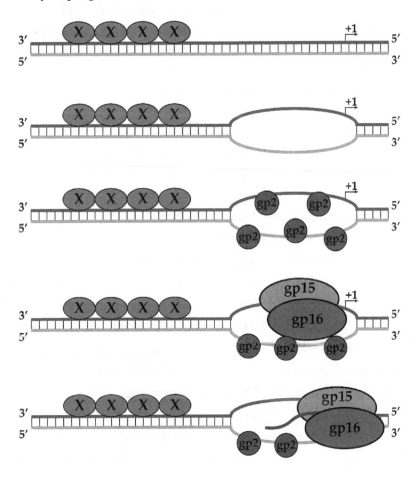

Figure 7. Model of middle transcription by N4 RNAPII. AT-rich double-stranded promoters are specifically recognized and melted by an unknown protein "X," allowing gp2 to bind to single-stranded DNA. Gp2 recruits N4 RNAPII to the promoter through direct interactions, allowing N4 RNAPII to recognize specific sequences in the template strand and initiate transcription. Template and non-template strands of promoter DNA are represented as purple and pink, respectively.

2.3. N4SSB Directs E. coli σ⁷⁰-RNAP to Synthesize N4 Late RNAs

Although N4 encodes two rifampicin-resistant transcribing activities, neither progeny nor virion structural proteins are produced in the presence of rifampicin, which suggests that the rifampicin-sensitive host RNAP is required for late transcription [10,11,19]. Progeny production in a rifampicin-resistant host was unaffected by presence of the drug, indicating that the host RNAP is responsible for N4 late transcription. Late transcripts localize to the right half of the phage genome and extend with opposite (leftward) polarity of early and middle transcripts [19]. A product of middle transcription, the N4 single-stranded DNA-binding protein (N4SSB), plays a dual role in N4 development. N4SSB is essential for DNA replication [58–60]. Genetic and biochemical analyses indicate that the N4SSB single-stranded DNA binding activity is dispensable while residues at the C-terminus are essential for late transcription activation. N4SSB interacts with a highly conserved region at the C-terminus of the β' subunit of *E. coli* σ⁷⁰-RNAP [61–63]. These findings reveal that bacteriophage N4 uses single-stranded DNA binding proteins *Eco*SSB, N4 gp2, N4SSB, as transcription activators. We surmise that the dependence on *Eco*SSB and gp2 derives from the interaction of vRNAP and N4 RNAPII with non-canonical DNA structures at the promoters, while N4SSB activation of late transcription couples phage morphogenesis to DNA replication.

3. N4 RNAPs Have Multiple Roles in Phage Development

Although we have focused on the roles of vRNAP and N4 RNAPII in transcription, their utility to N4 stretches across many developmental processes. The coding sequence of the vRNAP polypeptide (ORF50, 10.5 kbp) encompasses approximately 15% of the 72 kbp phage genome; its size and domain architecture suggest multiple functions during N4 development. vRNAP, present in four copies per virion, was localized above the portal protein by cryo-electron microscopy [64]. Based on this localization, injection of N4 genomic DNA must follow vRNAP out of capsids and into the host. Upon interaction of the N4 tail sheath protein (N4gp65) with *E. coli* outer membrane protein NfrA, a conformational change occurs in the phage tail that leads to the injection of the 3500 aa vRNAP through a 25 Å diameter tail tube and localization to the host cytoplasmic membrane [48,50,64–68]. Movement of vRNAP away from the portal allows the first 500 bp of the N4 genome to enter the host, where it is acted on by host DNA gyrase [14,64,69]. This enzyme introduces negative supercoils into N4 genomic DNA, leading to the extrusion of promoter hairpins [25]. vRNAP recognizes the promoter structure, which induces a structural change to activate the RNAP for transcription initiation [36,40]. Transcription of early genes leads to the injection of the genome, which is completed only upon the synthesis of the middle transcriptional machinery responsible for the synthesis of proteins required for N4 DNA replication [69]. One such product, N4SSB, interacts with the host σ⁷⁰-RNAP redirecting it to late promoters [62].

The developmental scheme outlined implicates a role of N4 transcriptional machineries in encapsidation, genome injection, transcription, and DNA replication. The encapsidation of its own early RNAP has distinct advantages for this phage in overcoming barriers of infection. N4, a member of the *Podoviridae* family, has a short, non-contractile tail incapable of traversing the host periplasm and injecting the DNA into the cytoplasm. vRNAP may have a role in overcoming this barrier, as suggested by its membrane association and role in genome injection [50,69]. Furthermore, the injection of genomic DNA in an inactive conformation provides a checkpoint for N4 phage to sample the energy state of the host by reading out the [ATP]/[ADP] ratio through the ATP-dependent activity of host DNA gyrase required for hairpin extrusion and early promoter activation [17,70,71]. The injection of vRNAP in an inactive conformation also precludes its interaction with host DNA, which is in greater abundance than the single copy of phage DNA. These mechanisms are but some of the many strategies employed by bacteriophage N4 to maximize progeny yield [72–74].

To our knowledge, N4 is the only bacteriophage that utilizes three different RNAPs to regulate the expression of its genes. Why does N4 use this strategy? The requirement for supercoil-dependent formation of the hairpin structure at the N4 vRNAP promoters concurrent with genome injection constrains their localization to the left end of the genome, which is injected into the host first. Moreover, the four vRNAP molecules injected from virions are insufficient for transcription from the large number of promoters. Therefore, to utilize all middle promoters and increase middle transcript abundance, additional polymerases must be synthesized. Synthesizing the 320 kDa vRNAP, with two large domains inessential for transcription, would waste resources where the minimal 80 kDa N4 RNAPII protein suffices. Why encode RNAPs required for early and middle transcription while relying on the host RNAP for late transcription in a pattern opposite of any other well-studied phage? Hijacking of host polymerase severely reduces host transcription, diverting resources to N4. Utilizing the N4SSB protein required for N4 DNA replication ensures that the transcription of late genes involved in virion assembly and lysis does not proceed until genome replication has begun. The ability to simultaneously transcribe the genes required for genome replication and morphogenetic proteins using two distinct transcriptional machineries allows for the production of 3000 progeny per infected bacterium [75].

4. Phylogenetic Analysis of N4-Like Phage Proteins Involved in Transcription

Original studies in 2002 suggested that vRNAP and N4 RNAPII are highly divergent members of the T7-like RNAP family with very little sequence homology to any other known phage RNAPs [34,49]. With the advent of next-generation sequencing technologies, there has been a large influx of phage genomes annotated and submitted in the past 10 years [76]. As a consequence, we are now able to identify a plethora of new phage and expand our knowledge of T7-like RNAP diversity.

Using this expanded dataset, homologs of vRNAP, N4 RNAPII, gp1, and gp2 were identified by DELTA-BLAST and subjected to maximum-likelihood neighbor-joining phylogenetic analyses using the software program MEGA6 [77]. We detected over 500 homologs of N4 RNAPII and 26 homologs of vRNAP. Of the >500 N4 RNAPII homologs, only 24 are heterodimers, while the remainders are single-subunit RNAPs. Sequence alignments show significantly greater homology to gp16 than gp15, which is not surprising given the localization of motifs A, B, and C in gp16 while gp15 is truncated and contains the more variable promoter recognition elements. Interestingly, all species possessing a virion-encapsidated RNAP homolog, except two, also possess a heterodimeric RNAP. The exceptions have sequences similar to gp16 split into two separate genes, highlighting the difficulty of using unannotated genomes. By parsimony, we infer that virion-encapsidated RNAPs predate heterodimeric T7-like RNAPs. Furthermore, 13 and 17 species contain annotated homologs of the middle transcription cofactors gp1 and gp2, respectively. Of these species, seven show absolute conservation of the early and middle transcriptional machineries (Figure 8). This suggests a phylogenetic relationship between virion-encapsidated and heterodimeric RNAP transcriptional paradigms.

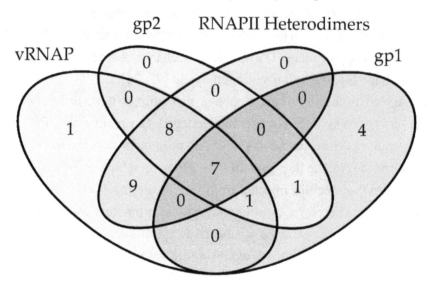

Figure 8. Venn diagram of species containing homologs of N4-encoded early and middle transcriptional machinery proteins vRNAP (blue), gp2 (green), gp1 (pink), and N4 RNAPII (purple) identified by DELTA-BLAST of NCBI RefSeq proteins. For simplicity, the category of species containing single-subunit RNAPs homologous to N4 RNAPII has been excluded. The number of species within each cross-section is indicated.

The origin and advantage of having a heterodimeric versus single-subunit RNAP remain unclear. Sequence alignment analysis of all heterodimeric RNAPs was performed and compared to the T7 RNAP sequence in an attempt to identify the dimerization interface. Several blocks of sequence homology in the C-terminus of gp15 and N-terminus of gp16 that do not overlap with motifs required for catalysis were identified. These blocks map to a three-helix bundle at the base of the T7 RNAP palm and may represent a dimerization domain required for heterodimer stability.

Finally, sequence analysis of the N- and C-terminal domains of vRNAP-like proteins revealed several blocks of sequence conservation. These sequence blocks may help to define the regions of each domain required for encapsidation and genome injection [78].

5. Conclusions

T7 RNAP, considered the founding member and prototype of the single-subunit family of RNAPs, catalyzes faithful initiation, elongation and termination of transcription independent of other phage or host proteins. However, biochemical and genetic characterization of other members of the family (vRNAP, N4 RNAPII, human and yeast mitochondrial RNAPs) show that these RNAPs rely on transcription factors for accurate initiation and/or elongation [35,79–81]. Therefore, T7 RNAP's property of transcriptional independence is the exception, rather than the rule (Table 1).

Table 1. T7-like RNAPs and their required cofactors.

Organism	RNAP	Cofactor(s)
T7	T7 RNAP	None
Saccharomyces cerevisiae mitochondria	Rpo41	Mtf1p
Homo sapiens mitochondria	POLRMT	TFAM and TFB2M
N4	N4 vRNAP	*Eco*SSB
N4	N4 RNAPII	gp1 and gp2

Through studying N4 RNAPs, we have expanded the diversity and knowledge of the T7-like "single-subunit" RNAPs, come to a new understanding of the importance of regulation of transcription through changes in DNA conformation, developed a model for understanding the role of cofactors in transcription initiation, expanded on the molecular basis of promoter recognition and transcription initiation, and discovered multiple mechanisms by which transcriptional machinery can be used to overcome barriers of infection.

Although the N4 transcriptional machinery provides a framework for understanding the importance of transcription in phage development, several open questions remain. Current work is focused on understanding the role that N4 RNAPII cofactors have on DNA conformation, how cofactors interact with the RNAP to initiate transcription, and whether these interactions are conserved among RNAPs such as POLRMT and Rpo41, which also rely on cofactors for transcription. We are also interested in exploring how conserved RNAP motifs involved in promoter recognition have evolved to recognize such a wide variety of promoter sequences and structures. Finally, the presence of the vRNAP polypeptide in virions and its involvement in genome injection provides a unique manipulable system to investigate the process of protein and genome injection into the host.

Acknowledgments

Work in our laboratory has been supported by NIH Genetics and Regulation Training Grant T32 GM007197 and NIH RO1 AI-12575. We wish to thank Alex Advani for advice on the phylogenetic analyses.

Author Contributions

Bryan R. Lenneman and Lucia B. Rothman-Denes co-wrote the manuscript.

References

1. Santangelo, T.J.; Artsimovitch, I. Termination and antitermination: RNA polymerase runs a stop sign. *Nat. Rev. Microbiol.* **2011**, *9*, 319–329.

2. Hinton, D.M. Transcriptional control in the prereplicative phase of T4 development. *Virol. J.* **2010**, doi:10.1186/1743-422X-7-289.

3. Losick, R.; Pero, J. Cascade of Sigma Factors. *Cell* **1981**, *25*, 582–584.

4. Monsalve, M.; Calles, B.; Mencía, M.; Salas, M.; Rojo, F. Transcription activation or repression by phage phi 29 protein p4 depends on the strength of the RNA polymerase-promoter interactions. *Mol. Cell* **1997**, *1*, 99–107.

5. Studier, F.W. Bacteriophage T7. *Science* **1972**, *176*, 367–376.

6. Bae, B.; Davis, E.; Brown, D.; Campbell, E.A.; Wigneshweraraj, S.; Darst, S.A. Phage T7 Gp2 inhibition of *Escherichia coli* RNA polymerase involves misappropriation of σ70 domain 1.1. *Proc. Natl. Acad. Sci. USA* **2013**, *110*, 19772–19777.

7. Liu, B.; Shadrin, A.; Sheppard, C.; Mekler, V.; Xu, Y.; Severinov, K.; Matthews, S.; Wigneshwerararaj, S. A bacteriophage transcription regulator inhibits bacterial transcription initiation by σ-factor displacement. *Nucleic Acids Res.* **2014**, *42*, 4294–4305.

8. Semenova, E.; Djordjevic, M.; Shralman, B.; Severinov, K. The tale of two RNA polymerases: Transcription profiling and gene expression strategy of bacteriophage Xp10. *Mol. Microbiol.* **2005**, *55*, 764–777.

9. Ceyssens, P.-J.; Minakhin, L.; van den Bossche, A.; Yakunina, M.; Klimuk, E.; Blasdel, B.; de Smet, J.; Noben, J.-P.; Bläsi, U.; Severinov, K.; Lavigne, R. Development of giant bacteriophage φKZ is independent of the host transcription apparatus. *J. Virol.* **2014**, *88*, 10501–10510.

10. Rothman-Denes, L.B.; Schito, G.C. Novel transcribing activities in N4-infected *Escherichia coli*. *Virology* **1974**, *60*, 65–72.

11. Vander Laan, K.; Falco, S.C.; Rothman-Denes, L.B. The program of RNA synthesis in N4-infected *Escherichia coli*. *Virology* **1977**, *76*, 596–601.

12. Falco, S.C.; Vander Laan, K.; Rothman-Denes, L.B. Virion-associated RNA polymerase required for bacteriophage N4 development. *Proc. Natl. Acad. Sci. USA* **1977**, *74*, 520–523.

13. Falco, S.C.; Zehring, W.; Rothman-Denes, L.B. DNA-dependent RNA polymerase from bacteriophage N4 virions. Purification and characterization. *J. Biol. Chem.* **1980**, *255*, 4339–4347.

14. Falco, S.C.; Zivin, R.; Rothman-Denes, L.B. Novel template requirements of N4 virion RNA polymerase. *Proc. Natl. Acad. Sci. USA* **1978**, *75*, 3220–3224.

15. Brody, E.N.; Geiduschek, E.P. Transcription of the bacteriophage T4 template. Detailed comparison of *in vitro* and *in vivo* transcripts. *Biochemistry* **1970**, *9*, 1300–1309.

16. Chamberlin, M.; Ring, J. Characterization of T7-specific ribonucleic acid polymerase. *J. Biol. Chem.* **1973**, *248*, 2235–2244.

17. Gellert, M.; Mizuuchi, K.; O'Dea, M.H.; Nash, H.A. DNA gyrase: An enzyme that introduces superhelical turns into DNA. *Proc. Natl. Acad. Sci. USA* **1976**, *73*, 3872–3876.

18. Gellert, M.; O'Dea, M. H.; Itoh, T.; Tomizawa, J. Novobiocin and coumermycin inhibit DNA supercoiling catalyzed by DNA gyrase. *Proc. Natl. Acad. Sci. USA* **1976**, *73*, 4474–4478.

19. Zivin, R.; Zehring, W.; Rothman-Denes, L.B. Transcriptional map of bacteriophage N4. Location and polarity of N4 RNAs. *J. Mol. Biol.* **1981**, *152*, 335–356.

20. Zivin, R.; Malone, C.; Rothman-Denes, L.B. Physical map of coliphage N4 DNA. *Virology* **1980**, *104*, 205–218.

21. Haynes, L.L.; Rothman-Denes, L.B. N4 virion RNA polymerase sites of transcription initiation. *Cell* **1985**, *41*, 597–605.

22. Glucksmann, M.A.; Markiewicz, P.; Malone, C.; Rothman-Denes, L.B. Specific sequences and a hairpin structure in the template strand are required for N4 virion RNA polymerase promoter recognition. *Cell* **1992**, *70*, 491–500.

23. Sinden, R.R.; Carlson, J.O.; Pettijohn, D.E. Torsional tension in the DNA double helix measured with trimethylpsoralen in living *E. coli* cells: Analogous measurements in insect and human cells. *Cell* **1980**, *21*, 773–783.

24. McClellan, J.A.; Boublíková, P.; Palecek, E.; Lilley, D.M. Superhelical torsion in cellular DNA responds directly to environmental and genetic factors. *Proc. Natl. Acad. Sci. USA* **1990**, *87*, 8373–8377.

25. Dai, X.; Greizerstein, M.B.; Nadas-Chinni, K.; Rothman-Denes, L.B. Supercoil-induced extrusion of a regulatory DNA hairpin. *Proc. Natl. Acad. Sci. USA* **1997**, *94*, 2174–2179.

26. Dai, X.; Kloster, M.; Rothman-Denes, L.B. Sequence-dependent extrusion of a small DNA hairpin at the N4 virion RNA polymerase promoters. *J. Mol. Biol.* **1998**, *283*, 43–58.

27. Dai, X.; Rothman-Denes, L.B. Sequence and DNA structural determinants of N4 virion RNA polymerase-promoter recognition. *Genes Dev.* **1998**, *12*, 2782–2790.

28. Davydova, E.K.; Santangelo, T.J.; Rothman-Denes, L.B. Bacteriophage N4 virion RNA polymerase interaction with its promoter DNA hairpin. *Proc. Natl. Acad. Sci. USA* **2007**, *104*, 7033–7038.

29. Markiewicz, P.; Malone, C.; Chase, J.W.; Rothman-Denes, L.B. *Escherichia coli* single-stranded DNA-binding protein is a supercoiled template-dependent transcriptional activator of N4 virion RNA polymerase. *Genes Dev.* **1992**, *6*, 2010–2019.

30. Glucksmann-Kuis, M.A.; Dai, X.; Markiewicz, P.; Rothman-Denes, L.B. *E. coli* SSB activates N4 virion RNA polymerase promoters by stabilizing a DNA hairpin required for promoter recognition. *Cell* **1996**, *84*, 147–154.

31. Chase, J. Single-stranded DNA binding proteins required for DNA replication. *Annu. Rev. Biochem.* **1986**, *55*, 103–136.

32. Davydova, E.K.; Rothman-Denes, L.B. *Escherichia coli* single-stranded DNA-binding protein mediates template recycling during transcription by bacteriophage N4 virion RNA polymerase. *Proc. Natl. Acad. Sci. USA* **2003**, *100*, 9250–9255.

33. Steitz, T.A. The structural basis of the transition from initiation to elongation phases of transcription, as well as translocation and strand separation, by T7 RNA polymerase. *Curr. Opin. Struct. Biol.* **2004**, *14*, 4–9.

34. Kazmierczak, K.M.; Davydova, E.K.; Mustaev, A.A.; Rothman-Denes, L.B. The phage N4 virion RNA polymerase catalytic domain is related to single-subunit RNA polymerases. *EMBO J.* **2002**, *21*, 5815–5823.

35. Cermakian, N.; Ikeda, T.M.; Miramontes, P.; Lang, B.F.; Gray, M.W.; Cedergren, R. On the evolution of the single-subunit RNA polymerases. *J. Mol. Evol.* **1997**, *45*, 671–681.

36. Murakami, K.S.; Davydova, E.K.; Rothman-Denes, L.B. X-ray crystal structure of the polymerase domain of the bacteriophage N4 virion RNA polymerase. *Proc. Natl. Acad. Sci. USA* **2008**, *105*, 5046–5051.

37. Jeruzalmi, D.; Steitz, T.A. Structure of T7 RNA polymerase complexed to the transcriptional inhibitor T7 lysozyme. *EMBO J.* **1998**, *17*, 4101–4113.

38. Cheetham, G.M.; Steitz, T.A. Structure of a transcribing T7 RNA polymerase initiation complex. *Science* **1999**, *286*, 2305–2309.

39. Cheetham, G.M.; Jeruzalmi, D.; Steitz, T.A. Structural basis for initiation of transcription from an RNA polymerase-promoter complex. *Nature* **1999**, *399*, 80–83.

40. Gleghorn, M.L.; Davydova, E.K.; Rothman-Denes, L.B.; Murakami, K.S. Structural basis for DNA-hairpin promoter recognition by the bacteriophage N4 virion RNA polymerase. *Mol. Cell* **2008**, *32*, 707–717.

41. Imburgio, D.; Rong, M.; Ma, K.; McAllister, W.T. Studies of promoter recognition and start site selection by T7 RNA polymerase using a comprehensive collection of promoter variants. *Biochemistry* **2000**, *39*, 10419–10430.

42. Rong, M.; He, B.; McAllister, W.T.; Durbin, R.K. Promoter specificity determinants of T7 RNA polymerase. *Proc. Natl. Acad. Sci. USA* **1998**, *95*, 515–519.

43. Chapman, K.A.; Burgess, R.R. Construction of bacteriphage T7 late promoters with point mutations and characterization by *in vitro* transcription properties. *Nucleic Acids Res.* **1987**, *15*, 465–475.

44. Li, T.; Ho, H.H.; Maslak, M.; Schick, C.; Martin, C.T. Major groove recognition elements in the middle of the T7 RNA polymerase promoter. *Biochemistry* **1996**, *35*, 3722–3727.

45. Kennedy, W.P.; Momand, J.R.; Yin, Y.W. Mechanism for *de novo* RNA Synthesis and initiating nucleotide specificity by T7 RNA polymerase. *J. Mol. Biol.* **2007**, *370*, 256–268.

46. Gleghorn, M.L.; Davydova, E.K.; Basu, R.; Rothman-Denes, L.B.; Murakami, K.S. X-ray crystal structures elucidate the nucleotidyl transfer reaction of transcript initiation using two nucleotides. *Proc. Natl. Acad. Sci. USA* **2011**, *108*, 3566–3571.

47. Basu, R.S.; Murakami, K.S. Watching the bacteriophage N4 RNA polymerase transcription by time-dependent soak-trigger-freeze X-ray crystallography. *J. Biol. Chem.* **2013**, *288*, 3305–3311.

48. Falco, S.C.; Rothman-Denes, L.B. Bacteriophage N4-induced transcribing activities in *Escherichia coli* I. Detection and characterization in cell extracts. *Virology* **1979**, *95*, 454–465.

49. Willis, S.H.; Kazmierczak, K.M.; Carter, R.H.; Rothman-Denes, L.B. N4 RNA polymerase II, a heterodimeric RNA polymerase with homology to the single-subunit family of RNA polymerases. *J. Bacteriol.* **2002**, *184*, 4952–4961.

50. Falco, S.C.; Rothman-Denes, L.B. Bacteriophage transcribing activities in *Escherichia coli* II. Association of the N4 transcriptional apparatus with the cytoplasmic membrane. *Virology* **1979**, *95*, 466–475.

51. Zehring, W.A.; Falco, S.C.; Malone, C.; Rothman-Denes, L.B. Bacteriophage N4-induced transcribing activities in *E. coli*. III. A third cistron required for N4 RNA polymerase II activity. *Virology* **1983**, *126*, 678–687.

52. Zehring, W.A.; Rothman-Denes, L.B. Purification and characterization of coliphage N4 RNA polymerase II activity from infected cell extracts. *J. Biol. Chem.* **1983**, *258*, 8074–8080.

53. Abravaya, K.; Rothman-Denes, L.B. *In vitro* requirements for N4 RNA polymerase II-specific Initiation. *J. Biol. Chem.* **1989**, *264*, 12695–12699.

54. Abravaya, K.; Rothman-Denes, L.B. N4 RNA polymerase II sites of transcription initiation. *J. Mol. Biol.* **1990**, *211*, 359–372.

55. Hammer, M.; Lenneman, B.R.; Rothman-Denes, L.B. Determinants of bacteriophage N4 RNAPII promoter specificity. Unpublished data, 2015.

56. Sousa, R. Structural and mechanistic relationships between nucleic acid polymerases. *Trends Biochem. Sci.* **1996**, *21*, 186–190.

57. Carter, R.H.; Demidenko, A.A.; Hattingh-Willis, S.; Rothman-Denes, L.B. Phage N4 RNA polymerase II recruitment to DNA by a single-stranded DNA-binding protein. *Genes Dev.* **2003**, *17*, 2334–2345.

58. Guinta, D.; Stambouly, J.; Falco, S.C.; Rist, J.K.; Rothman-Denes, L.B. Host and phage-coded functions required for coliphage N4 DNA replication. *Virology* **1986**, *150*, 33–44.

59. Lindberg, G.K.; Rist, J.K.; Kunkel, T.A.; Sugino, A.; Rothman-Denes, L.B. Purification and characterization of bacteriophage N4-induced DNA polymerase. *J. Biol. Chem.* **1988**, *263*, 11319–11326.

60. Lindberg, G.; Kowalczykowski, S.C.; Rist, J.K.; Sugino, A.; Rothman-Denes, L.B. Purification and characterization of the coliphage N4-coded single-stranded DNA binding protein. *J. Biol. Chem.* **1989**, *264*, 12700–12708.

61. Choi, M.; Miller, A.; Cho, N.Y.; Rothman-Denes, L.B. Identification, cloning, and characterization of the bacteriophage N4 gene encoding the single-stranded DNA-binding protein: A protein required for phage replication, recombination, and late transcription. *J. Biol. Chem.* **1995**, *270*, 22541–22547.

62. Cho, N.Y.; Choi, M.; Rothman-Denes, L.B. The bacteriophage N4-coded single-stranded DNA-binding protein (N4SSB) is the transcriptional activator of *Escherichia coli* RNA polymerase at N4 late promoters. *J. Mol. Biol.* **1995**, *246*, 461–471.

63. Miller, A.; Dai, X.; Choi, M.; Glucksmann-Kuis, M.A.; Rothman-Denes, L.B. Single-stranded DNA-binding proteins as transcriptional activators. *Methods Enzymol.* **1996**, *274*, 9–20.

64. Choi, K.H.; Mcpartland, J.; Kaganman, I.; Bowman, V.D.; Rothman-Denes, L.B.; Rossmann, M.G. Insight into DNA and protein transport in double-stranded DNA viruses: The structure of bacteriophage N4. *J. Mol. Biol.* **2009**, *378*, 726–736.

65. Kiino, D.R.; Rothman-Denes, L.B. Genetic analysis of bacteriophage N4 adsorption. *J. Bacteriol.* **1989**, *171*, 4595–4602.

66. Kiino, D.R.; Singer, M.S.; Rothman-Denes, L.B. Two overlapping genes encoding membrane proteins required for bacteriophage N4 adsorption. *J. Bacteriol.* **1993**, *175*, 7081–7085.

67. Kiino, D.R.; Licudine, R.; Wilt, K.; Yang, D.H.C.; Rothman-Denes, L.B. A cytoplasmic protein, NfrC, is required for bacteriophage N4 adsorption. *J. Bacteriol.* **1993**, *175*, 7074–7080.

68. McPartland, J.; Rothman-Denes, L.B. The tail sheath of bacteriophage N4 interacts with the *Escherichia coli* receptor. *J. Bacteriol.* **2009**, *191*, 525–532.

69. Demidenko, A.A.; Rothman-Denes, L.B. Bacteriophage N4 genome injection is driven by the phage transcription systems. Unpublished data, 2015.

70. Hsieh, L.S.; Rouviere-Yaniv, J.; Drlica, K. Bacterial DNA supercoiling and [ATP]/[ADP] ratio: Changes associated with salt shock. *J. Bacteriol.* **1991**, *173*, 3914–3917.

71. Hsieh, L.S.; Burger, R.M.; Drlica, K. Bacterial DNA supercoiling and [ATP]/[ADP]. Changes associated with a transition to anaerobic growth. *J. Mol. Biol.* **1991**, *219*, 443–450.

72. Yano, S.T.; Rothman-Denes, L.B. A phage-encoded inhibitor of *Escherichia coli* DNA replication targets the DNA polymerase clamp loader. *Mol. Microbiol.* **2011**, *79*, 1325–1338.

73. Stojković, E.A.; Rothman-Denes, L.B. Coliphage N4 N-acetylmuramidase defines a new family of murein hydrolases. *J. Mol. Biol.* **2007**, *366*, 406–419.

74. Khuong, N.; Loose, M.; Yano, S.; Rothman-Denes, L.B. Coliphage N4 inhibits cell division by targeting FtsZ and FtsA. *Proc. Natl. Acad. Sci. USA* **2015**, submitted for publication.

75. Schito, G.C. Development of coliphage N4: Ultrastructural studies. *J. Virol.* **1974**, *13*, 186–196.

76. Hatfull, G.F. Bacteriophage genomics. *Curr. Opin. Microbiol.* **2008**, *11*, 447–453.

77. Tamura, K.; Stecher, G.; Peterson, D.; Filipski, A.; Kumar, S. MEGA6: Molecular evolutionary genetics analysis version 6.0. *Mol. Biol. Evol.* **2013**, *30*, 2725–2729.

78. Kaganman, I.; Davydova, E.K.; Kazmierczak, K.M.; Rothman-Denes, L.B. The N-terminal and C-terminal domains of bacteriophage N4 vRNAP are required for genome injection and encapsidation respectively. Unpublished data, 2015.

79. Morozov, Y.I.; Agaronyan, K.; Cheung, A.C.M.; Anikin, M.; Cramer, P.; Temiakov, D. A novel intermediate in transcription initiation by human mitochondrial RNA polymerase. *Nucleic Acids Res.* **2014**, *42*, 3884–3893.

80. Masters, B.S.; Stohl, L.L.; Clayton, D.A. Yeast mitochondrial RNA polymerase is homologous to those encoded by bacteriophages T3 and T7. *Cell* **1987**, *51*, 89–99.

81. Deshpande, A.P.; Patel, S.S. Mechanism of transcription initiation by the yeast mitochondrial RNA polymerase. *Biochim. Biophys. Acta Gene Regul. Mech.* **2012**, *1819*, 930–938.

4

Bacteriophages and Biofilms

David R. Harper [1,*]**, Helena M. R. T. Parracho** [1]**, James Walker** [2]**, Richard Sharp** [2]**, Gavin Hughes** [3]**, Maria Werthén** [4,5]**, Susan Lehman** [1] **and Sandra Morales** [1]

[1] AmpliPhi Biosciences, Glen Allen, VA 23060, USA;
   E-Mails: hparracho@thenativeantigencompany.com (H.M.R.T.P.);
   sml@ampliphibio.com (S.L.); spm@ampliphibio.com (S.M.)
[2] Public Health England, Porton Down, Salisbury SP4 0JG, UK;
   E-Mails: jimmy.walker@phe.gov.uk (J.W.); richard.sharp@phe.gov.uk (R.S.)
[3] Gavin Hughes—The Surgical Materials Testing Laboratory, Bridgend, South Wales CF31 1RQ, UK;
   E-Mail: gavin@smtl.co.uk
[4] Maria Werthén, Mölnlycke Health Care AB, SE-402 52 Gothenburg, Sweden; E-Mail:
   maria.werthen@molnlycke.com
[5] Department of Biomaterial Science, University of Gothenburg, SE-405 30 Gothenburg, Sweden

* Author to whom correspondence should be addressed; E-Mail: drh@ampliphibio.com

Abstract: Biofilms are an extremely common adaptation, allowing bacteria to colonize hostile environments. They present unique problems for antibiotics and biocides, both due to the nature of the extracellular matrix and to the presence within the biofilm of metabolically inactive persister cells. Such chemicals can be highly effective against planktonic bacterial cells, while being essentially ineffective against biofilms. By contrast, bacteriophages seem to have a greater ability to target this common form of bacterial growth. The high numbers of bacteria present within biofilms actually facilitate the action of bacteriophages by allowing rapid and efficient infection of the host and consequent amplification of the bacteriophage. Bacteriophages also have a number of properties that make biofilms susceptible to their action. They are known to produce (or to be able to induce) enzymes that degrade the extracellular matrix. They are also able to infect persister cells, remaining dormant within them, but re-activating when they become metabolically active. Some cultured biofilms also seem better able to support the replication of bacteriophages than comparable planktonic systems. It is perhaps unsurprising that bacteriophages, as the natural predators of bacteria, have the ability to target this common form of bacterial life.

Keywords: bacteriophage; biofilm; antibiotic resistance; phage therapy; depolymerase; persister cells

1. Introduction: Biofilms

Biofilms are aggregations of cells, which may be eukaryotic or prokaryotic in nature, surrounded by a matrix of extracellular polymeric substance (EPS) produced, at least in part, by the cells within the biofilm [1]. This EPS consists of long chain sugars, DNA and other biological macromolecules [2], the precise nature of which can be highly variable.

Bacteria within a biofilm can show high levels of resistance to agents, such as biocides and antibiotics. The level of such agents that is needed to produce antibacterial effects can be over 1000 times greater than the level required for free-living (planktonic) bacteria [3]. Suggestions that the biofilm matrix forms an impermeable barrier to all such agents are now thought to be an oversimplification. Some agents may be at least partially blocked, for example the sequestration of tobramycin reported by Mah *et al.* [4]. In contrast, others can readily penetrate the biofilm matrix, as reported for ofloxacin and peracetic acid by Spoering and Lewis [5].

The presence of different conditions (including gaseous, as well as nutrient stratifications) leads to different cell states and, thus, to the existence of distinct zones within the biofilm. Quorum sensing, where cells communicate by releasing small chemical molecules as a signaling process to other cells, results in the biofilm acting in many ways as a community rather than simply a cluster of independent cells [2].

It is now thought that much of the apparent resistance or regrowth of bacteria within the biofilm arises from the presence within the matrix of metabolically inactive "persister" cells. The slow growth and severely limited metabolic activity of these persister bacteria may prevent the action of many antibiotics. Such cells can reactivate after such stress, leading to the regrowth of the biofilm after treatment [6].

Given the ubiquity and advantages of biofilm growth, it is perhaps not unexpected that biofilms appear to be associated with the majority of bacterial infections [2]. Biofilms are thought to underlie much of the reported resistance to antibiotics in clinical use. Such resistance can be apparent *in vivo*, even when the cells themselves, outside the biofilm, such as in the suspended or planktonic state, are susceptible to such agents and, thus, show up as sensitive in assays that do not permit biofilm formation.

An outline of the life cycle of bacterial biofilms is presented in Figure 1, exemplified by the motile bacterium, *Pseudomonas aeruginosa*, which has been used in many studies in this area [7,8]. It should be noted that non-motile bacteria, such as *Staphylococcus aureus*, also form biofilms.

Figure 1. The life cycle of bacteria in biofilms.

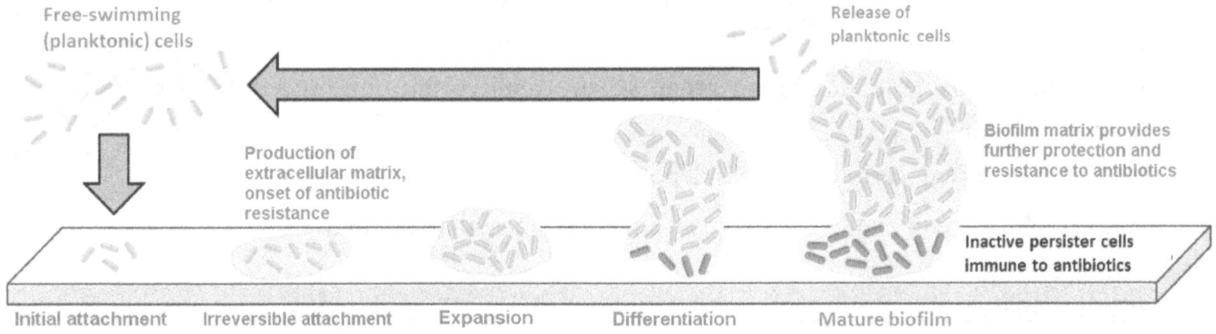

Free-swimming
(planktonic) cells

Release of
planktonic cells

Production of
extracellular matrix,
onset of antibiotic
resistance

Biofilm matrix provides
further protection and
resistance to antibiotics

Inactive persister cells
immune to antibiotics

Initial attachment   Irreversible attachment   Expansion   Differentiation   Mature biofilm

2. Bacteriophage and Their Effects on Biofilms

Bacteriophages (often known simply as phages) are naturally occurring viruses that infect bacteria. As such, they are unaffected by antibiotic resistance and (unlike many antibiotics) are able to target bacteria within biofilms [1]. They can either coexist with their host by inserting themselves into the bacterial genome (lysogenic bacteriophages) or destroy them (lytic bacteriophages; the type most suited to therapeutic use). Lytic bacteriophages replicate inside their hosts, then release many new bacteriophages able to infect more bacteria.

It has often been assumed that biofilms confer resistance to bacteriophages, due to the impermeability of the biofilm matrix. However, although they are far larger than chemical antibiotics, bacteriophages are still far smaller than their bacterial hosts, and many bacteriophages can and do infect bacteria within biofilms.

Bacteriophages act differently on bacteria contained within biofilms than do chemical antibiotics or biocides. Indeed, one can make the argument that phage have co-evolved with bacterial biofilms, and thus, their infection of adherent bacterial populations would be expected. There are at least four mechanisms underlying this difference (Figure 2).

Figure 2. Destruction of a biofilm by bacteriophages.

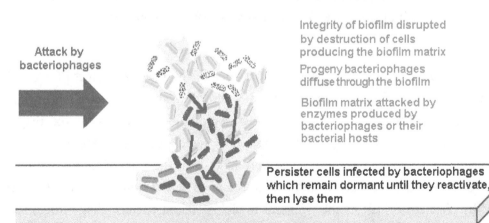

Attack by
bacteriophages

Integrity of biofilm disrupted
by destruction of cells
producing the biofilm matrix

Progeny bacteriophages
diffuse through the biofilm

Biofilm matrix attacked by
enzymes produced by
bacteriophages or their
bacterial hosts

Persister cells infected by bacteriophages
which remain dormant until they reactivate,
then lyse them

Biofilm attacked by bacteriophages

1. Bacteriophages replicate within their host cells, resulting in localized increases in bacteriophage numbers (amplification). This releases increasing numbers of infectious progeny bacteriophages into the biofilm. By spreading through the biofilm, eliminating the bacteria producing the EPS material, bacteriophages can progressively remove the biofilm and reduce the potential for regeneration.

2. Bacteriophages can carry or express depolymerizing enzymes that degrade the EPS.

3. Bacteriophages can induce depolymerizing enzymes that degrade the EPS from within the host genome.

4. Persister cells can be infected by bacteriophages; although bacteriophage cannot replicate within and destroy inactive cells, they can remain within these bacteria until they reactivate and then commence a productive infection, which then destroys the cells.

Bacteriophages can kill their host cells without replicating when they greatly outnumber their bacterial targets, a process called "lysis from without" [9], but such numbers are rarely achieved outside the laboratory. Use of lower numbers of bacteriophages kills the host cells by replication within them followed by lysis during the release of progeny bacteriophages, which then go on to destroy yet more bacteria. Such an infection results in the localized amplification of bacteriophages and can result in self-sustaining infections, with localized spread and ongoing amplification, where the host bacteria are present at levels high enough to support this. However, where insufficient host bacteria are present, this may interrupt the infectious cycle. Given that biofilms are both very common and the location of high levels of bacteria, it is likely that the effective targeting of bacteria within biofilms by bacteriophages represents an evolutionary adaptation to use this rich resource. The mechanisms that they use for this are likely to be based on their need to deal with bacterial capsular polysaccharides during the normal course of infection. This form of infectious spread within a biofilm, with lysis of bacteria as the key element, has been referred to as "active penetration" [10].

It is known that many bacteriophage genomes contain genes for enzymes capable of breaking down elements of the biofilm matrix [1,11,12]. In many cases, these are soluble enzymes that target the host bacterial cell wall during the release from the host cell, but these also have the potential to degrade the biofilm EPS when released from lysing host cells. These cells also release DNA, which can contribute to the biofilm matrix, but can also release DNase enzymes, which are present in the host cells as part of normal replication.

In addition to this, many bacteriophages, such as the T4 and HK620 bacteriophages of *Escherichia coli* have enzymes present on the tail of the virus particle, where they aid the penetration of the bacterial cell wall. While they could theoretically play a role in degrading the biofilm matrix, they are often masked until the tail reconfigures during infection and, thus, have a very localized action [11,12]. The requirements for these proteins are precise, since they have to fit into and function within the virus structure. The presence of such enzymes within the tail seems to be a common feature of bacteriophage infection, noted by Yan *et al.* [12] as the "general model of tailed bacteriophage infection". In this model, components of the bacteriophage tail recognize and digest the capsular polysaccharide, thus allowing the tail to contact the cell membranes through which it then injects the bacterial genome.

Yan *et al.* [12] further notes that "polysaccharide depolymerase protein is a common constituent of the tail structure of a bacteriophage" and that "many tailspike proteins have endoglycosidase activity, hydrolyzing their polysaccharide receptors". However, this activity is restricted as noted above.

As well as the carriage of the genes coding for such activity, it has long been known that some bacteriophages can induce the expression of depolymerase enzymes in their host bacteria [13]. However, until genome sequencing became possible, it was very difficult to confirm whether this was encoded by the bacteriophage itself.

With bacteriophage GH4, the expression of alginase activity was seen in assays *in vitro* (Figure 3). Despite early suggestions that an alginase was encoded within the bacteriophage genome [14], sequencing showed no such gene, indicating that this alginase is induced from within the host cell. It has been suggested that this could be either a bacteriophage-induced method of rendering the biofilm matrix more porous, thus aiding infection by progeny bacteriophage, or, alternatively, a flight response by infected bacteria, seeking to facilitate movement away from the focus of infection.

Figure 3. Expression of alginase in bacteriophage-infected *Pseudomonas aeruginosa*.

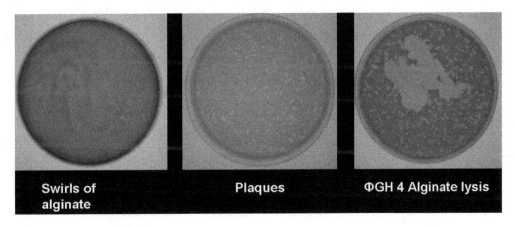

The ability to infect persister cells and to remain within them until they reactivate, then to initiate a productive, lytic infection was reported by Pearl *et al.* [15], who stated that "Intriguingly, we found that, whereas persistent bacteria are protected from prophage induction, they are not protected from lytic infection. Quantitative analysis of gene expression reveals that the expression of lytic genes is suppressed in persistent bacteria. However, when persistent bacteria switch to normal growth, the infecting phage resumes the process of gene expression, ultimately causing cell lysis". This is likely to be a significant element in the bacteriophage-mediated destruction of biofilms.

Supporting this, studies with a static phase culture of *Pseudomonas aeruginosa* embedded in a collagen matrix modeling biofilm wound infection [16] showed no killing of bacteria, even during prolonged incubation (up to five days), until the matrix was disrupted and bacterial growth resumed. Killing by bacteriophages was then readily apparent in the growing cultures [17]. Indeed, such activity can actually be problematical in assessing bacterial numbers after bacteriophage treatment.

3. Depolymerases

A large number of enzymes exist that are capable of depolymerizing components of the bacterial biofilm EPS. Among bacteriophages, these include both some of the enzymes produced to release bacteriophages from the host cell (these also include endopeptidases) and tailspike proteins that aid infection, but which are usually restricted to highly localized activity. However, it has been suggested that such proteins, while restricted in activity within the virus particle, can be released from lysing cells in a more generally active form, which can affect the biofilm matrix [18].

It is important to note that different species of bacteria produce different EPS components. Thus, a depolymerase active against the polysaccharides produced by one species may not digest that produced by other bacteria. However, depolymerases are likely to have broader activity than their parent bacteriophages among closely related bacteria, since the complexity (and, hence, the variability) in the EPS is lower than that of the host bacteria. Son *et al.* [19] observed this by comparing the activity of a bacteriophage of *Staphylococcus aureus* with that of the depolymerase that it produced. However, neither would affect bacteria other than *Staphylococci*, suggesting that multiple depolymerases would be required for targeting mixed biofilms.

Where an active depolymerase is produced and released, distinctive "haloes" may be seen around the bacteriophage plaques formed on bacterial lawns, where the bacterial polysaccharide has been damaged or destroyed. Gutiérrez *et al.* [20] used this approach to detect such activity in two bacteriophages infecting *Staphylococcus epidermidis*, both of which were then confirmed by sequencing to contain genes for pectin lyases, while Glonti *et al* [21] identified haloes in cultures of a bacteriophage infecting *Pseudomonas aeruginosa* and purified a depolymerase protein from the bacteriophage using electrophoresis.

Bacteriophage polysaccharide depolymerases have been classified by Yan [12] as endorhamnosidases, alginate lyases, endosialidases and hyaluronidases (glycoside hydrolases). All are known to be carried by multiple bacteriophages. Other relevant enzymes also exist. For example, Broudy *et al.* [22] identified a secreted DNase enzyme produced by a bacteriophage infecting *Streptococcus pyogenes*.

4. Bacteriophage Growth in Biofilms

Experimental data indicates that bacteriophages grow well in *Pseudomonas aeruginosa* biofilms [23], at least in the early stages of biofilm growth. Using two-day-old biofilms [3], it was found that of 17 strains of *Pseudomonas aeruginosa* that were insensitive to a test set of bacteriophages in a plaque assay (thus, using planktonic bacterial hosts), 8/17 strains supported the growth of the same bacteriophages in the biofilm model. These early-stage biofilms are, however, able to block the activity of antibiotics, with Amikacin at 100× the MIC actually enhancing bacterial growth in this system (Figure 4), even though the bacterium was shown to be sensitive at the MIC by disc testing when growing planktonically. This accords with the findings of Gupta *et al.* [8], who also found that the onset of antibiotic resistance occurred in the early stages of biofilm development. Thus, bacteriophages were shown to be able to kill bacteria in situations where conventional antibiotics cannot do so.

Figure 4. The effect of high dose antibiotic (Amikacin at 100× MIC) on an early-stage biofilm of *Pseudomonas aeruginosa*.

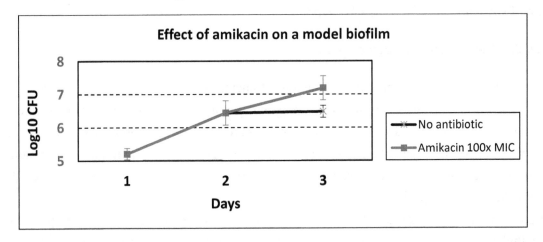

In early studies that helped to demonstrate the potential of bacteriophages for biofilm control, Hanlon *et al.* [24] found that *Pseudomonas aeruginosa* bacteriophages could destroy bacteria in a mature (20-day-old) biofilm and (perhaps surprisingly, given their size) could diffuse through even the thickest (12%) alginate gel studied, although diffusion was slower than through thinner alginate gels. Hanlon also observed that the bacteriophages studied could degrade the alginate polymer directly, apparently via a bacteriophage-carried enzymatic activity, although this was not identified. Whatever the activity, it was clearly different from the highly-restricted tailspike proteins.

Sillankorva *et al.* [1] used bacteriophages of both *Pseudomonas fluorescens* and *Staphylococcus lentus* and demonstrated the effective reduction of single species and mixed biofilms with these agents. Both bacteriophages were fully sequenced, and it was demonstrated that neither coded for a polysaccharide depolymerase (although the *Pseudomonas fluorescens* bacteriophage encoded an endopeptidase). Similarly, Doolittle *et al.* [25] showed that *Escherichia coli* bacteriophage T4 spread efficiently through a biofilm, although it does not code for any polysaccharide depolymerases other than a very restricted tailspike protein, which is only broken out of the bacteriophage tail during the penetration of the host cell. However, Doolittle *et al.* [25] also worked with the E79 bacteriophage of *Pseudomonas aeruginosa* and showed that this was less effective than T4 at penetrating biofilms.

Although it is clear that naturally occurring bacteriophages can penetrate biofilms even when they do not produce polysaccharide depolymerases (or when these are of a very restricted function), not all studies have shown efficient infection within biofilms and some workers continue to believe that EPS-degrading enzymes are necessary for biofilm applications [18].

Tait *et al.* [26] reported that a mixture of three bacteriophages could completely eliminate a single species biofilm, but that this was less effective when other, insensitive bacterial species were present. Kay *et al.* [27] also showed that mixed biofilms could ablate the efficacy of bacteriophages. Despite this, Sillankorva *et al.* [1] showed that efficiency in model biofilms could be high, even with a bacteriophage targeting a single bacterial species, stating that "phages can be adopted as a method to kill a specific bacterium even when its host resides in mixed consortium". Sillankorva *et al.* [1] also showed that mature (seven-day-old) biofilms could be targeted effectively using bacteriophages

Thus, it is clear that natural bacteriophages can and often do express enzymes capable of disrupting biofilms, but that these do not seem to be essential for infectivity in this situation. The potential for the induction of such enzymes from the host genome is, of course, far harder to identify.

5. Enhancing the Activity of Bacteriophages

One concern over the commercial development of bacteriophages as control agents has historically been the opinion that it was not possible to patent natural bacteriophages. This is incorrect as demonstrated by patents awarded to multiple groups protecting such bacteriophages, based, as for all patents, on novelty and "surprising" activity [14,28,29]. However, a number of companies have attempted to use engineered bacteriophages, which has the effect of enhancing the novelty inherent in their intellectual property position.

One approach to this is the addition of virulence-enhancing factors to the bacteriophage genome. This was the approach taken by Lu and Collins [30], who inserted a gene for Dispersin B (a glycoside hydrolase known to have biofilm-dispersing activity) into *Escherichia coli* bacteriophage T7. This was then expressed at high levels during infection, being released into the extracellular matrix from lysed cells. Lu and Collins [30] found that this substantially improved biofilm removal compared to the parent T7 (which was itself a recombinant, containing a gene from bacteriophage T3 to permit replication in this system). T7 naturally produces an endopeptidase, but does not produce a polysaccharide depolymerase. This appears to have represented an experimental system selected to highlight the effect of the expressed enzyme, since Lu and Collins [30] noted that "E. coli, which produces the K1 polysaccharide capsule, is normally resistant to infection by T7 phage" but that this contrasts with Escherichia coli bacteriophage T4, "which can infect and replicate within Escherichia coli biofilms and disrupt biofilm morphology by killing bacterial cells". This approach was then entered into commercial development, targeted at use in biofilm removal in industrial processes.

The value of this type of approach, with the additional regulation inherent in the use of genetically modified organisms (GMOs), has of course to be considered alongside the demonstrated ability of at least some natural bacteriophages to target biofilms effectively. Clearly, generating novel recombinant bacteriophages in this way does enhance patentability, but this has to be considered alongside the potential for developing natural bacteriophages for this purpose.

It should be noted that Lu and Collins [30] used a soluble enzyme, released from the infected cell by lysis. This is relatively straightforward compared to engineering expression within the virus particle, where there are far more constraints on activity unless more permissive methods, such as phage display, are used.

The alternative approach of co-administering bacteriophages with biofilm dispersing agents, such as alginate lyases or even DNase enzymes (such as the commercial product Pulmozyme), has also been suggested [14]. However, this would not permit such agents to be present for succeeding generations of bacteriophage, unlike those that are expressed from within the bacteriophage genome.

6. Biotic and Abiotic Systems

Bacteriophages are replicating biological control agents and are thus inherently different from chemical agents, as noted above. Given the replicating nature of the agent, when the levels of host cells fall below critical (low) levels, bacteriophage amplification is no longer sustainable. However, the work of Marza *et al.* [31] showed that even a picogram-range dose (400 PFU) of a single bacteriophage type is capable of resolving a chronic infection *in vivo*. This single case study showed cycles of improvement and deterioration in the clinical condition, albeit with an overall positive trend, leading to successful resolution after several months. This aligns with the expectations of predator-prey cycles in such a system, as illustrated by Parracho *et al.* [32] (Figure 5).

Figure 5. Predator-prey cycles as applied to bacteriophages [30].

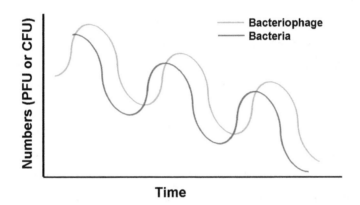

**Predator-prey population dynamics of bacteria (prey; CFU)
during "active" therapy involving bacteriophage (predator; PFU)
replication following a single dose of a bacteriophage therapeutic.
Bacteriophage multiplies *in vivo*, and clears when infection resolves**

In an *in vivo* system, other factors are acting alongside bacteriophages, which act to eliminate an infection. As well as the effects of the immune system, infected material may be cleared by biological effectors (for example, by ciliary action in the lung or ear), mechanically (by wound debridement or physical removal from the upper regions of the lung), chemically (by the disinfection of burns) or even by inherent properties of the infection itself (for example, by the shutdown of pathogenic effectors via quorum sensing when bacterial numbers drop below critical levels). Bacteriophages are a form of biological control [33] and, as such, act in concert with other effectors. This is illustrated by the observation that the efficacy of bacteriophage therapy *in vivo* is reduced when the host animal is immunosuppressed [34].

When bacteriophages are used therapeutically *in vivo*, they have been shown to be able to clear infections, even where these are characterized by biofilm formation, for example in work with chronic infections of the ear in both animal and human systems [31,35,36]. Successful clearance is likely to arise from a combination of direct antibacterial effects and other effectors, as noted above.

In an abiotic system, such as industrial fouling by biofilms, there is a lack of such additional factors, and this may be why enhancing the activity of bacteriophages has attained greater emphasis in such systems. Many of the systems used to study the effects of bacteriophages on biofilms are simple, *in vitro* systems where such effects might be anticipated [26]. However, Sillankorva *et al.* [1] has shown the efficient clearing of both single and multispecies biofilms cultured on steel plates by bacteriophages, so this is by no means an absolute.

7. Combination with Other Agents

Verma [37] noted that lytic bacteriophages could make mature biofilms more amenable to antibiotic treatment, which aligns with findings from the few clinical trials of bacteriophage efficacy reported to date [28,38]. Based on this, the combined or sequential use of antibiotics and bacteriophages has been identified as having potential for therapeutic applications. Supporting this, Yilmaz et al. [39] showed increased effects when bacteriophages and antibiotics were used in combination, including clearance of *Staphylococcus aureus* biofilms. Other combinations are also possible. Sharp et al. [14] proposed the use of a polysaccharide lyase and of DNase enzymes to degrade the biofilm matrix, to be administered alongside bacteriophages. This is also discussed by Abedon [40], although differential diffusion of bacteriophages and co-administered enzymes is noted as an issue. Physical cleaning of wounds can also be combined with bacteriophage use. Seth et al. [41] found using a rabbit ear model that, while neither wound debridement nor phage treatment alone produced an effect in this system, a combination of both was effective.

Similarly, for industrial biofilms, Sillankorva et al. [1] noted that "Although phages can decrease bacterial populations present in biofilms, these biological agents alone are most likely not efficient enough to be applied to control industrial biofilms. Commonly used cleaning procedures remove not only microorganisms, but also all undesirable materials (e.g., foreign bodies, cleaning chemicals, soil, *etc.*), and phages are not capable of this task. However, phages could have a similar function to nowadays used biocides and sanitizers, and be used after the major cleaning processes, to kill specific bacterium on the remainder biofilms." Similarly, Ganegama Arachchi et al. [42] showed efficient clearance of *Listeria monocytogenes* biofilms from steel surfaces by a mixture of three bacteriophages, even when these were contaminated with fish protein. However, Ganegama Arachchi et al. [42] also found that this was accelerated by mechanical clearing, stating that "Phages were more effective on biofilm cells dislodged from the surface compared with undisturbed biofilm cells. Therefore, for short-term phage treatments of biofilm it should be considered that some disruption of the biofilm cells from the surface prior to phage application will be required". As with biological systems, other combinations are possible. Liao et al. [43] observed synergistic effects on silicone catheter segments in the prevention of biofilm formation when bacteriophages were combined with commensal bacteria, while Zhang and Hu [44] observed increased effects on filters when combining bacteriophages with a biocide (chlorine).

8. Engineered and Natural Bacteriophages

As with any biological system, it is possible to optimize bacteriophages for specific functions by either GM approaches [30] or, more recently, by the use of synthetic biology. In doing so, one has to be aware that there are likely to be additional regulatory requirements related to such products [40]. For example, the French company, Pherecydes Pharma, was established to explore the use of such approaches [45]. However, in 2011, they reported that the regulatory complexities surrounding the use of GM agents had refocused their activities on natural bacteriophages [46]. They are currently a main participant in the multicenter EU-sponsored PhagoBurn clinical trial, using natural (non-recombinant) bacteriophages [47].

Bacteriophages are thought to be the most numerous form of life, with around 10^{31} bacteriophages present in the global biosphere [48]. Of these, only a few thousand have been sequenced. Rohwer [49] estimated that "less than 0.0002% of the global phage metagenome has been sampled". As a result, most bacteriophage genes are unknown. Even where the genomes have been sequenced, many of the genes are ORFans; that is, unique genes with no known homologues in other organisms [50]. This results from the highly diverse nature of bacteriophages and means that it is extremely difficult to map the functional elements of bacteriophage replication. To support attempts to enhance or design bacteriophages, there needs to be a far greater understanding of the basic genetics of the systems under study, accompanied by expanding knowledge of just what is present in existing ecosystems. There are signs that technology is beginning to deliver such an understanding [51], but this remains at an early stage.

As a result of the massive and often readily available diversity of bacteriophages, sampling from the environment can be used to identify and isolate naturally evolved bacteriophages for almost any target or application, including biofilms. It is further possible to optimize these bacteriophages for their intended applications either in the initial selection or by serial passage or other standard techniques [52]. This approach can also be used to identify multiple agents, allowing for the possibility of substitution into an approved product [53], which could be more difficult to attain with highly developed GM or other products based around a small number of agents.

9. Conclusions

Bacteriophages possess unique properties and show considerable promise in the control of biofilms. However, such applications are still evolving, and large-scale uses are still under development. Thus, the identification of the most effective approaches has to be, at present, speculative in nature. In time and as more results are published, best practices for such uses will, of course, emerge.

Author Contributions

DRH was involved with performing the work, analyzing the data, and writing and editing the manuscript.

HMRTP, JW, RS, GII, and MW were involved with performing the work, analyzing the data, and editing the manuscript.

SL and SM were involved with analyzing the data and writing and editing the manuscript.

Conflicts of Interest

D. R. Harper is a director and officer of Biocontrol Limited, the U.K. subsidiary of AmpliPhi Biosciences, and an officer of AmpliPhi Biosciences, and holds both shares and stock options in AmpliPhi Biosciences.

S. Lehman is an employee of AmpliPhi Biosciences.

S. Morales is an employee of AmpliPhi, Australia, the Australian subsidiary of AmpliPhi Biosciences.

H. M. R. P. Parracho is a former employee of Biocontrol Limited.

M. Werthén is an employee of Mölnlycke Health Care AB.

Biocontrol Limited holds a licensed patent from Public Health England, which is subject under certain conditions to both milestone and royalty payments.

The views expressed in this publication are those of the authors and not necessarily those of Public Health England or any other government agency.

References

1. Sillankorva, S.; Neubauer, P.; Azaredo, J. *Use of Bacteriophages to Control Biofilms*; LAP Lambert Academic Publishing: Saarbrücken, Germany, 2011.

2. Flemming, H.-C. Biofilms. In *The Encyclopedia of Life Sciences*; John Wiley and Sons: Chichester, UK, 2008.

3. Ceri, H.; Olson, M.E.; Stremick, C.; Read, R.R.; Buret, A. The Calgary biofilm device: New technology for rapid determination of antibiotic susceptibilities of bacterial biofilms. *J. Clin. Microbiol.* **1999**, *37*, 1771–1776.

4. Mah, T.-F.; Pitts, B.; Pellock, B.; Walker, G.C.; Stewart, P.S.; O'Toole, G.A. A genetic basis for *Pseudomonas aeruginosa* biofilm antibiotic resistance. *Nature* **2003**, *426*, 306–310.

5. Spoering, A.L.; Lewis, K. Biofilms and planktonic cells of *Pseudomonas aeruginosa* have similar resistance to killing by antimicrobials. *J. Bacteriol.* **2001**, *183*, 6746–6751.

6. Fauvart, M.; de Groote, V.N.; Michiels, J. Role of persister cells in chronic infections: Clinical relevance and perspectives on anti-persister therapies. *J. Med. Microbiol.* **2011**, *60*, 699–709.

7. Stoodley, P.; Sauer, K.; Davies, D.G.; Costerton, J.W. Biofilms as complex differentiated communities. *Annu. Rev. Microbiol.* **2002**, *56*, 187–209.

8. Gupta, K.; Marques, C.N.H.; Petrova, O.E.; Sauer, K. Antimicrobial tolerance of *Pseudomonas aeruginosa* biofilms is activated during an early developmental stage and requires the two-component hybrid SagS. *J. Bacteriol.* **2013**, 195, 4975–4981.

9. Abedon, S.T. Lysis from without. *Bacteriophage* **2011**, *1*, 46–49.

10. Abedon, S.T.; Thomas-Abedon, C. Phage therapy pharmacology. *Curr. Pharm. Biotechnol.* **2010**, *11*, 28–47.

11. Leiman, P.G.; Chipman, P.R.; Kostyuchenko, V.A.; Mesyanzhinov, V.V.; Rossman, M.G. Three-dimensional rearrangement of proteins in the tail of bacteriophage T4 on infection of its host. *Cell* **2004**, *118*, 419–429.

12. Yan, J.; Mao, J.; Xie, J. Bacteriophage polysaccharide depolymerases and biomedical applications. *BioDrugs* **2013**, *28*, 265–274.

13. Bartell, P.F.; Orr, T.E. Origin of polysaccharide depolymerase associated with bacteriophage infection. *J. Virol.* **1969**, *3*, 290–296.

14. Sharp, R.; Hughes, G.; Hart, A.; Walker, J.T. Bacteriophage for the treatment of bacterial biofilms. U.S. Patent 7758856 B2, 2006.

15. Pearl, S.; Gabay, C.; Kishony, R.; Oppenheim, A.; Balaban, N.Q. Nongenetic Individuality in the host-phage interaction. *PLoS Biol.* **2008**, *5*, e120.

16. Werthén, M.; Henriksson, L.; Jensen, P.Ø.; Sternberg, C.; Givskov, M.; Bjarnsholt, T. An *in vitro* model of bacterial infections in wounds and other soft tissues. *APMIS* **2010**, *118*, 156–164.

17. Monk, A.; Parracho, H.; Cass, J.; McConville, M.; Harper, D.; Werthén, M.; Erikson, K. Bacteriophages and biofilms: A waiting game? In Proceedings of the 18th Biennial Evergreen International Phage Biology Meeting, Olympia, WA, USA, 9–14 August 2009.

18. Cornelissen, A.; Ceyssens, P.J.; T'Syen, J.; van Praet, H.; Noben, J.P.; Shaburova, O.V.; Krylov, V.N.; Volckaert, G.; Lavigne, R. The T7-related *Pseudomonas putida* phage φ15 displays virion-associated biofilm degradation properties. *PLoS One* **2011**, *6*, e18597.

19. Son, J.S.; Lee, S.J.; Jun, S.Y.; Yoon, S.J.; Kang, S.H.; Paik, H.R.; Kang, J.O.; Choi, Y.J. Antibacterial and biofilm removal activity of a podoviridae *Staphylococcus aureus* bacteriophage SAP-2 and a derived recombinant cell-wall-degrading enzyme. *Appl. Microbiol. Biotechnol.* **2010**, *86*, 1439–1449.

20. Gutiérrez, D.; Martínez, B.; Rodríguez, A.; García, P. Genomic characterization of two *Staphylococcus epidermidis* bacteriophages with anti-biofilm potential. *BMC Genomics* **2012**, *13*, e228.

21. Glonti, T.; Chanishvili, N.; Taylor, P.W. Bacteriophage-derived enzyme that depolymerizes the alginic acid capsule associated with cystic fibrosis isolates of *Pseudomonas aeruginosa*. *J. Appl. Microbiol.* **2010**, *108*, 695–702.

22. Broudy, T.A.; Pancholi, V.; Fischetti, V.A. The *in vitro* Interaction of *Streptococcus pyogenes* with human pharyngeal cells induces a phage-encoded extracellular DNase. *Infect. Immun.* **2002**, *70*, 2805–2811.

23. Sharp, R.; Walker, J.T.; Riley, P.; Budge, C.; West, K.; Hughes, G. Bacteriophage therapy to control biofilms of *Pseudomonas aeruginosa* in the lungs of patient with cystic fibrosis. In *Biofilm Communities—Order from Chaos*; McBain, A., Allison, D., Brading, M., Rickard, A., Verran, J., Walker, J., Eds.; Bioline: Cardiff, UK, 2003; pp. 237–245.

24. Hanlon, G.W.; Denyer, S.P.; Olliff, C.J.; Ibrahim, L.J. Reduction in exopolysaccharide viscosity as an aid to bacteriophage penetration through *Pseudomonas aeruginosa* biofilms. *Appl. Environ. Microbiol.* **2001**, *67*, 2746–2753.

25. Doolittle, M.M.; Cooney, J.J.; Caldwell, D.E. Tracing the interaction of bacteriophage with bacterial biofilms using fluorescent and chromogenic probes. *J. Ind. Microbiol.* **1996**, *16*, 331–341.

26. Tait, K.; Skillman, L.C.; Sutherland, I.W. The efficacy of bacteriophage as a method of biofilm eradication. *Biofouling* **2002**, *18*, 305–311.

27. Kay, M.K.; Erwin, T.C.; McLean, R.J.; Aron, G.M. Bacteriophage ecology in *Escherichia coli* and *Pseudomonas aeruginosa* mixed-biofilm communities. *Appl. Environ. Microbiol.* **2011**, *77*, 821–829.

28. Soothill, J.S.; Hawkins, C.; Harper, D.R. Bacteriophage-containing therapeutic agents. U.S. Patent 8105579 B2, 2011.

29. Sulakvelidze, A.; Pasternack, G.R. *Pseudomonas aeruginosa* bacteriophage and uses thereof. U.S. Patent 7622293 B2, 2008.

30. Lu, T.K.; Collins, J.J. Dispersing biofilms with engineered enzymatic bacteriophage. *Proc. Natl. Acad. Sci.* **2007**, *27*, 11197–11202.

31. Marza, J.A.S.; Soothill, J.S.; Boydell, P.; Collyns, T.A. Multiplication of therapeutically administered bacteriophages in *Pseudomonas aeruginosa* infected patients. *Burns* **2006**, *32*, 644–646.

32. Parracho, H.M.R.T.; Burrowes, B.H.; Enright, M.C.; McConville, M.L.; Harper, D.R. The role of regulated clinical trials in the development of bacteriophage therapeutics. *J. Mol. Genet. Med.* **2012**, *6*, 279–286.

33. Harper, D.R. Biological control by microorganisms. In *The Encyclopedia of Life Sciences*; John Wiley and Sons: Chichester, UK, 2013.

34. Tiwari, B.; Kim, S.; Rahman, M.; Kim, J. Antibacterial efficacy of lytic *Pseudomonas* bacteriophage in normal and neutropenic mice models. *J. Microbiol.* **2011**, *49*, 994–999.

35. Hawkins, C.; Harper, D.R.; Burch, D.; Anggard, E., Soothill, J. Topical treatment of *Pseudomonas aeruginosa* otitis of dogs with a bacteriophage mixture: A before/after clinical trial. *Vet. Microbiol.* **2010**, *146*, 309–313.

36. Wright, A.; Hawkins, C.; Anggard, E.A.; Harper, D.R. A controlled clinical trial of a therapeutic bacteriophage preparation in chronic otitis due to antibiotic-resistant *Pseudomonas aeruginosa*; a preliminary report of efficacy. *Clin. Otolaryngol.* **2009**, *34*, 349–357.

37. Verma, V.; Harjai, K.; Chhibber, S. Structural changes induced by a lytic bacteriophage make ciprofloxacin effective against older biofilm of *Klebsiella pneumonia*. *Biofouling* **2010**, *26*, 729–737.

38. Harper, D.R. Beneficial effects of bacteriophage treatment. U.S. Patent 8475787 B2, 2010.

39. Yilmaz, C.; Colak, M.; Yilmaz, B.C.; Ersoz, G.; Kutateladze, M.; Gozlugol, M. Bacteriophage therapy in implant-related infections: An experimental study. *J. Bone Jt. Surg. Am.* **2013**, *95*, 117–125.

40. Abedon, S.T. *Bacteriophages and Biofilms*; Nova Science Publishers: New York, NY, USA, 2011.

41. Seth, A.K.; Geringer, M.R.; Nguyen, K.T.; Agnew, S.P.; Dumanian, Z.; Galiano, R.D.; Leung, K.P.; Mustoe, T.A.; Hong, S.J. Bacteriophage therapy for *Staphylococcus aureus* biofilm-infected wounds: A new approach to chronic wound care. *Plast. Reconstr. Surg.* **2013**, *131*, 225–234.

42. Ganegama Arachchi, G.J.; Cridge, A.G.; Dias-Wanigasekera, B.M.; Cruz, C.D.; McIntyre, L.; Liu, R.; Flint S.H.; Mutukumira, A.N. Effectiveness of phages in the decontamination of *Listeria monocytogenes* adhered to clean stainless steel, stainless steel coated with fish protein, and as a biofilm. *J. Ind. Microbiol. Biotechnol.* **2013**, *40*, 1105–1116.

43. Liao, K.S.; Lehman, S.M.; Tweardy, D.J.; Donlan, R.M.; Trautner, B.W. Bacteriophages are synergistic with bacterial interference for the prevention of *Pseudomonas aeruginosa* biofilm formation on urinary catheters. *J. Appl. Microbiol.* **2012**, *113*, 1530–1539.

44. Zhang, Y.; Hu, Z. Combined treatment of *Pseudomonas aeruginosa* biofilms with bacteriophages and chlorine. *Biotechnol. Bioeng.* **2013**, *110*, 286–295.

45. Pouillot, F.; Blois, H.; Iris, F. Genetically engineered virulent phage banks in the detection and control of emergent pathogenic bacteria. *Biosecur. Bioterror.* **2010**, *8*, 155–169.

46. Gabard, J. Natural *versus* engineered phages: Benefits and drawbacks for industrial applications. In Proceedings of the Phages 2011 Meeting, Oxford, UK, 19–21 September 2011.

47. PhagoBurn. Available online: http://www.phagoburn.eu (accessed on 18 January 2014).

48. Rohwer, F.; Edwards, R. The phage proteomic tree: A genome-based taxonomy for phage. *J. Bacteriol.* **2002**, *184*, 4529–4535.

49. Rohwer, F. Global phage diversity. *Cell* **2003**, *113*, 141.

50. Yin, Y.; Fischer, D. Identification and investigation of ORFans in the viral world. *BMC Genomics* **2008**, *9*, e24.

51. Kristensen, D.M.; Cai, X.; Mushegian, A. Evolutionarily conserved orthologous families in phages are relatively rare in their prokaryotic hosts. *J. Bacteriol.* **2011**, *193*, 1806–1814.

52. Kelly, D.; McAuliffe, O.; Ross, R.P.; O'Mahony, J.; Coffey, A. Development of a broad-host-range phage cocktail for biocontrol. *Bioeng. Bugs* **2011**, *2*, 31–37.

53. Verbeken, G.; Pirnay, J.-P.; de Vos, D.; Jennes, S.; Zizi, M.; Lavigne, R.; Casteels, M.; Huys, I. Optimizing the European regulatory framework for sustainable bacteriophage therapy in human medicine. *Arch. Immunol. Ther. Exp.* **2012**, *60*, 161–172.

5

Bacteriophage 434 Hex Protein Prevents RecA-Mediated Repressor Autocleavage

Paul Shkilnyj, Michael P. Colon and Gerald B. Koudelka *

Department of Biological Sciences, University at Buffalo, Buffalo, NY 14260, USA;
E-Mails: shkilnyj@gmail.com (P.S.); mpcolon@buffalo.edu (M.P.C.)

* Author to whom correspondence should be addressed; E-Mail: koudelka@buffalo.edu

Abstract: In a λ^{imm434} lysogen, two proteins are expressed from the integrated prophage. Both are encoded by the same mRNA whose transcription initiates at the P_{RM} promoter. One protein is the 434 repressor, needed for the establishment and maintenance of lysogeny. The other is Hex which is translated from an open reading frame that apparently partially overlaps the 434 repressor coding region. In the wild type host, disruption of the gene encoding Hex destabilizes λ^{imm434} lysogens. However, the *hex* mutation has no effect on lysogen stability in a *recA⁻* host. These observations suggest that Hex functions by modulating the ability of RecA to stimulate 434 repressor autocleavage. We tested this hypothesis by identifying and purifying Hex to determine if this protein inhibited RecA-stimulated autocleavage of 434 repressor *in vitro*. Our results show that *in vitro* a fragment of Hex prevents RecA-stimulated autocleavage of 434 repressor, as well as the repressors of the closely related phage P22. Surprisingly, Hex does not prevent RecA-stimulated autocleavage of phage lambda repressor, nor the *E. coli* LexA repressor.

Keywords: bacteriophage; RecA; lysogeny

1. Introduction

Upon infection of a host cell, the lambdoid phages choose between two developmental fates, opting either to replicate and lyse the cell, or to enter the latent, or lysogenic phase in which the phages' chromosome is integrated into that of the host and is replicated along with it [1,2]. Maintenance of the lysogenic state requires the activity of the phage's cI repressor protein. The repressor functions to maintain lysogeny by repressing transcription from promoters P_R and P_L, thereby preventing synthesis of proteins needed for phage lytic development. At the same time, the repressor activates transcription of its own gene by stimulating transcription from P_{RM} and thereby maintains appropriate repressor concentrations inside the cell [1].

The survival of the integrated prophage is linked to the fitness and survival of its host. Thus to ensure their survival, the lambdoid phages employ mechanisms that allow prophages to enter the lytic growth pathway and thus escape from hosts whose survival is in doubt. The best understood regulator of lambdoid prophage induction is the RecA protein [3–5]. As the master regulator of the SOS response, RecA inactivates LexA protein, the repressor of the SOS response genes, by stimulating its nascent autocleavage activity [5] thereby eliminating LexA protein's DNA binding activity. Similarly, RecA also stimulates autoproteolysis of the lambdoid phage repressors [6,7]. When the repressor is cleaved into its constituent carboxyl terminal and amino terminal domains, it is incapable of binding DNA with high affinity DNA binding or forming higher order oligomers (dimers, tetramers, *etc.*) Consequently transcription from P_R and P_L are no longer repressed, and the prophage enters the lytic growth pathway.

Transcription from P_{RM} requires the stimulatory activity of DNA bound repressor. Hence this promoter is only active in the presence of repressor, *i.e.*, in the lysogenic state. In addition to the repressor gene many lambdoid bacteriophages encode additional proteins on the transcript initiated from P_{RM}. Since these genes are not essential for lysogenic or lytic growth, they are considered accessory [8]. However in some cases they apparently provide an advantage to both the host and the phage. For example, the RexA and RexB proteins, which are produced by bacteriophage λ lysogens, excludes superinfection by bacteriophage T4rII$^-$ [9]. In bacteriophage 933W a eukaryotic-like tyrosine kinase is cotranscribed with cI repressor [10] and its activity aborts infection of the 933W lysogen by a superinfecting HK97 phage [11]. Further examples of proteins that are expressed during lysogeny of temperate bacteriophages are found in the tripartite immunity system of phages P1 and P7 [12], and phage K139 [13]. These proteins also prevent superinfection by other phages.

In a lysogen, bacteriophage 434 expresses, in addition to cI repressor, a second protein Hex, which is encoded on the mRNA transcript initiated from P_{RM}. Susskind and Botstein demonstrated that *S. typhimurium* lysogenized with $λ^{imm434}$ excluded growth of superinfecting phage P22 [14]. However, *S. Typhimurium* lysogenized with $λ^{imm434:Hex-}$ did not exclude the growth of superinfecting phage P22. The name Hex, therefore, stands for heterologous exclusion.

Since the heterologous phage exclusion functions of each of the proteins encoded along with repressor is apparently highly specific for particular phage(s), we wondered whether these proteins may have a more general role in bacteriophage biology. Our results show that deletion of the open reading frame of Hex increases the spontaneous induction frequency of $λ^{imm434}$ phage. We also show that in a RecA$^-$ strain of *E. coli*, the spontaneous induction of $λ^{imm434:hex-}$ and wild type $λ^{imm434}$ are identically low. The results of genetic and biochemical experiments suggest that, when present, Hex increases the stability of bacteriophage 434 by interfering with RecA-mediated autocleavage of the phage 434 repressor. Consistent with this finding, we found that purified Hex blocks RecA-mediated stimulation of 434 repressor autocleavage. We found that Hex also prevents RecA stimulated autoproteolysis of the repressor encoded by bacteriophage P22, but not the repressor of bacteriophage λ or *E. coli* LexA. All of which are susceptible to RecA stimulated autoproteolysis. This finding suggests that Hex does not block RecA activity by directly interacting with it.

2. Results

Computer analysis of the three reading frames in the mRNA initiated from P_{RM} of the bacteriophage 434 identified a a potential open reading frame with a capacity to code for a 158 amino acid protein located 3' to the 434 repressor gene (Figure 1). The predicted amino acid sequence of the identified ORF is shown in Figure 2, beginning with a methionine encoded by the 5'-most AUG of the transcript. To ensure that the transcript initiated at P_{RM} extends through the putative ORF downstream from the repressor gene, we isolated total bacterial RNA from MG1655(λ^{imm434}) lysogens, reverse transcribed this RNA using primer 1 (Figure 1) and the resulting cDNA was analyzed by PCR using a series of primers complementary to this region (Figure 1). Since the RNA was isolated from a phage lysogen, RNA encoding the repressor must be present. As expected, a product of ~627 bp, the predicted size, was obtained when primer set complementary to this region (Primers 2 + 3) of the RNA was used (Figure 3, lane 1). When PCR was performed using primers 1 + 3, a DNA product of about 1100 base pairs was obtained (Figure 3, lane 2). This observation shows that in the MG1655::λ^{imm434} lysogen, transcripts initiated at P_{RM} continue through the predicted Hex ORF.

Figure 1. Schematic representation of the immunity region of bacteriophage 434. The N- and C-terminal domains of the cI repressor dimers are represented by black and gray circles, respectively bound to the operator sites. In above configuration shown, transcription of the promoters P_R and P_L are repressed and transcription from P_{RM} is activated on. The Hex gene is located on the same transcript initiated at P_{RM}. Primers designated 1, 2, and 3 were used to analyze the P_{RM} transcript (see text).

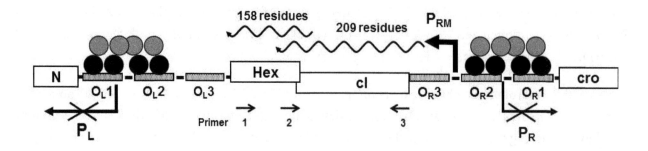

Figure 2. The putative amino acid sequence of Hex. The predicted amino acid sequence starts from the 5'-most AUG, in the hex transcript. Additional potential methionine initiators highlighted. The red amino acids highlight the beginning and end of the sHex polypeptide (see also Figure 5).

```
                10                  20                  30
    M A Y D S Y Q R E L Q D Y R C C R G S E G K I R M I R I A A
                40                  50                  60
    L L S I L L T T S A N S E C W I V T N L H G Y G A M N G D R
                70                  80                  90
    Y E F T K D S T E D S V F N V T I N G D K S S V Y E S V S G
                100                 110                 120
    V Y P E M K Y T A L S S N T M V G E Y Q S G G G I T V E T W
                130                 140                 150
    S I T T D K K A L Y S K V M N I P G M Q Q L T S T K S F V G

    D V V G T C N Q
```

Figure 3. RT-PCR analysis of total bacterial RNA isolated from MG1655::λ^{imm434} lysogens. Total RNA was reverse transcribed using Primer 1 (see Figure 1). The cDNA was PCR separately amplified with primers 2 + 3 (lane 1) or 1 + 3 (lane 2), and products separated on an agarose gel. The numbers represent sizes of molecular weight standards.

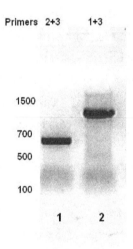

2.1. Effect of Hex Mutation on λ^{imm434} Prophage Induction

E. coli strain MG1655 was lysogenized with λ$^{imm434hex-}$ and the amount of phage produced spontaneously by *E. coli* strain MG1655 separated lysogenized with λ^{imm434} and λ$^{imm434hex-}$ was determined as described in the Experimental Section. We find that the MG1655::λ$^{imm434hex-}$ lysogens produced 30-fold more phage during a five hour incubation then did the wild type (MG1655::λ$^{imm434)}$ lysogens (Figure 4). This finding suggests that Hex may function in regulating the lysis-lysogeny of integrated λ^{imm434} prophage. Alternatively, the increased in number of phage produced during the incubation may be a result of an effect of *hex* on the number of phage released/cell. We tested this idea by measuring phage burst sizes for both of these bacteriophages (see Experimental Section). We find that the burst sizes of the λ^{imm434} and λ$^{imm434hex-}$ phages are indistinguishable from each other; under the conditions of our measurements, each cell infected by these phages produces 104 phage.

Figure 4. Effect of *hex* mutation on stability of MG1655::λ^{imm434} lysogens. MG1655(λ^{imm434}) and MG1655($\lambda^{\text{imm434hex}-}$) were grown to saturation, washed and resuspended in fresh media for 4 hours. The amount of phage release into the culture was determined by plating various amounts of supernatant on MG1655 cells that did not contain λ^{imm434} prophage. The concentration of lysogen cells was determined by plating various dilutions of cells on LB agar plates. Error bars are standard deviation of twenty four separate repeats.

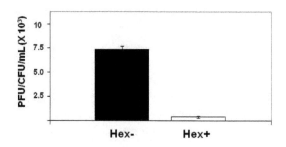

2.2. Role of RecA and Hex in Regulating Prophage Stability

Since RecA affects the level of spontaneous phage induction in a RecA$^+$ host [15], we tested the idea that Hex may affect spontaneous induction by influencing RecA's co-protease anti-repressor activity. To test this idea, we compared the spontaneous induction frequency of λ^{imm434} and $\lambda^{\text{imm434hex}-}$ prophages in wild type and RecA$^-$ MG1655 strains. The spontaneous induction frequency of MG1655recA938::cat ($\lambda^{\text{imm434hex}-}$) is $1.8 \pm 0.08 \times 10^{-5}$, which is nearly identical to the spontaneous induction rate of MG1655recA938::cat::(λ^{imm434}) (Table 1). In contrast, the spontaneous induction frequency of $\lambda^{\text{imm434hex}-}$ and λ^{imm434} in wild type MG1655 are remarkably different. In the wild type host, the amount of induction of MG1655($\lambda^{\text{imm434hex}-}$) is nearly 30-fold higher than the induction frequency of MG1655(λ^{imm434}) (Table 1) Although the frequency of spontaneous induction in a RecA$^-$ host is much lower than in the wild type MG1655 host, the nearly identical frequency of induction in RecA$^-$ host of both $\lambda^{\text{imm434hex}-}$ and wild type λ^{imm434} suggests that Hex increases the stability of λ^{imm434} by interfering with RecA mediated autocleavage of the cI repressor.

Table 1. Effect of *hex* Deletion on Spontaneous Induction of RecA$^-$ MG1655(λ^{imm434}) Lysogens. The values (+/− standard deviation) were determined as described in the Experimental Section. Data is an average based on twenty four separate repeats.

Strain		PFU/mL/viable cell (X1000)		
Wild-type		0.275 ± 0.06		
*rec*A-		$1.8 \pm 0.08 \times 10^{-5}$		
hex-		7.34 ± 0.1		
*rec*A-/*hex*-		$1.8 \pm 0.23 \times 10^{-5}$		

Figure 5. Expression and purification of Hex polypeptide. Full length *hex* gene sequence was inserted into plasmid pMAL-c2X to generate a fusion protein between the genes encoding MBP and Hex. *E. coli* extracts containing the fusion peptide were passed through an amylase column and the eluate digested by Factor Xa (see Experimental Section). Shown is a Coomassie stain of the products of the Factor Xa digest of the eluate from amylase column, fractionated by SDS-PAGE (left panel). Bacterial strain BL21(DE3)[pLysS] was transformed with plasmid psHex containing the gene coding for the identified sHex (see Figure 2). The sHex was purified from cell extracts using ammonium sulfate, GuHCl, followed by size exclusion chromatography. The right panel shows Coomassie stain of an SDS-PAGE of the the pooled protein-containing fractions from the sizing column. The positions of sHex, maltose binding protein (MBP) and molecular weight standards are indicated.

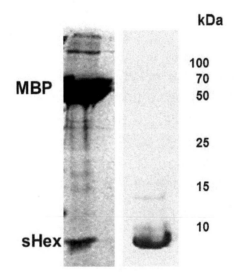

2.3. Expression and Purification of Hex

We placed the *hex* gene under control of the T7 RNA polymerase promoter and attempted to express this full length gene product. We repeatedly observed that upon induction of Hex expression, cultures bearing plasmids encoding full-length Hex stopped growing, and subsequently showed a decrease in optical density at 600 nm. Therefore we initially attempted to purify Hex as a fusion with maltose binding protein (MBP). We reasoned that as a fusion with MBP (Maltose Binding Protein), the Hex peptide may be less detrimental to cell viability. When a plasmid encoding a MBP-Hex fusion protein was transformed into *E. coli* strain BL21(DE3)[pLysS], over-expression of the MBP-Hex protein was induced upon addition of IPTG. The expressed peptide was purified on an amylose resin and fractions containing the MPB-Hex were pooled and digested with factor Xa to separate the two proteins and then fractioned on SDS-PAGE (Figure 5). Sequence analysis shows that the MBP-Hex fusion protein contains a single Factor Xa target sequence, located at the fusion site between MBP and Hex. Thus we would have anticipated this treatment would have released a ~17.5 kDa protein-the size of full-length Hex. Instead, treatment of the MBP-Hex protein with Factor Xa produced two fragments, one the approximate size of MBP, and the other roughly 6 kDa (Figure 5). Repeated attempts to generate the full Hex (MW ~16 kDa) from the MBP-Hex fusion by cleaving the fused peptide with the Factor Xa under various conditions, or sources of Factor Xa failed. We are unsure as to why we are unable to obtain full length Hex by Factor Xa treatment. However we hypothesized that this 6 kDa protein fragment represented a stable fragment of Hex and proceeded to analyze this Hex fragment.

2.4. Characterization of sHex

To determine whether the 6 kDa protein fragment that results from factor Xa cleavage of the MBP-Hex fusion is a fragment of Hex, we cut this fragment out of a 15% SDS-polyacrylamide Tris-Tricine gel, and determined its sequence by MALDI-TOF mass spectrometry. The resulting sequence confirmed this peptide derives from full length Hex. The shortened Hex polypeptide (hereafter referred to as sHex) sequence begins with glutamic acid 43 and ends with asparagine 103 (see Figure 1).

Figure 6. Effect of sHex expression on the stability of MG1655::$\lambda^{imm434\ hex-}$ lysogens in MG1655. MG1655::$\lambda^{imm434\ hex-}$ lysogens were transformed with a control plasmid (black bars) or a plasmid encoding the sHex only (hatched bar). For comparison, the amount of spontaneously released from MG1655(λ^{imm434}) lysogens (denoted as WT) under these conditions is also shown (white bar). The amount of phage spontaneously released was determined as described in the Experimental Section. Error bars are standard deviation of twenty four separate experiments.

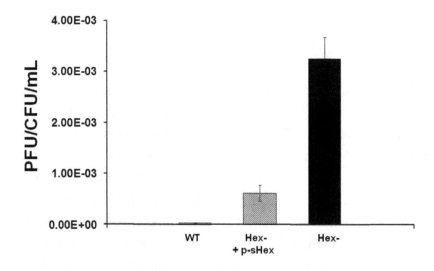

We determined whether sHex is capable of complementing the effects of Hex deletion. To accomplish this, we amplified DNA encoding the sHex sequence and placed expression of sHex gene under the control of the T7 RNA polymerase promoter in pET17b [16], creating the plasmid p-sHex. The sequence of this protein is as given in Figure 2, except it contains an initiator methonine precedingE43. MG1655($\lambda^{imm434hex-}$) lysogens were transformed with p-sHex together with pGP1-2 which encodes a T7 RNA polymerase under the control of the temperature sensitive *cI857* mutant λ repressor [16]. As a control, MG1655($\lambda^{\text{imm434hex-}}$) cells were transformed with pGP1-2, and pET17b lacking the sHex insert. Cultures of these cells were grown to saturation overnight at 32 °C, washed, and resuspended in fresh medium and incubated at 37 °C for four hours. After 4 hours, the spontaneous induction frequencies of $\lambda^{\text{imm434hex-}}$ prophages in MG1655 with or without the sHex encoding plasmid were determined as described (Experimental Section). Expression of sHex reduces the spontaneous induction frequency of Hex⁻ lysogens by ~7-fold compared to cells transformed with control plasmid (Figure 6). This finding indicates that sHex can act in *trans* to partially complement the *hex-* defect in $\lambda^{\text{imm434hex-}}$. This observation suggests that sHex constitutes the 'active' fragment of Hex.

2.5. sHex Inhibits RecA-Mediated Autocleavage of 434 Repressor in Vitro

The results in Table 1 suggest that Hex may alter the spontaneous induction frequency of λ^{imm434} prophages by interfering with RecA mediated autocleavage of the 434 repressor, thereby reducing the amount bacteriophage 434 produced. To test this hypothesis we purified sHex as described in the Experimental Section and examined the effect of this protein on RecA-mediated autocleavage of 434 repressor. Mixing sHex (5 µM) with 300 nM 434 repressor does not result in the formation of any 434 repressor cleavage products (Figure 7A, lane 2), whereas mixing 434 repressor with RecA (see Experimental Section) causes the formation of two lower molecular weight antibody reactive species (Figure 7A, lane 3). The molecular weights of these products correspond to the 434 repressor N- and C-terminal domain fragments, consistent with the previous observation that RecA stimulates 434 repressor autocleavage [17]. When increasing concentrations of sHex are added into mixtures of 434 repressor and RecA, the amount of repressor cleavage products observed progressively decreases (Figure 7A, lanes 5–8). In an identical set of experiments in which increasing concentrations of BSA instead of sHex were added into the mixtures of 434 repressor and RecA, BSA did not interfere with the reaction (Figure 7B) Therefore the findings in Figure 7 indicate that sHex interferes with the ability of RecA to catalyze autoproteolysis of 434 repressor *in vitro*.

2.6. Effect of sHex on RecA-Mediated Autocleavage of Related Repressors P22, λ, and the Bacterial LexA Repressor

To determine whether or not the sHex inhibition of RecA stimulated autoproteolysis is specific to the 434 cI repressor, we examined the effect of sHex on RecA stimulated autocleavage of related cI repressors from lambdoid phages P22 and λ, as well as the SOS repressor LexA, which also undergoes RecA stimulated autocleavage. In an identical procedure as described above, we examined the effect of adding sHex on RecA stimulated autocleavage of P22 repressor (Figure 8A), λ repressor (Figure 8B) and LexA repressor (Figure 8C). Similar to the case with 434 repressor, added sHex interferes with the RecA stimulated autocleavage of P22 repressor (compared Figure 8A with Figure 7A). In contrast, added sHex does not inhibit RecA mediated autocleavage of the λ repressor (Figure 8B) under these conditions. Similar to the λ repressor case, sHex is also incapable of blocking RecA stimulated autocleavage of the SOS repressor LexA (Figure 8C) under these conditions.

Figure 7. Effect of Hex on RecA-stimulated autocleavage of 434 Repressor. (**A**) 434 repressor [300 nM] (lane1) was incubated with sHex (lane2), 8 μM RecA (lanes 3) or 8 μM RecA plus the indicated sHex concentrations (lanes 5–8). (**B**) In identical procedure as in (A), except BSA was substituted for sHex. The figure is a composite immunoblot of the reaction products reacted with anti-434R antibodies and visualized by chemiluminescence.

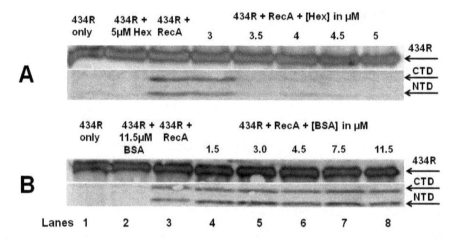

Figure 8. Effect of Hex on RecA-stimulated autocleavage of bacteriophage and bacterial repressors. Bacteriophage repressors from phage P22 (**A**), λ (**B**), and the bacterial SOS repressor LexA (**C**) were incubated in the absence (lane1) or presence of indicated concentrations of sHex and/or 5 μM RecA. The figure is a composite immunoblot of the reaction products reacted with anti-protein antibodies and visualized by chemiluminescence.

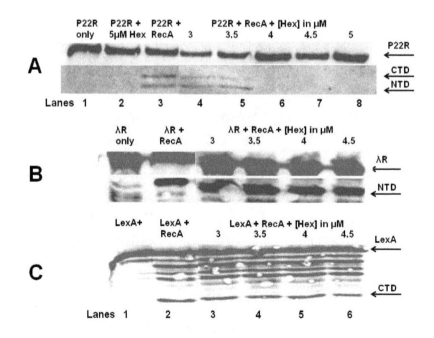

3. Discussion

Our results show that the Hex protein of bacteriophage 434 reduces spontaneous induction of the 434 prophage. Based on the results of *in vitro* biochemical data, Hex apparently reduces prophage spontaneous induction by inhibiting RecA mediated autoproteolysis of the cI repressor. The *hex* gene is cotranscribed along with that of the cI repressor gene on mRNA that is initiated at the promoter for repressor maintenance (P_{RM}) in a P_{RM}-*cI-Hex* operon (Figures 1 and 3). Many temperate prophages produce accessory proteins that benefit the phage, the *E. coli* host, or both, from the same message as their repressor. The most common advantage conferred by many accessory proteins is exclusion of other superinfecting phages [9,12,13,18]. Our data show that the Hex protein of bacteriophage 434, in addition to conferring immunity to phage P22 when expressed in *S. typhimurium*, also reduces induction of λ^{imm434} by interfering with RecA stimulated autocleavage of phage 434 repressor. When the RecA protein is not present in *E. coli* host, $\lambda^{imm434Hex^-}$ induces with same frequency as the wild type phage (Table 1), however in a wild type *E. coli* host, $\lambda^{imm434Hex^-}$ induction is 30 fold higher. Producing sHex from a plasmid in $\lambda^{imm434hex^-}$ lysogens reduces the spontaneous induction about 7 fold (Figure 6). These results suggest that Hex is an advantageous adaptation to bacteriophage 434 in that it "fine tunes" the phage lysis-lysogeny decision.

The host SOS response to DNA damage causes the lysogenic phage to enter lytic growth; once this cycle initiated, it is irreversible. Under normal conditions, (aerated rich media and lack of DNA damaging agents) SOS genes are expressed at a very low basal level [12]. Furthermore, under normal conditions there is enough repressor made to reduce the transcription of P_R roughly 1,000 fold [1]. However, since phage 434 cI repressor's cooperatively binds DNA, a 5 fold drop in repressor concentration causes repressor to only partially occupy O_R1 and O_R1, and thus increase the activity of P_R by 50% [19]. The small decrease in repressor concentration, combined with a small amount of active RecA, which may lead to further autocleavage of the repressor. This situation could result in unwanted spontaneous induction. Bacteriophage 434 may use Hex to prevent this "false alarm" induction.

The *in vitro* autocleavage experiments shown in Figures 7 and 8 demonstrate that sHex blocks RecA mediated autocleavage of the phage 434 repressor and P22 repressor, but not λ repressor or *E. coli* LexA repressor (a protein central in the SOS response). Although the 434 repressor, P22 repressor, λ repressor and LexA are structurally similar, they exhibit differences in their mechanisms of RecA-mediated autocleavage. For example, 434 repressor undergoes autocleavage most efficiently as a DNA-bound dimer [17], as opposed to lambda and LexA repressors which can be efficiently autocleaved as monomers [20,21]. P22 repressor autocleavage increases with increase repressor concentrations, which suggests that P22 repressor is a better substrate for RecA-catalyzed inactivation as a dimer [4]. The different mechanisms of intra or intermolecular cleavage are also prevalent in other RecA targets such as UmuD [22]. These differences in RecA mediated autocleavage of various repressors may account for the selectivity of Hex for the repressors of phages 434 and P22.

RecA function can be inhibited by competing for DNA binding by DNA binding proteins, filament formation, or preventing RecA from hydrolyzing ATP [23]. For example, RecA dependent proteolysis of P22 repressor is blocked when P22 antirepressor-repressor complex is formed, which prevents access of RecA to the P22 repressor [24]. Phage P22 *virA* produces the antirepressor protein [24]. The specific manner in which Hex prevents RecA mediated autocleavage of 434 and P22 repressors suggests that Hex does not function by directly competing with RecA for DNA, does not prevent ATP hydrolysis by RecA, nor does it directly interact with RecA active site for repressor interaction. Rather, Hex may interact with 434 repressor, and its close homolog the P22 repressor (and with repressor-DNA complexes), thereby preventing RecA mediated autocleavage. Regardless of the precise mechanism, it is clear that Hex reduces RecA-mediated repressor autocleavage. We further speculate that by being selective only to phages 434 & P22 repressors, Hex does not interfere with bacterial SOS repair. A number of DNA repair cycles may be necessary throughout the bacterial life cycles in order to repair minor DNA damage. These repairs will require a small amount of SOS gene products, active RecA being one of them. However the phage must be able to distinguish life threatening DNA damage from more minor effects, and Hex may function in this capacity.

Outside of the biological function of Hex, for puzzling, and as of yet unidentified reasons, overexpression of full-length Hex is lethal to cells. We were also unable to isolate the MBP-Hex fusion. Treatment of the fusion protein with factor Xa, surprisingly, yielded a much shorter peptide then the full length Hex. Neither long nor short Hex contain factor Xa cleavage sites, nor are there any basic residues present at the ends of the sHex which sometimes can be cleaved by factor Xa. However a presence of cryptic Xa sites cannot be excluded. The possibility that full length Hex is processed to sHex by the activity of residual bacterial proteases unrelated to factor Xa present during the purification process also cannot be excluded. A more intriguing possibility is that full length Hex has an intrinsic autoprotease activity, and undergoes spontaneous proteolytic cleavage. Regardless of the precise mechanism of the full length Hex modification, we conclude that we have identified the active Hex peptide, and demonstrated its role in bacteriophage 434 lysis-lysogeny modulation.

4. Experimental Section

4.1. Media and Growth Conditions

Liquid cell cultures were grown in Luria broth or M9 minimal media [25], each of which were supplemented with 100 μg/mL ampicillin and 30 μg/mL kanamycin, where needed, to select for the presence of plasmids.

4.2. Bacterial Strains, Phages, and Plasmids

The host strain used for all spontaneous induction analysis was MG1655 [26,27]. The MG1655 *recA* mutants were created by P1 transduction using a lysate from *E. coli* GW4212, which bears the *recA938::cat* allele [28] (a gift from Mark Sutton, University at Buffalo, Buffalo, NY, USA). Both wild type and the *recA* mutant MG1655 were lysogenized with λ^{imm434} or $\lambda^{imm434:\ hex-}$ as previously described [8]. The *hex* mutant version of λ^{imm434} was a gift from Lynn Thomason, Center for Cancer Research, National Cancer Institute at Frederick, Frederick, MD, USA. This phage was generated completely replacing the Hex coding sequence with that encoding the gene for chloramphenicol acyltransferase flanked by its promoter and transcription terminator. The replacement was done by recombineering.

The Hex DNA sequence ATGGCCTATGACTCCTATCAACGGGAACTGCAAGATTATCG GTGTTGTCGTGGAAGCGAGGGTAAAATTCGTATGATCAGGATTGCGGCGCTACTCTCAAT ACTCTTAACTACCAGCGCCAATTCTGAATGCTGGATTGTCACAAACCTGCACGGGTACGG GGCAATGAATGGCGATCGTTACGAGTTTACAAAAGACAGCACGGAAGATTCCGTTTTCAA CGTAACAATAAATGGCGATAAATCATCAGTTTATGAATCAGTTTCTGGCGTCTATCCAGA GATGAAATACACTGCTTTGTCATCGAACACTATGGTAGGAGAATACCAGTCTGGAGGAGG CATAACCGTTGAAACTTGGTCAATCACTACAGACAAAAAAGCTCTTTACTCCAAAGTAAT GAACATCCCAGGTATGCAACAACTTACATCAACCAAATCCTTTGTTGGTGATGTAGTCGG AACCTGCAATCAGTAAT was PCR amplified and cloned into XbaI and HindIII sites of pMAL-c2X (New England Biolabs), generating a MBP-Hex fusion. This construct encodes a Factor Xa cleavage site in the linker between Hex and MBP. Expression of this fusion protein is under control of the IPTG (isopropyl-β-ᴅ-thiogalactopyranoside) inducible *lac* promoter. A truncated version of Hex (short hex—sHex) was created by PCR amplifying the following sequence from the central portion of the hex gene: GAATGCTGGATTGTCACAAACCTGCACGGGTACGGGGCAATGAATGGCGATCGTTACGA GTTTACAAAAGACAGCACGGAAGATTCCGTTTTCAACGTAACAATAAATGGCGATAAATC ATCAGTTTATGAATCAGTTTCTGGCGTCTATCCAGAGATGAAATACACTGCTTTGTCATCG AAC using forward primer 5'-GGCCATCATATGGAATGCTGGATTGTC-3', and a reverse primer 5'-CCAGCGCTCGAGTTAGTTCGATGACAAAGC-3' (IDT Technologies). This DNA fragment was cleaved with NdeI and XhoI (New England Biolabs), purified and ligated into pET17b (Novagen) which was cleaved with the same enzymes.

4.3. Analysis of RNA Originating from P_{RM}

RNA was isolated from MG1655$^{\lambda imm434}$ as described in reference [29]. Briefly, cell cultures were grown to OD$_{600}$ = 0.6, collected by centrifugation at 12,000 × *g* at 4 °C, and resuspended in protoplasting buffer (15 mM Tris-HCl pH 8.0, 0.45 M Sucrose, 8 mM EDTA) containing 50 mg/mL lysozyme, and incubated 15 minutes on ice. Protoplasts were collected by centrifugation for 5min at 5,900 × *g* 4 °C, resuspended in 0.5 mL gram-negative lysing buffer containing DEPC, and incubated for 5 min at 37 °C. The RNA was precipitated from the preparation by adding saturated NaCl (34 g/100 mL of water) and 100% ethanol, in an overnight incubation at −20 °C. The RNA was collected by centrifugation and washed with ice-cold 70% ethanol, dried, and dissolved in DEPC-treated water. The RNA was digested for 10 minutes with RNAse-free DNase I (Promega Madison, WI, USA) which was subsequently deactivated at 65 °C for 10 minutes. The RNA was immediately used in RT-PCR reactions. Synthesis of first-strand cDNA was performed using AffinityScript™ QPCR cDNA Synthesis Kit (Qiagen Valencia, CA, USA). The primer 5'-CGAGATTGAGGTGGGGATTAC-3' was incubated with 3 µg total RNA, first strand master mix,

AffinityScript RT/RNase block enzyme mixture, and incubated at 25 °C for 5 minutes, followed by 15 minutes in 42 °C. The reaction was terminated by placing the mixture at 95 °C for 5 minutes, chilled on ice, and immediately used in PCR with the following primer combinations: 5'-CGCGTCCAGTATCTACTAACA-3' with 5'-CAGCTTGGACTTAACCAGGCT-3', or 5'-CGAGATTGAGGTGGGGATTAC-3' with 5'-CAGCTTGGACTTAACCAGGCT-3'. The PCR products were fractionated by electrophoresis on agarose gels.

4.4. Analysis of Spontaneous Induction

Measurements of spontaneous induction frequency were performed as described previously [30]. Briefly, cultures of MG1655 lysogenized with λ^{imm434} or $\lambda^{imm434\ hex-}$ were grown to saturation overnight in LB or M9 minimal media at 37 °C. Phage that spontaneously produced overnight were removed from the stationary cells by washing three times with fresh medium. During the last wash the cells were resuspended in a volume of fresh medium that was equal to the starting culture volume. The cultures were incubated with shaking at 37 °C for 5 hours. Cell cultures were then centrifuged at 8,000 × g, and the phage-containing supernatant was sterilized by addition of chloroform. To determine the number of CFU (colony forming units), an aliquot of the remaining phage culture was evenly spread over Luria broth (LB) solidified with Luria agar (LA) in a Petri dish. The amount of phage released into the supernatant was determined by evenly distributing the released phage on nonlysogenic MG1655 as described previously [8]. The amount of phage released per cell in culture was calculated as the plaque forming units PFU/CFU ratio.

4.5. Measurement of Burst Size

Standard methodologies were used for determining the mean number of phage particles per bacterium [31,32] Briefly, phage particles at a concentration of 1×10^7 pfu mL^{-1} were mixed with 1×10^8 cfu mL^{-1} MG1655 in LB supplemented with 10 mM MgSO$_4$ and 5 mM CaCl$_2$ for 5 minutes at 37 °C. The mixture was diluted 10,000-fold with pre-warmed LB to give a final culture volume of 10 mL. At intervals up to 75 minutes, two 0.2 mL samples were acquired to determine amount of unadsorbed (U) phage particles and total (T) phage particles. The U phage particles were additionally treated with 2% CHCl$_3$. The burst size is calculated as $b = \frac{F}{(T-U)}$ where F is the final phage count at time point 75 minutes and T and U are total and unadsorbed phage, respectively.

4.6. Analysis of Complementation of the Hex Mutation

The MG1655 $\lambda^{imm434\ hex-}$ strain was transformed with pET17b bearing the short Hex gene under the T7 RNA polymerase promoter, together with plasmid pGP1-2 which contains temperature inducible T7 RNA polymerase [16]. Control MG1655 $\lambda^{imm434\ hex-}$ lysogens were transformed with plasmids pGP1-2 and pET17b that does not express any Hex derivatives. The cultures were grown overnight at 32 °C, and the cells collected by centrifugation as described above, washed three times and resuspended in a volume that was equal to the starting culture volume, and incubated at 37 °C for three hours. Cell cultures were then centrifuged at 8,000 × g and the phage containing supernatant was sterilized by addition of chloroform. To determine the number of CFU, an aliquot of the remaining phage culture was evenly spread on LB/LA Petri dish as described above. The amount of phage released into the supernatant was determined by infecting nonlysogenic MG1655 as described previously [8]. The amount of phage released per cell in culture was calculated as the PFU/CFU ratio.

4.7. RecA-Mediated in Vitro Autoproteolysis

The standard buffer used in all assays contained 50 mM KCl, 15 mM Tris (pH 7.5), 2 mM $MgCl_2$, 0.1 mM EDTA, 2 mM dithiothreitol. Concentrations of repressors from phage 434, λ, P22, and LexA in each reaction was 300 nm. λ repressor and anti-lambda repressor antibodies were a gift from Ann Hochschild (Harvard University Medical School, Cambridge MA, USA). LexA and anti-LexA antibodies were a gift from John Little, (University of Arizona, Tuscon, AZ, USA). Reaction mixtures were incubated 10 minutes at room temperature with different repressors, 0.5 mM γ-S-ATP, 2 mM DTT, and 3 μM oligo(dT_{20}). After incubation, RecA protein was added to a final concentration of 8 μM, and the mixture was then incubated for 2 hours at 37 °C. Following incubation, Tris/Tricine loading dye (0.1 M Tris, pH 6.8, 1% sodium dodecyl sulfate (SDS), 20 mM 2-mercaptoethanol, 20% glycerol, 0.01% bromophenol blue) were added, and the samples were boiled for 5 min. The products were fractionated by electrophoresis on 15% SDS-polyacrylamide Tris-Tricine gels [33]. The repressor and its cleaved products were visualized by chemiluminescent detection of Western blots by a Storm Imager using an ECL-Plus kit (both obtained from GE Life Sciences, Piscataway, NJ, USA) with horseradish peroxidase secondary antibody.

4.8. Expression and Purification of Hex and sHex

A saturated, overnight culture of BL21(DE3)::pLysS with pMAL-c2X (New England Biolabs, Beverly, MA, USA), bearing the full Hex sequence, was diluted 1:50 in three liters of LB broth supplemented with 100 μg of ampicillin/mL and 20 μg of chloramphenicol/ml. The culture was grown for 2 hours at 37 °C. Cultures were induced to produce Hex by the addition of 0.5 mM IPTG. After additional growth for 4 hours at 37 °C, the cells were harvested by centrifugation at 10,000 × g for 10 minutes, and the cell pellet was resuspended in 30 mL of lysis buffer (100 mM Tris [pH 7.5], 200 mM NaCl, and 10 mM EDTA). All subsequent procedures were performed at 4 °C. The cells were lysed by using a French press, and the cell lysate was diluted to 100 mL with lysis buffer. The lysate was centrifuged at 10,500 × g for 20 minutes to remove cellular debris. The resulting supernatant was passed through a 2 mL Amylose resin column, washed three times with 5 bed volumes of resin buffer (100 mM Tris [pH 7.5], 500 mM NaCl), and eluted with resin buffer plus 10mM maltose in 500 μL fractions. The purification progress was monitored by analyzing fractions collected at each step by SDS-PAGE. MBP-Hex fractions were pooled and digested with factor Xa (Qiagen, Valencia, CA, USA) as specified by this manufacturer. Expected size of Hex peptide is ~17.5 kDa. The products were analyzed on 15% SDS-polyacrylamide Tris-Tricine gels. Unexpectedly, no full length Hex protein was observed. Instead, a ~6 kDa product was observed to accumulate. This product was cut out from the gel and its amino acid sequence was determined using MALDI-TOF MS analysis. Subsequently, DNA encoding this sHex (short Hex) was PCR amplified, cloned and expressed as described above. Purification of sHex was performed as described for the MBP-Hex fusion protein, except after breaking the cells and centrifuging the lysate at 10,500 × g for 20 minutes to remove cellular debris, sHex was precipitated from solution with addition of ammonium sulfate to the final concentration of 10%, and harvested by centrifugation at 10,000 × g for 10 minutes. The insoluble pellet was dissolved in final concentration of 4 M Guanidine-HCl (GuHCl) in 100 mM Tris [pH 7.5], 100 mM NaCl, and passed through 20 cm Sephadex G50 column equilibrated with the same buffer. The sHex containing fractions were pooled, concentrated, and dialyzed in buffers of successively lower concentrations of GuHCl in order to remove GuHCl. Purity of sHex was monitored by SDS-PAGE.

Acknowledgments

The work was supported in part by the College of Arts and Sciences at the University at Buffalo and by a grant from the National Science Foundation (MCB-0956454) to GBK. The authors also wish to thank Lynn Thomason and Donald Court of the National Cancer Institute Frederick National Laboratory, Frederick, Maryland, USA for generously providing reagents and for helpful discussions and comments on the manuscript.

References

1.   Ptashne, M. *A Genetic Switch*; Blackwell PRESS: Palo Alto, CA, USA, 1986.

2.   Pirrotta, V. Operators and promoters in the OR region of phage 434. *Nucleic Acids Res.* **1979**, *6*, 1495–1508.

3.   Janion, C. Inducible SOS response system of DNA repair and mutagenesis in *Escherichia coli*. *Int. J. Biol. Sci.* **2008**, *4*, 338–344.

4.   Phizicky, E.M.; Roberts, J.W. Induction of SOS functions: Regulation of proteolytic activity of *E. coli* RecA protein by interaction with DNA and nucleoside triphosphate. *Cell* **1981**, *25*, 259–267.

5.   Butala, M.; Zgur-Bertok, D.; Busby, S.J. The bacterial LexA transcriptional repressor. *Cell. Mol. Life Sci.* **2009**, *66*, 82–93.

6.   Phizicky, E.M.; Roberts, J.W. Kinetics of RecA protein-directed inactivation of repressors of phage lambda and phage P22. *J. Mol. Biol.* **1980**, *139*, 319–328.

7.   Sauer, R.T.; Nelson, H.C.; Hehir, K.; Hecht, M.H.; Gimble, F.S.; DeAnda, J.; Poteete, A.R. The lambda and P22 phage repressors. *J. Biomol. Struct. Dyn.* **1983**, *1*, 1011–1022.

8.   Arber, W.; Enquist, L.; Hohn, B.; Murray, N.E.; Murray, K.; Hendrix, R.W.; Roberts, J.W.; Stahl, F.W.; Weisberg, R.A. *Lambda II*; Cold Spring Harbor Laboratory: Cold Spring Harbor, NY, USA, 1983; pp. 433–466.

9.   Slavcev, R.A.; Hayes, S. Stationary phase-like properties of the bacteriophage lambda Rex exclusion phenotype. *Mol. Genet. Genom.* **2003**, *269*, 40–48.

10. Tyler, J.S.; Mills, M.J.; Friedman, D.I. The operator and early promoter region of the Shiga toxin type 2-encoding bacteriophage 933W and control of toxin expression. *J. Bacteriol.* **2004**, *186*, 7670–7679.

11. Friedman, D.I.; Mozola, C.C.; Beeri, K.; Ko, C.-C.; Reynolds, J.L. Activation of a prophage-encoded tyrosine kinase by a heterologous infecting phage results in a self-inflicted abortive infection. *Mol. Microbiol.* **2011**, *82*, 567–577.

12. Heinrich, J.; Velleman, M.; Schuster, H. The tripartite immunity system of phages P1 and P7. *FEMS Microbiol. Rev.* **1995**, *17*, 121–126.

13. Nesper, J.; Blass, J.; Fountoulakis, M.; Reidl, J. Characterization of the major control region of Vibrio cholerae bacteriophage K139: Immunity, exclusion, and integration. *J. Bacteriol.* **1999**, *181*, 2902–2913.

14. Susskind, M.M.; Botstein, D. Superinfection exclusion by lambda prophage in lysogens of Salmonella typhimurium. *Virology* **1980**, *100*, 212–216.

15. Livny, J.; Friedman, D.I. Characterizing spontaneous induction of Stx encoding phages using a selectable reporter system. *Mol. Microbiol.* **2004**, *51*, 1691–1704.

16. Tabor, S.; Richardson, C.C. A bacteriophage T7 RNA polymerase/promoter system for controlled exclusive expression of specific genes. *Proc. Natl. Acad. Sci. USA* **1985**, *82*, 1074–1078.

17. Pawlowski, D.R.; Koudelka, G.B. The preferred substrate for RecA-mediated cleavage of bacteriophage 434 repressor is the DNA-bound dimer. *J. Bacteriol.* **2004**, *186*, 1–7.

18. Kliem, M.; Dreiseikelmann, B. The superimmunity gene sim of bacteriophage P1 causes superinfection exclusion. *Virology* **1989**, *171*, 350–355.

19. Bushman, F.D. The bacteriophage 434 right operator. Roles of O R 1, O R 2 and O R 3. *J. Mol. Biol.* **1993**, *230*, 28–40.

20. Little, J.W. Autodigestion of lexA and phage lambda repressors. *Proc. Natl. Acad. Sci. USA* **1984**, *81*, 1375–1379.

21. Slilaty, S.N.; Rupley, J.A.; Little, J.W. Intramolecular cleavage of LexA and phage lambda repressors: Dependence of kinetics on repressor concentration, pH, temperature, and solvent. *Biochemistry* **1986**, *25*, 6866–6875.

22. McDonald, J.P.; Frank, E.G.; Levine, A.S.; Woodgate, R. Intermolecular cleavage by UmuD-like mutagenesis proteins. *Proc. Natl. Acad. Sci. USA* **1998**, *95*, 1478–1483.

23. Cox, M.M. Regulation of bacterial RecA protein function. *Crit. Rev. Biochem. Mol. Biol.* **2007**, *42*, 41–63.

24. Klinge, S.; Voigts-Hoffmann, F.; Leibundgut, M.; Arpagaus, S.; Ban, N. Crystal structure of the eukaryotic 60S ribosomal subunit in complex with initiation factor 6. *Science* **2011**, *334*, 941–948.

25. Sambrook, J.; Fritsch, E.F.; Maniatis, T. *Molecular Cloning — A Laboratory Manual*; Cold Spring Harbor Press: Cold Spring Harbor, NY, USA, 1989.

26. Guyer, M.S.; Reed, R.E.; Steitz, T.; Low, K.B. Identification of a sex-factor-affinity site in *E. coli* as gamma delta. *Cold Spr. Harb. Symp. Quant. Biol.* **1981**, *45*, 135–140.

27. Blattner, F.R.; Plunkett, G., III; Bloch, C.A.; Perna, N.T.; Burland, V.; Riley, M.; Collado-Vides, J.; Glasner, J.D.; Rode, C.K.; Mayhew, G.F.; *et al.* The complete genome sequence of *Escherichia coli* K-12. *Science* **1997**, *277*, 1453–1474.

28. Winans, S.C.; Elledge, S.J.; Krueger, J.H.; Walker, G.C. Site-directed insertion and deletion mutagenesis with cloned fragments in *Escherichia coli*. *J. Bacteriol.* **1985**, *161*, 1219–1221.

29. Engberg, J.; Nielsen, H.; Lenaers, G.; Murayama, O.; Fujitani, H.; Higashinakagawa, T. Comparison of primary and secondary 26S rRNA structures in two Tetrahymena species: Evidence for a strong evolutionary and structural constraint in expansion segments. *J. Mol. Evol.* **1990**, *30*, 514–521.

30. Shkilnyj, P.; Koudelka, G.B. Effect of salt shock on the stability of λ^{imm434} lysogens. *J. Bacteriol.* **2007**, *189*, 3115–3123.

31. Delbrück, M. The growth of bacteriophage and lysis of the host. *J. Gen. Physiol.* **1940**, *23*, 643–660.

32. Wang, I.N. Lysis timing and bacteriophage fitness. *Genetics* **2006**, *172*, 17–26.

33. Schagger, H.; Von Jagow, G. Tricine-sodium dodecyl sulfate-polyacrylamide gel electrophoresis for the separation of proteins in the range from 1 to 100 kDa. *Anal. Biochem.* **1987**, *166*, 368–379.

6

Bacteriophages with the Ability to Degrade Uropathogenic *Escherichia Coli* Biofilms

Andrew Chibeu [1,*], Erika J. Lingohr [1], Luke Masson [2,3], Amee Manges [4], Josée Harel [5], Hans-W. Ackermann [6], Andrew M. Kropinski [1,7,*] and Patrick Boerlin [8]

[1] Laboratory for Foodborne Zoonoses, Public Health Agency of Canada, Guelph, ON, N1G 3W4, Canada; E-Mail: erika.lingohr@phac-aspc.gc.ca

[2] Biotechnology Research Institute, National Research Council of Canada, 6100 Royalmount Avenue, Montréal, QC H4P 2R2, Canada; E-Mail: luke.masson@cnrc-nrc.gc.ca

[3] Département de microbiologie et immunologie, Université de Montréal, 2900, boul. Édouard-Montpetit, Montréal, QC H3T 1J4, Canada

[4] Department of Epidemiology, Biostatistics and Occupational Health, McGill University, 1020 avenue des Pins Ouest, Montréal, QC H3A 1A2, Canada; E-Mail: amee.manges@mcgill.ca

[5] Groupe de Recherche sur les Maladies Infectieuses du Porc (GREMIP) and Centre de Recherche en infectiologie porcine (CRIP), Université de Montréal, Faculté de médecine vétérinaire, Saint-Hyacinthe, QC J2S 7C6, Canada; E-Mail: josee.harel@umontreal.ca

[6] Felix d'Herelle Reference Center for Bacterial Viruses, Department of Microbiology, Immunology and Infectionlogy, Faculty of Medicine, Laval University, QC G1K 4C6, Canada; E-Mail: ackermann@mcb.ulaval.ca

[7] Department of Molecular and Cellular Biology, University of Guelph, ON N1G 2W1, Canada

[8] Department of Pathobiology, Ontario Veterinary College, University of Guelph, ON N1G 2W1, Canada; E-Mail: pboerlin@uoguelph.ca

* Authors to whom correspondence should be addressed; E-Mails: andrew.chibeu@phac-aspc.gc.ca (A.C.); kropinsk@queensu.ca (A.M.K.)

Abstract: *Escherichia coli*-associated urinary tract infections (UTIs) are among the most common bacterial infections in humans. UTIs are usually managed with antibiotic therapy, but over the years, antibiotic-resistant strains of uropathogenic *E. coli* (UPEC) have emerged. The formation of biofilms further complicates the treatment of these infections by making them resistant to killing by the host immune system as well as by antibiotics. This has encouraged research into therapy using bacteriophages (phages) as a supplement or substitute for antibiotics. In this study we characterized 253 UPEC in terms of their biofilm-forming capabilities, serotype, and antimicrobial resistance. Three phages were then isolated (vB_EcoP_ACG-C91, vB_EcoM_ACG-C40 and vB_EcoS_ACG-M12) which were able to lyse 80.5% of a subset (42) of the UPEC strains able to form biofilms. Correlation was established between phage sensitivity and specific serotypes of the UPEC strains. The phages' genome sequences were determined and resulted in classification of vB_EcoP_ACG-C91 as a *SP6likevirus*, vB_EcoM_ACG-C40 as a *T4likevirus* and vB_EcoS_ACG-M12 as *T1likevirus*. We assessed the ability of the three phages to eradicate the established biofilm of one of the UPEC strains used in the study. All phages significantly reduced the biofilm within 2–12 h of incubation.

Keywords: UPEC; bacteriophage; biofilms

---

1. Introduction

Urinary tract infections (UTIs) are among the most common bacterial infections in humans. They account for more than seven million visits to physicians' offices per year in the United States of America [1]. Approximately 20% of women develop UTIs sometime during their lifetime. Above the age of 50, men and women have a similar incidence of UTIs [1–5]. UTIs have tremendous economic impact on health care systems in both direct and indirect costs associated with treatment [6,7]

Uropathogenic strains of *Escherichia coli* (UPECs) account for about 75%–85% of UTIs [4,8]. There has been an evolution toward antibiotic resistance in UPECs, with decreasing susceptibility to first-line agents such as ampicillin, nitrofurantoin, sulphamethoxazole/trimethoprim (SXT) and fluoroquinolones [4,9–13]. An alternative to antimicrobial treatment would be of great relevance to treat multiresistant UTIs [5], as well as to avoid the continuous selection of resistant pathogens and to safeguard the efficacy of antimicrobial agents for cases of emergency and life-threatening infections.

In hospital-acquired infections as well as in geriatrics patients, most UTI are associated with indwelling catheters which act as foci for biofilm formation [14]. Biofilms play a significant role in the ability of bacteria to withstand killing by host immune responses and antibiotics. There is a need to develop new therapeutic strategies to eradicate biofilm infections [15].

The development of biofilms by bacteria does not protect cells from bacteriophage killing. These viruses can penetrate the extracellular matrix that binds macromolecules and prevents their diffusion into the biofilm [16–18] and kill cells [19–21]. Furthermore, certain phages have evolved to deal with capsules by possessing virion-associated polysaccharide depolymerases [22–24] leading us to believe that they are ideal agents to help reduce biofilm-associated infections as well as to kill planktonic cells.

In this paper, we describe a subset of a collection of 253 *E. coli* isolates from UTIs specifically selected on their ability to form biofilms and the activity of three environmentally isolated bacteriophages on these strains. To assess their efficacy and safety as therapeutic agents against UPEC, the phages were characterized in detail with regards to their spectrum of activity, genetic structure, as well as activity on biofilms.

2. Results and Discussion

2.1. Biofilm Forming Capabilities, Serotype and Antimicrobial Resistance

Each of the 253 UTI *E. coli* isolates was screened for static biofilm formation in 96-well untreated polystyrene microtiter plates following growth in artificial urine medium as described in the Experimental section. Isolates were considered positive for biofilm formation if the crystal violet-stained biofilm had an OD_{600} equal to or greater than 3-fold the value obtained in the well containing bacteria-free medium. The UTI *E. coli* strain CTF073 was used as a positive control and 12 wells of this strain were included with each plate of samples of UPEC assayed for biofilm formation. The plotted results were the average of three independent experiments, each with six replicates per isolate. Only 42 out of the 253 *E. coli* isolates tested in this study produced biofilms in microtiter plate wells with the levels of biofilm formation varying among isolates (Figure 1). This is in contrast to a previous study [25] where a majority of the UTI strains tested produced biofilms *in vitro*. A possible reason for the different outcomes in the two studies is the fact that UPEC do not form biofilm well on polystyrene plates in urine or artificial urine medium as compared to M9 minimal medium used by [25]. Among the 42 biofilm-forming isolates, Can67 showed the highest level of biofilm formation (OD_{600} above 1.0) whereas MSHS94 and Can72 resulted in weak biofilms formation with an OD_{600} below 0.2.

The biofilm-forming UTI *E. coli* isolates totalized 22 different somatic O antigens and 14 different H antigens (Table 1). We then investigated their susceptibility profiles against 15 antimicrobials. The investigated isolates were frequently resistant with 32 (78%) isolates displaying resistance to antimicrobial agents, while 16 (39%) isolates displayed multidrug resistance phenotypes. These results are in agreement with previous studies that have shown a high frequency of antimicrobial resistance among UPEC [4,9–13]. However, inspection of the different levels of biofilm formation on microtiter plates (Figure 1) and comparing this to the results on antimicrobial susceptibility (Table 1) showed that there was no apparent direct correlation between antimicrobial resistance and the amount of biofilm formed by each isolate.

Figure 1. Urinary tract infection (UTI) *E. coli* isolates positive for biofilm formation on 96-well microtiter plates. Isolates were considered as biofilm formers if the OD_{600} for the crystal violet-stained biofilm was equal to or greater than 3-fold the OD_{600} for a bacteria free medium. Data points represent an average of three independent experiments, each with 6 replicate wells for each isolate tested. Isolate CTF073 is a positive control for biofilm formation. Error bars indicate standard error of means.

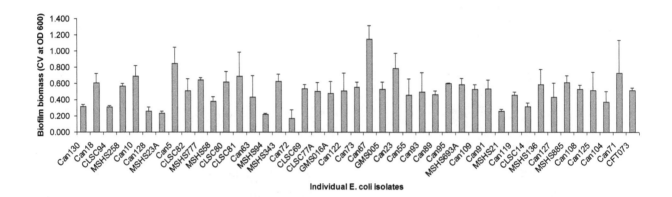

Table 1. List of UTI *E. coli* forming biofilms, their serotypes, respective antimicrobials resistances and sensitivity to the three isolated phages ACG-C91, ACG-C40 and ACG-M12. AMC = Amoxicillin/Clavulanic acid; AMP = Ampicillin; FOX = Cefoxitin; TIO = Ceftiofur; CRO = Ceftriaxone; CHL = Chloramphinicol; CIP = Ciprofloxacin; KAN = Kanamycin; NAL = Nalidixic acid; STR = Streptomycin; SOX = Sulfisoxazole, TCY = Tetracycline; SXT = Sulphamethoxazole/Trimethoprim. + symbolises sensitivity to phage; - resistance to phage; nd, not determined.

Strain	Serotype	Antimicrobials to which isolates are resistant	ACG-C40	ACG-C91	ACG-M12
Can130	O1:NM		+	+	-
Can18	O2:H4		-	+	-
CLSC94	O2:H7		+	+	-
MSHS258	O2:H18		+	+	-
Can10	O8:H10		-	-	-
Can128	O8:NM	SOX, TCY	+	-	-
MSHS23A	O6:H1	AMP, STR, SOX,	+	-	+
Can5	O6:H1	NAL	+	-	+
CLSC82	O6:H1	AMP	+	-	+
MSHS777	O6:H1	STR	+	-	+
MSHS58	O6:H25	KAN	-	-	-
CLSC80	O11:H18	STR, SOX	+	-	-
CLSC81	O14:H4		+	-	+
Can63	O14:H31	TCY	-	-	+
MSHS94	O18ac:H7	AMP, STR, SOX, TCY	+	+	+

Table 1. *Cont.*

Strain	Serotype	Antimicrobials to which isolates are resistant	ACG-C40	ACG-C91	ACG-M12
MSHS343	O18ac:H7		+	+	+
Can72	O18ac:NM	AMP	+	-	-
CLSC69	O21:H14		-	-	-
CLSC77A	O22:H1		-	-	-
GMS016A	O25:H1		+	+	+
Can122	O25:H4	AMP, TIO, CRO, CIP, NAL, SOX, TCY, SXT	-	-	-
Can73	O25:H4	AMP, CIP, NAL	-	-	-
Can67	O25:H4	AMC, AMP, CIP, NAL	-	-	-
GMS005A	O35:H10		+	-	-
Can23	O68:H18		+	-	-
Can55	O75:H7	AMC, AMP, FOX, TIO, CRO	+	+	-
Can93	O75:H7		-	+	-
Can89	O78:H5		+	-	-
Can95	O106:H18	CHL, TCY	-	-	-
MSHS693A	O117:H5		+	-	-
Can109	O134:H31	NAL	+	+	-
Can91	O135:H6		+	+	+
MSHS21	O135:H11		+	-	-
Can119	O153:H18	AMP, TIO, CRO, NAL, STR, SOX, SXT	+	+	-
CLSC14	O153:NM	AMP, STR, SOX, TCY, SXT	+	-	-
MSHS136	O166:H15	AMC, AMP	-	-	-
Can127	OR:H4	AMP,CHL, TCY	-	+	+
MSHS885	OR:H4	AMP, CHL, STR, SOX, TCY	+	+	-
Can108	OR:H4	SOX, TCY, STX	+	+	-
Can125	OR:H4	AMP, CIP, NAL, SOX, SXT	+	-	+
Can104	OR:H7		-	+	+
Can71	OR:H40		+	-	+
CFT073	nd	nd	+	+	+

It is important to note that the ability of biofilms to withstand antibiotic killing is a function of their mode of growth. Antimicrobial agents have been shown to penetrate biofilms at different rates depending on the particular agent and the biofilm [26]. The antimicrobial susceptibility tests carried out in this study using the broth microdilution method did not test the susceptibility of the bacteria to the antimicrobial agent after biofilm formation. The antimicrobials were added to cell suspensions at time of media inoculation and not after the biofilms had formed. The results of [27] demonstrated that older biofilms of *E. coli* resisted ampicillin treatment to a greater extent than their younger counterparts. However, in our study, a majority of the biofilm forming *E. coli* UTI isolates (78%) were resistant to at least one antimicrobial agent. This compares to 53.1% of the non-biofilm forming *E. coli*

UTI isolates that were resistant to at least one out of the 15 antimicrobials tested (results not shown). This is consistent with what is known about antimicrobial resistance among biofilm forming bacterial strains [28].

2.2. Phage Morphology

UPEC strains Can 91, Can 40 and MSHS1210 were used as propagating strains for the isolation of phages from serially diluted CsCl purified phage concentrate. All three isolates were from midstream urine specimens with Can 40 and Can 91 being part of the CANWARD study whereas MSHS1210 was from the McGill University student health service. The single unique phages isolated from UPEC strains Can 91, Can 40 and MSHS1210 were vB_EcoP_ACG-C91, vB_EcoM_ACG-C40 and vB_EcoS_ACG-M12, respectively.

Transmission electron microscopy revealed that phage ACG-C40 has an elongated head, a neck, and a contractile tail with tail fibers (Figure 2a). Its head is 110 × 82 nm, while the extended tail is 114 × 8 nm. Phage ACG-M12 (Figure 2b) has an isometric head of about 157 nm in diameter between opposite apices and a relatively flexible tail of 172 × 7 nm, which terminate in 1–2 fibers of 12 nm in length. Phage ACG-C91 (Figure 2c) has an isometric head of 65–68 nm and a short tail of 12 × 8 nm, which carries fibers of 13 nm in length with a terminal swelling. Based on their morphology, ACG-C40 is classified as a member of the family *Myoviridae*, ACG-M12 is part of the *Siphoviridae* and ACG-C91 belongs to the family *Podoviridae*.

Figure 2. Negative staining of phages ACG-C91 (**a**); ACG-M12 (**b**) and ACG-C40 (**c**) with 2% uranyl acetate or 2% phosphotungstate. Final magnification is × 297,000. Bars indicate 100 nm.

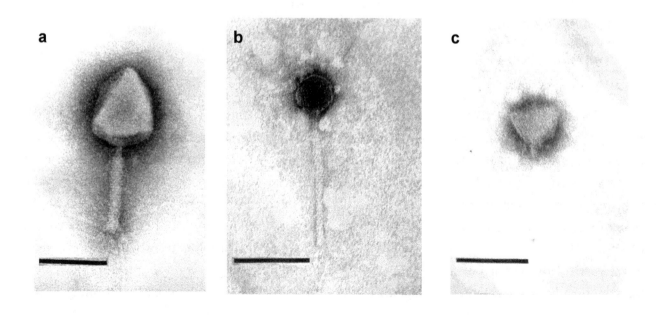

2.3. Lytic Spectra of Phages against Biofilm Forming UTI Isolates and Relation to Isolates' Serotypes

The lytic activity of ACG-C91, ACG-C40 and ACG-M12 tested against the biofilm forming *E. coli* UTI isolates. In the spot tests, 33 (80.5%) of the biofilm forming *E. coli* UTI isolates were infected by at least one of the three phages (Table 1).

Out of the three phages isolated in this study, phage ACG-C40 had the broadest host range, with the ability to lyse 28 (66.67%) of the biofilm forming UTI *E. coli* isolates tested (Table 1). It is not unusual for phages in the family *Myoviridae* to have such a large host range. A previous study by [29], showed that bacteriophage T4 lysed 41.1% of the 69 clinical isolates of *E. coli* tested. Other T4-like phages such as AR1 have been observed to have similar broad host ranges [30]. The ability of phage ACG-C40 to cause lysis in a wide range of *E. coli* isolates with different serotypes makes it an important candidate for phage therapy applications. The broad host range of UTI *E. coli* isolates is an indicator of how well adapted phage ACG-C40 is to infecting UPEC, which in itself is a desirable property for a phage to be used in phage therapy against such strains. The criteria of choosing phages which are well adapted for infection of targeted hosts for therapeutic purposes have been applied in similar previous studies [31–33].

The correlation between host serotype and phage sensitivity is seen in the fact that among the biofilm forming strains, all the O25:H4 strains were resistant to the three tested phages. The O6:H1 strains were sensitive to phages ACG-C40 and ACG-M12. Further in depth studies to identify the receptors of the tested phages and the receptors spatial orientation on the surface of the host strains under tested conditions would give insight as to why the phages are resistant or susceptible to all strains of a given serotype.

2.4. Salient Bacteriophage Genome Features

Genome sequencing and bioinformatics analysis of the three isolated phages revealed that all three are non-temperate and that none carries any known bacterial virulence genes for humans or animals. None of the three phages possessed a demonstrable extracellular polysaccharide or exopolysaccharide (EPS) depolymerase gene. The importance of phage-associated depolymerases in biofilm eradication has been recognized in previous studies [22–24,34]. Depolymerase-producing phages were found to be more effective in eradicating mature biofilms than non-depolymerase producing phages [34]. Phages producing depolymerases have thus been used in concert with antimicrobials to facilitate deeper penetration of antimicrobials by degrading the EPS [23,34]. However, phages that do not produce EPS depolymerases have also found use in biofilm degradation. Such phages include naturally occurring phages such as T4 [16,17,35] and phages such as T7 engineered to express recombinant dispersin B (DsbB) [36].

2.4.1. Phage ACG-C40

The sequence of the phage genome consisted of 167,396 bp (G+C content 35%) which is close to the genome size of other "T4-like viruses" [37]. It was predicted to encode 282 ORFs and 10 tRNAs. The latter clustered between 68, 043 and 68, 989 bp on the phage's genome.

The DNA of this virus was resistant to digestion by all restriction endonucleases tested. An analysis of the genome of this phage (Additional file: Figure A1; Table A1) reveals that it does, like coliphage T4, encode, a glucosyl transferase (orf 060). Bacteriophage T4 contains glucosylated 5-hydroxymethylcytosine (hmdCyt) instead of cytosine which makes its DNA resistant to most restriction enzymes except for EcoRV and NdeI [38]. Based on the presence of the glucosyl transferase gene and the resistance to enzymes, we conclude that bacteriophage ACG-C40 also contains hypermodified bases, probably glucosylated hmdCyt.

The genome of this phage revealed that the phage potentially encodes two versions of the Hoc protein; a shorter and a longer version which may result from translational frameshifting generating Hoc protein with C termini of different lengths (supplementary file: Table S1).

2.4.2. Phage ACG-C91

Phage ACG-C91 has a 43,731 bp genome with G+C content of 45% and predicted 55 ORFs (supplementary file: Table S2). The phage DNA was sensitive to restriction enzymes BglI, XbaI, EcoRV, EcoRI, NdeI and SalI.

2.4.3. Phage ACG-M12

Phage ACG-M12 has a 46, 054 bp genome with a G+C content of 44% and 77 ORFs (supplementary file: Table S3). The phage DNA was sensitive to all restriction enzymes tested except NaeI and SmaI.

2.5. Comparative Genomics

Pairwise comparisons using CoreGenes 3.0 [39] at default stringency setting ("75") revealed that at the protein level, ACG-C40, ACG-C91 and ACG_M12 were 85.9% ,78.8% and 77.3% similar to bacteriophages T4, SP6 and RTP, respectively. Bacteriophage ACG-C40 genome encodes 239 proteins homologous to T4 proteins, ACG-C91 genome has 41 encoded proteins homologous to those of SP6 and ACG-M12 genome encodes 58 proteins with homology to RTP proteins (supplementary files: Table A1; Table A2; Table A3).

A Mauve alignment [40,41] of phage ACG-C91 DNA against the genome of phage SP6 shows regions of sequence similarity (coloured red) with two regions where there are no DNA sequence similarities (coloured white). The regions of non-sequence similarity correspond to the regions encoding phage ACG-C91 internal virion protein proteins (orf 38), putative tail protein (orf 41) and the endosialidase (orf 53). The genome region encoding the endosialidase is lacking altogether in phage SP6 (supplimentary file: Table A1; supplementary file: Figure A2a). This protein is found in K1-specific phages, such as K1F and K1-5, that encode virion-associated endosialidases hydrolyzing the K1 polysialic acid structure of the K1 capsule-producing *E.coli* strains. Mauve alignment of ACG-M12 against the genome of phage RTP was also performed and it revealed regions of sequence similarity (red) interspersed with four regions where no sequence similarity exists (white). The regions of genome homology disparity include phage ACG-M12 genome regions encoding: hypothetical proteins orf2, orf3 and orf4 (1025 bp–1588 bp), the major tail protein orf26 (12435 bp–13091 bp), putative tail fibre orf47 (29255–32074) and conserved hypothetical proteins orf62 and orf63 (supplementary file: Figure A2b).

2.5. Bacteriophage Eradication of Established Biofilms

UTI *E. coli* isolate Can 91 was selected to determine the effectiveness of the isolated phages to degrade established biofilms because of the isolate's ability to form strong biofilms in microtiter wells (OD_{600} 0.538) after 48 hrs and its sensitivity to all the three phages isolated. Phages ACG-C40, ACG-C91 and ACG-M12 treatment of preformed UTI *E. coli* Can 91 biofilms yielded reductions (based on OD_{600} measurement of CV-stained cells) in biofilm mass compared to untreated controls (Figure 3). All of the three phages displayed, at low concentrations (10^5 PFU/mL), a steady reduction of biofilm biomass over time with the largest percentage reduction of biofilm biomass being realized after 8 h incubation. The same pattern was also observed at higher phage concentrations (10^7 PFU/mL and 10^9 PFU/mL). At all phage concentrations, it was evident that the biofilm started re-establishing itself after 24 h (Figure 3).

These results indicate that under the conditions tested, the biofilm reduction by the phages is not dose-dependent, a fact that could be taken advantage of if the phages are to be applied *in vivo* to remove established biofilms. It would mean that low titers of the phages were as effective as using a higher titer in eradicating established biofilms. Re-establishment of biofilms after 24 h exposure to the phages *in vitro* may be attributed to development of bacterial resistance against the phages. This could be overcome by using a cocktail of phages. It may also be a good strategy to augment phage treatment with chemical antimicrobials which will be effective in prevention of the re-establishment of biofilms by the phage resistant cells.

Further studies need to be carried out to assess the efficacy of phage ACG-C40, ACG-C91 and ACG-M12 in the prevention of biofilm formation for comparison with this study where we have assayed their efficacy in the eradication of established biofilms.

Figure 3. Phage disruption of established *E. coli* strain Can 91 biofilm. Biofilms grown in polystyrene microtiter plate wells for 48 h, were initially inoculated with 10^5, 10^7 and 10^9 pfu of phages (**a**) ACG-C40 and (**b**) ACG-C91 (**c**) ACG-M12. After 2, 4, 8 and 24 h of phage treatment at 37 °C, average biofilm biomass in corresponding microtiter plate wells, were scored relatively to untreated control samples (100%) and represented on the Y-axes. Three independent experiments were performed, each starting from a different overnight culture and each with six repeats for each parameter combination. Error bars indicate standard error of means.

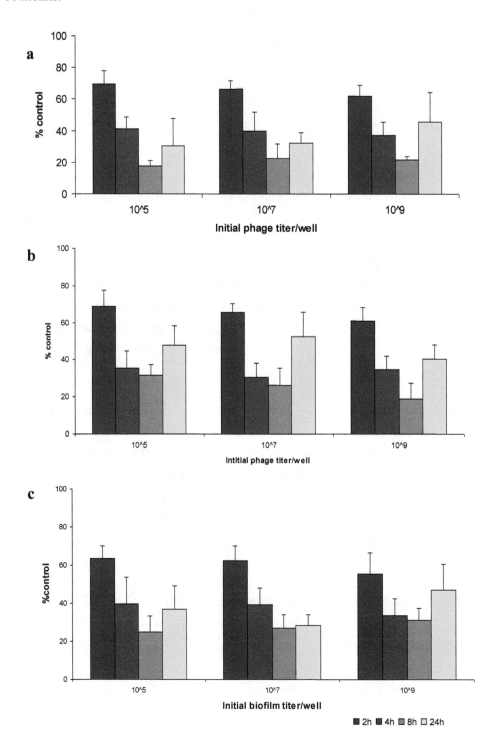

3. Experimental Section

3.1. Sampling and E. coli Strain Isolation

A group of 123 strains of UTI *E. coli* were isolated from women with community-acquired UTI either from a student health service (McGill University, Montréal, QC, Canada) or at a community health center also in Montréal. To ensure no bias in the selection of isolates due to recurrent UTI or treatment failures, only one isolate per patient was used.

An additional 130 isolates were part of the ongoing Canadian national surveillance study (CANWARD) for testing urinary culture pathogens for antimicrobial susceptibilities. The study involved out-patients attending hospital clinics and emergency rooms, and in patients on medical and surgical wards, and in intensive care units [42]. Each study site was asked to submit only clinically significant isolates, as defined by local site criteria, from out- and in-patients with urinary infections. Fifty (2009) to 100 (2007, 2008) consecutive urinary tract isolates per year per medical center site (one isolate per patient) were collected. Primary isolate identification was performed by the submitting medical center site and confirmed by the coordinating laboratory, as required, based on morphological characteristics and spot tests [43]. If an isolate identification made by the coordinating laboratory did not match that provided by the submitting site, the isolate was removed from the study. All isolates were stored at −70 °C in skim milk.

At the Department of Epidemiology, Biostatistics and Occupational Health of McGill University, the isolates from the McGill University student health service and the community health center, also in Montréal, as well as those from the CANWARD study were recovered on MacConkey and CLED agar (Uricult dipslides: Orion Diagnostica, Oy, Finland). A single representative colony was picked from the MacConkey side of the Uricult slide and grown in LB broth and archived in 15% glycerol at −80 °C. When needed, further testing, based upon indole production and the presence of lysine or ornithine decarboxylase, was also performed to confirm the bacterial identification.

3.2. Culture Conditions

For all experiments, bacteria were grown in LB medium or synthetic urine medium. The latter medium was based on that devised by [44], and its composition and method of sterilization have been described previously [45]. Briefly the medium was composed of $CaCl_2$, 0.65 g/L; $MgCl_2$, 0.65 g/L; NaCl, 4.6 g/L; Na_2SO_4, 2.3 g/L; Na_3-citrate, 0.65 g/L; Na_2-oxalate, 0.02 g/L; KH_2PO_4, 2.8 g/L; KCl, 1.6 g/L; NH_4Cl, 1.0 g/L; urea, 25 g/L; creatinine, 1.1 g/L; Luria-Bertani (LB) broth (Difco, Sparks, MD, USA), 10 g/L. The pH of the medium is adjusted to pH 5.8 prior to sterilization by filtration through a 0.2-μm filter.

3.3. E. coli Serotyping

Identification of somatic (O) and flagellar (H) antigens was performed by standard agglutination methods [46]. The isolates were serotyped at the Public Health Agency of Canada, Laboratory for Foodborne Zoonoses, Guelph, ON, Canada.

3.4. Susceptibility Testing

The broth microdilution method was used (Sensititre System; Trek Diagnostics, Cleveland, OH, USA), following the protocols of the Canadian Integrated Program for Antimicrobial Resistance Surveillance [47,48]. Each *E. coli* isolate was tested for the following antimicrobial agents (breakpoints are indicated in parentheses): amikacin ($\geq$64 µg/mL), amoxicillin-clavulanic acid ($\geq$32 and $\geq$16 µg/mL, respectively), ampicillin ($\geq$32 µg/mL), cefoxitin ($\geq$32 µg/mL), ceftiofur ($\geq$8 µg/mL), ceftriaxone ($\geq$2 µg/mL) (11), chloramphenicol ($\geq$32 µg/mL), ciprofloxacin ($\geq$4 µg/mL), gentamicin ($\geq$16 µg/mL), kanamycin ($\geq$64 µg/mL), naladixic acid ($\geq$32 µg/mL), streptomycin ($\geq$64 µg/mL), sulfisoxazole ($\geq$512 µg/mL), tetracycline ($\geq$16 µg/mL), and trimethoprim-sulfamethoxazole ($\geq$4 and $\geq$76 µg/mL, respectively).

3.5. Bacteriophage Isolation

Three liters of preliminary treated sewage sample from the Guelph water treatment plant were centrifuged at 3,000 RPM for 20 min to eliminate solid debris. The clarified supernatant was concentrated to 400 mL using the tangential flow system as described in [49]. NaCl (20.5 g) was then added to the concentrate and stirred until it dissolved before 40 g of polyethylene glycol was added to precipitate the phage present in the concentrate. This was left overnight while being stirred gently with a magnetic stirrer at 4 °C. Precipitated phage was recovered by centrifugation at 3,000 RPM for 10 min and resuspended in 3 mL SM buffer [50]. This was subjected to two rounds of CsCl gradient purification steps and the resulting viral band isolated.

E. coli isolates Can 91, Can 40 and MSHS1210 were used as propagating strains for the isolation of single plaques from serially diluted CsCl purified phage concentrate. All three isolates were from midstream urine specimens with Can 40 and Can 91 being part of the CANWARD study whereas MSHS1210 was from the McGill University student health service.

LB agar (Difco) plugs from standard agar overlay method [51] were resuspended in 500 µL SM buffer, diluted and used for further single plaque isolation on the respective *E. coli* propagation strains. This was repeated three times for each strain to ensure unique phages for each UPEC isolate. The single unique phages isolated from *E. coli* isolates Can 91, Can 40 and MSHS1210 were named vB_EcoP_ACG-C91, vB_EcoM_ACG-C40 and vB_EcoS_ACG-M12, respectively, following the naming convention of [52].

3.6. Transmission Electron Microscopy

Phages were pelleted at 25,000 × g for 1 h, using a Beckman high-speed centrifuge and a JA-18.1

fixed-angle rotor (Beckman, Palo Alto, CA, USA). Phage pellets were washed twice under the same conditions in neutral 0.1 M ammonium acetate and resuspended. The phages were then deposited on copper grids with carbon-coated Formvar films and stained with 2% uranyl acetate (pH 4) or 2% phosphotungstate (pH 7.2). They were examined in a Philips EM electron microscope. Magnification was calibrated using T4 phage tails as size standards.

3.7. Bacteriophage DNA Isolation, Restriction Analysis and Sequencing

To separate phage from bacterial debris, crude phage lysates were centrifuged at $10,000 \times g$ for 15 min at 4 °C. Bacterial nucleic acids in the supernatants were digested with pancreatic DNase 1 and RNase A, each to a final concentration of 10 μg/mL (Sigma-Aldrich Canada Ltd., Oakville, ON, USA). Phage particles were then precipitated in the presence of 10% w/v (final concentration) polyethylene glycol (PEG-8000; Sigma-Aldrich) at 4 °C overnight. The precipitated phage particles were recovered by centrifugation, resuspended in TM buffer (10 mM Tris-HCl, pH 7.8, 1 mM $MgSO_4$). The DNA was extracted from a portion of the precipitated viral particles using the SDS/proteinase K method modified from [50], followed by extraction with phenol:chloroform:isoamyl alcohol (25:24:1, vol/vol), ethanol precipitation and resolution in 10 mM Tris-HCl (pH 7.5). The DNA was characterized spectrophotometrically.

Phage DNA was digested with restriction enzymes BamHI, BglI, EcoRV, HindIII, NaeI, XbaI, NdeI, PstI, and SmaI (New England Biolabs, Hertfordshire, UK), according to the manufacturer's instructions. Phage Lambda DNA (Fermentas Inc., Hanover, MD, USA) digested by the same enzymes was used as a positive control for the restriction digest analysis.

The DNA was subjected to pyrosequencing (454 technology) at the McGill University and Génome Québec Innovation Centre (Montréal, QC, Canada) to between 195 and 930 fold coverage.

3.8. Bioinformatic Analysis

The sequences were rearranged to resemble their GenBank homologs and annotated using MyRAST [53]. The generated gbk file was incorporated into Kodon (Applied Maths, Austin, TX, USA) and proof-read. Proteins were generated using gbk2faa program [54] and screened for homologs using BLAST algorithm [55]. In addition they were searched for conserved motifs in the Pfam database [56] and transmembrane helix predictions using TMHMM [57] and Phobius [58]. The sequence was also screened for tRNAs using tRNAscan-SE program [59]. The final sequences were converted to Sequin format [60] using gbk2sqn [61] before deposition into GenBank. The accession numbers are ACG-C91 JN986844, ACG-M12 JN986845 and ACG-C40 JN986846.

3.9. Biofilm Assay

Each of the 253 E. coli isolates was screened for biofilm formation using the 96-well plate assay as described in [62], with minor modifications. Briefly, *E. coli* isolates were used to inoculate 5 mL of synthetic urine medium and grown for 16 h at 37 °C. A 1:100 dilution of each of the cultures was made in synthetic urine media and 200 µL of each diluted culture was added to six wells in an untreated 96-well polystyrene flat-bottomed, 96-well microtiter plates (Costar; Corning Inc., Corning, NY, USA). The plate was covered and incubated 37 °C for 48 h without agitation. Negative control wells that contained 200 µL of sterile synthetic urine media only were included. Following incubation, planktonic bacteria were removed by pipetting out the culture and washing the wells twice with PBS to remove loosely bound cells, using a multichannel pipettor. A 1% (wt/vol; 210 µL) crystal violet (CV) solution was added to the wells for 10 min. The CV stain was then decanted and the unbound stain removed by washing the plate in a water tray. The plates were left to dry following which 200 µL of 95% ethanol added in each well to solubilize the bound CV from the stained *E. coli* biofilms. The absorbance of CV at 600 nm was measured in a Wallac-Victor^2 1420 Multilabel Counter (Perkin-Elmer, Boston, MA, USA).

3.10. Bacteriophage Host Range on Biofilm Forming UPEC Isolates

The host range of phages ACG-C91, ACG-C40 and ACG-M12 against biofilm forming *E. coli* UTI isolates was determined by standard spot tests [63]. Briefly, this involved mixing 200 µL overnight culture of each isolate with 3 mL molten top LB agar (0.6% agar) and pouring on 96 mm diameter plates containing bottom agar. This was left to solidify then 10 µL of phage to be tested was spotted in the middle of the plate and left to dry before incubation inverted overnight at 37 °C.

3.11. Bacteriophage Activity against Established Biofilms

E. coli isolate CLSC 94 was used to test the effect of the two environmentally isolated phages ACG-C40 and ACG-C91 on established biofilms. The method of [64] with minor modifications was used. Briefly, stationery-phase culture of *E. coli* CLSC94 was diluted 1:100 and 200 µL of the inoculum added to 12 wells of an untreated 96-well polystyrene microtiter plate. This was incubated for 24 h at 37 °C without agitation. Negative control wells that contained 200 µL of sterile synthetic urine media only were included. Following incubation period, planktonic bacteria were removed by pipetting out the culture, and washing the wells twice with PBS to remove loosely bound cells. The wells were then treated with 200 µL of ACG-C40 and ACG-C91 phage lysate at titers of 10^5, 10^7 and 10^9 PFU/mL added to each of the 12 wells. Phages were also added to the negative control wells which contained only UPEC cells grown in sterile synthetic urine media. After incubation periods of 2 h, 4 h, 8 h and 24 h, planktonic bacteria were removed and wells were washed twice with PBS to remove loosely bound cells. The remaining biofilm was stained as described above.

4. Conclusions

We were able to isolate phages that are effective against UPECstrains and are able to degrade biofilms formed by the UPEC strains. The phages' DNA sequences were determined. They were screened and found to be devoid of undesirable laterally transferable virulence and antimicrobial resistance determinants on the basis of homologies with known virulence and resistant genes available in GenBank.

The study resulted in detailed characterization of a large set of UPEC with regards to biofilm forming capabilities and antimicrobial resistance. The candidate phages with adequate lytic spectrum for therapeutic purposes (including efficacy for disruption of existing biofilms) were characterized in detail. The phages have a promising potential for phage therapy of UTIs caused by biofilm forming UPEC.

Acknowledgments

AC holds a Natural Science and Engineering Research Council (NSERC) postdoctoral fellowship. AMK was supported by a Discovery Grant from NSERC. This research was supported by a Government of Canada Genomics R&D Initiative to PB and AMK.

References

1. Foxman, B. Epidemiology of urinary tract infections: Incidence, morbidity, and economic costs. *Dis.-a-Mon.* **2003**, *49*, 53_70.

2. Sanford, J.P. Urinary tract symptoms and infections. *Ann. Rev. Med.* **1975**, *25*, 485.

3. Nicolle, L.E.; Ronald, A.R. Recurrent urinary tract infection in adult women: Diagnosis and treatment. *Infect. Dis. Clin. N. Am.* **1987**, *1*, 793–806.

4. Nicolle, L.E. Epidemiology of urinary tract infections. *Clin. Microbiol. Newslett.* **2002**, *24*, 135–140.

5. Czaja, C.A.; Hooton, T.M. Update on acute uncomplicated urinary tract infection in women. *Postgrad. Med.* **2006**, *119*, 39–45.

6. Foxman, B.; Barlow, R.; D'Arcy, H.; Gillespie, B.; Sobel, J.D. Urinary tract infection: Self-reported incidence and associated costs. *Ann. Epidemiol.* **2000**, *10*, 509–515.

7. Russo, T.A.; Johnson, J.R. Medical and economic impact of extraintestinal infections due to *Escherichia coli*: Focus on an increasingly important endemic problem. *Microbes Infect.* **2003**, *5*, 449–456.

8. Bacheller, C.D.; Bernstein, J.M. Urinary tract infections. *Med. Clin. N. Am.* **1997**, *3*, 719–730.

9. Gupta, K.; Hooton, T.M.; Stamm, W.E. Increasing antimicrobial resistance and the management of uncomplicated community-acquired urinary tract infections. *Ann. Intern. Med.* **2001**, *135*, 41–50.

10. Gupta, K. Addressing antibiotic resistance. *Am. J. Med.* **2002**, *113*, 29S⁻34S.

11. Karlowsky, J.A.; Kelly, L.J.; Thornsberry, C.; Jones, M.E.; Sahm, D.F. Trends in antimicrobial resistance among urinary tract infection isolates of *Escherichia coli* from female outpatients in the United States. *Antimicrob. Agents Chemother.* **2002**, *46*, 2540–2545.

12. Karlowsky, J.A.; Thornsberry, C.; Jones, M.E.; Sahm, D.F. Susceptibility of antimicrobial-resistant urinary *Escherichia coli* isolates to fluoroquinolones and nitrofurantoin. *Clin. Infect. Dis.* **2003**, *36*, 183–187.

13. Zhanel, G.G.; Hisanaga, T.L.; Laing, N.M.; DeCorby, M.R.; Nichol, K.A.; Palatnik, L.P.; Johnson, J.; Noreddin, A.; Harding, G.K.; Nicolle, L.E.; *et al.* NAUTICA Group. Antibiotic resistance in outpatient urinary isolates: Final results from the North American Urinary Tract Infection Collaborative Alliance (NAUTICA). *Int. J. Antimicrob. Agents* **2005**, *26*, 380–388.

14. Morris, N.S.; Stickler, D.J.; McLean, R.J. The development of bacterial biofilms on indwelling urethral catheters. *World J. Urol.* **1999**, *17*, 345–350.

15. Stewart, P.S.; Costerton, J.W. Antibiotic resistance of bacteria in biofilms. *Lancet* **2001**, *358*, 135–138.

16. Doolittle, M.M.; Cooney, J.J.; Caldwell, D.E. Lytic infection of *Escherichia coli* biofilms by bacteriophage T4. *Can. J. Microbiol.* **1995**, *42*, 12–18.

17. Doolittle, M.M.; Cooney, J.J.; Caldwell, D.E. Tracing the interaction of bacteriophage with bacterial biofilms using fluorescent and chromogenic probes. *J. Indust. Microbiol.* **1996**, *16*, 331–341.

18. Lacroix-Gueu, P.; Briandet, R.; Leveque-Fort, S.; Bellon-Fontaine, M.N.; Fontaine-Aupart, M.P. *In situ* measurements of viral particles diffusion inside mucoid biofilms. *C. R. Biol.* **2005**, *328*, 1065–1072.

19. Wood, H.L.; Holden, S.R.; Bayston, R. Susceptibility of *Staphylococcus epidermidis* biofilm in CSF shunts to bacteriophage attack. *Eur. J. Pediatr. Surg.* **2001**, *11*, S56–S57.

20. Sillankorva, S.; Oliveira, R.; Vieira, M.J.; Sutherland, I.W.; Azeredo, J. Bacteriophage φS1 infection of *Pseudomonas fluorescens* planktonic cells *versus* biofilms. *Biofouling* **2004**, *20*, 133–138.

21. Curtin, J.J.; Donlan, R.M. Using bacteriophages to reduce formation of catheter-associated biofilms by *Staphylococcus epidermidis*. *Antimicrob. Agents Chemother.* **2006**, *50*, 1268–1275.

22. Hughes, K.A.; Sutherland, I.W.; Jones, M.V. Biofilm susceptibility to bacteriophage attack: The role of phage-borne polysaccharide depolymerise. *Microbiology* **1998**, *144*, 3039–3047.

23. Hughes, K.A.; Sutherland, I.W.; Clark, J.; Jones, M.V. Bacteriophage and associated polysaccharide depolymerases-novel tools for study of bacterial biofilms. *J. Appl. Microbiol.* **1998**, *85*, 583–590.

24. Hanlon, G.W.; Denyer, S.P.; Ollif, C.J.; Ibrahim, L.J. Reduction in exopolysaccharide viscosity as an aid to bacteriophage penetration through *Pseudomonas aeruginosa* biofilms. *Appl. Environ. Microbiol.* **2001**, *67*, 2746–2753.

25. Rijavec, M.; Müller-Premru, M.; Zakotnik, B.; Zgur-Bertok, D. Virulence factors and biofilm production among *Escherichia coli* strains causing bacteraemia of urinary tract origin. *J. Med. Microbiol.* **2008**, *57*, 1329–1334.

26. Hoyle, B.D.; Alcantara, J.; Costerton, J.W. *Pseudomonas aeruginosa* biofilm as a diffusion barrier to piperacillin. *Antimicrob. Agents Chemother.* **1992**, *36*, 2054–2056.

27. Ito, A.; Taniuchi, A.; May, T.; Kawata, K.; Okabe, S. Increased antibiotic resistance of *Escherichia coli* in mature biofilms. *Appl. Environ. Microbiol.* **2009**, *75*, 4093–4100.

28. Gilbert, P.; Maira-Litran, T.; McBain, A.J.; Rickard, A.H.; Whyte, F.W. The physiology and collective recalcitrance of microbial biofilm communities. *Adv. Microbial. Physiol.* **2002**, *46*, 203–256.

29. Sillankorva, S.; Oliveira, D.; Moura, A.; Henriques, M.; Faustino, A.; Nicolau, A.; Azeredo, J. Efficacy of a broad host range lytic bacteriophage against *E. coli* adhered to urothelium. *Curr. Microbiol.* **2011**, *62*, 1128–1132.

30. Goodridge, L.; Gallaccio, A.; Griffiths, M.W. Morphological, host range, and genetic characterization of two coliphages. *Appl. Environ. Microbiol.* **2003**, *69*, 5364–5371.

31. Chibani-Chennoufi, S.; Sidoti, J.; Bruttin, A.; Dillmann, M.L.; Kutter, E.; Qadri, F.; Sarker, S.A.; Brüssow, H. Isolation of *Escherichia coli* bacteriophages from the stool of pediatric diarrhea patients in Bangladesh. *J. Bacteriol.* **2004**, *186*, 8287–8294.

32. Zuber, S.; Ngom-Bru, C.; Barretto, C.; Bruttin, A.; Brüssow, H.; Denou, E. Genome analysis of phage JS98 defines a fourth major subgroup of T4-like phages in *Escherichia coli. J. Bacteriol.* **2007**, *189*, 8206–8214.

33. Nishikawa, H.; Yasuda, M.; Uchiyama, J.; Rashel, M.; Maeda, Y.; Takemura, I.; Sugihara, S.; Ujihara, T.; Shimizu, Y.; Shuin, T.; *et al.* T-even-related bacteriophages as candidates for treatment of *Escherichia coli* urinary tract infections. *Arch. Virol.* **2008**, *153*, 507–515.

34. Verma, V.; Harjai, K.; Chibber, S. Structural changes induced by a lytic bacteriophage make ciprofloxacin effective against older biofilm of *Klebsiella pneumoniae. Biofouling* **2010**, *26*, 729–737.

35. Corbin, B.D.; McLean, R.J.; Aron, G.M. Bacteriophage T4 multiplication in a glucose-limited *Escherichia coli* biofilm. *Can. J. Microbiol.* **2001**, *47*, 680–684.

36. Lu, T.K.; Collins, J.J. Dispersing biofilms with engineered enzymatic bacteriophage. *Proc. Natl. Acad. Sci. USA* **2007**, *104*, 11197–11202.

37. Petrov, V.M.; Ratnayaka, S.; Nolan, J.M.; Miller, E.S.; Karam, J.D. Genomes of the T4-related bacteriophages as windows on microbial genome evolution. *Virol. J.* **2010**, *7*, 292.

38. Carlson, K.; Raleigh, E.A.; Hattman, S. Restriction and Modification. In *Bacteriophage T4*; Karam, J.D., Ed.; ASM Press: Washington, DC, USA, 1994; pp. 369–381.

39. Mahadevan, P.; King, J.F.; Seto, D. CGUG: *In silico* proteome and genome parsing tool for the determination of "core" and unique genes in the analysis of genomes up to ca. 1.9 Mb. *BMC Res. Notes* **2009**, *2*, 168.

40. Darling, A.C.; Mau, B.; Blattner, F.R.; Perna, N.T. Mauve: Multiple alignment of conserved genomic sequence with rearrangements. *Genome Res.* **2004**, *14*, 1394–1403.

41. Kropinski, A.M.; Borodovsky, M.; Carver, T.J.; Cerdeño-Tárraga, A.M.; Darling, A.; Lomsadze, A.; Mahadevan, P.; Stothard, P.; Seto, D.; Van Domselaar G.; *et al. In silico* identification of genes in bacteriophage DNA. *Methods Mol. Biol.* **2009**, *502*, 57–89.

42. Canadian Antimicrobial Resistance Alliance. Available online: http://www.can-r.com/ (accessed on 26 March 2012).

43. Zhanel, G.G.; Adam, H.J.; Low, D.E.; Blondeau, J.; Decorby, M.; Karlowsky, J.A.; Weshnoweski, B.; Vashisht, R.; Wierzbowski, A.; Hoban, D.J. Canadian antimicrobial resistance alliance (CARA). Antimicrobial susceptibility of 15,644 pathogens from Canadian hospitals: results of the CANWARD 2007–2009 study. *Diagn. Microbiol. Infect. Dis.* **2011**, *69*, 291–306.

44. Griffith, D.P.; Musher, D.M.; Itin, C. Urease: The primary cause of infection-induced urinary stones. *Investig. Urol.* **1976**, *13*, 346–350.

45. Stickler, D.J.; Morris, N.S.; Winters, C. Simple physical model to study the formation and physiology of biofilms on urethral catheters. *Methods Enzymol.* **1999**, *310*, 494–501.

46. Ewing, E.H. Genus *Escherichia*. In *Identification of Enterobacteriaceae*, 4th ed.; Edwards, P.R., Ewing, W.H., Eds.; Elsevier Science: New York, NY, USA, 1986; pp. 96–134.

47. Government of Canada. Canadian Integrated Program for Antimicrobial Resistance Surveillance (CIPARS) 2005. Public Health Agency of Canada: Guelph, Ontario, Canada, 2007.

48. Cockerill, F.R.; Wikler, M.A.; Bush, K. Performance standards for antimicrobial susceptibility testing; twentieth informational supplement M100-S20; Clinical and Laboratory Standards Institute: Wayne, PA, USA, 2010.

49. Casas, V.; Rohwer, F. Phage metagenomics. *Methods Enzymol.* **2007**, *421*, 259–268.

50. Sambrook, J.; Fritsch, E.F.; Maniatis, T. *Molecular Cloning: A Laboratory Manual*; Cold Spring Harbor: New York, NY, USA, 1989; Volume 2.

51. Adams, M.H. Bacteriophages. Interscience Publishers: New York, NY, USA, 1959.

52. Kropinski, A.M.; Prangishvili, D.; Lavigne, R. Position paper: The creation of a rational scheme for the nomenclature of viruses of Bacteria and Archaea. *Environ. Microbiol.* **2009**, *11*, 2775–2777.

53. myRAST- The SEED Server. Available online: http://blog.theseed.org/servers/presentations/t1/drast-overview.html (accessed on 9 February 2012).

54. GBKFAA. Available online: http://lfz.corefacility.ca/gbk2faa/ (accessed on 9 February 2012).

55. Basic Local Alignment Search Tool (BLAST). Available online: http://blast.ncbi.nlm.nih.gov/ (accessed on 9 February 2012).

56. Finn, R.D.; Mistry, J.; Tate, J.; Coggill, P.; Heger, A.; Pollington, J.E.; Gavin, O.L.; Gunasekaran P.; Ceric, G.; Forslund, K.; *et al*. The Pfam protein families database. *Nucleic Acids Res.* **2010**, *38*, D211–D222.

57. Sonnhammer, E.L.; Von, H.G.; Krogh, A. A hidden Markov model for predicting transmembrane helices in protein sequences. In Proceedings of the International Conference on Intelligent Systems for Molecular Biology, Montreal, QC, Canada, 28 June–1 July 1998; pp. 175–182.

58. Kall, L.; Krogh, A.; Sonnhammer, E.L. A combined transmembrane topology and signal peptide prediction method. *J. Mol. Biol.* **2004**, 338, 1027–1036.

59. Lowe, T.M.; Eddy, S.R. tRNAscan-SE: A program for improved detection of transfer RNA genes in genomic sequence. *Nucleic Acids Res.* **1997**, *25*, 955–964.

60. Benson, D.A.; Karsch-Mizrachi, I.; Lipman, D.J.; Ostell, J.; Sayers, E.W. GenBank. *Nucleic Acids Res.* **2011**, *39*, D32–D37.

61. Genbank to Sequin File Converter. Available online: http://lfz.corefacility.ca/gbk2sqn/ (accessed on 9 February 2012).

62. O'Toole, G.A.; Pratt, L.A.; Watnick, P.I.; Newman, D.K.; Weaver, V.B.; Kolter, R. Genetic approaches to study of biofilms. *Methods Enzymol.* **1999**, *310*, 91–109.

63. Kutter, E. Phage Host Range and Efficiency of Plating. In *Bacteriophages: Methods and Protocols*; Clokie, M.R.J., Kropinski, A.M., Eds.; Humana Press: New York, NY, USA, 2009; Volume 1, pp. 141–151.

64. Soni, K.A.; Nannapaneni, R. Removal of *Listeria monocytogenes* biofilms with bacteriophage P100. *J. Food Prot.* **2010**, *73*, 1519–1524.

A Genetic Approach to the Development of New Therapeutic Phages to Fight *Pseudomonas Aeruginosa* in Wound Infections

Victor Krylov *, Olga Shaburova, Sergey Krylov and Elena Pleteneva

Laboratory for Bacteriophages Genetics. Mechnikov Research Institute of Vaccines and Sera, RAMS, 5a, Maliy Kazenniy per., Moscow 105064, Russia; E-Mails: oshabs@mail.ru (O.S.); sergeykrylovv@gmail.com (S.K.); f2600@yandex.ru (E.P.)

*   Author to whom correspondence should be addressed; E-Mail: krylov.mech.inst@mail.ru

Abstract: *Pseudomonas aeruginosa* is a frequent participant in wound infections. Emergence of multiple antibiotic resistant strains has created significant problems in the treatment of infected wounds. Phage therapy (PT) has been proposed as a possible alternative approach. Infected wounds are the perfect place for PT applications, since the basic condition for PT is ensured; namely, the direct contact of bacteria and their viruses. Plenty of virulent ("lytic") and temperate ("lysogenic") bacteriophages are known in *P. aeruginosa*. However, the number of virulent phage species acceptable for PT and their mutability are limited. Besides, there are different deviations in the behavior of virulent (and temperate) phages from their expected canonical models of development. We consider some examples of non-canonical phage-bacterium interactions and the possibility of their use in PT. In addition, some optimal approaches to the development of phage therapy will be discussed from the point of view of a biologist, considering the danger of phage-assisted horizontal gene transfer (HGT), and from the point of view of a surgeon who has accepted the Hippocrates Oath to cure patients by all possible means. It is also time now to discuss the possible approaches in international cooperation for the development of PT. We think it would be advantageous to make phage therapy a kind of personalized medicine.

Keywords: *Pseudomonas aeruginosa* bacteriophages diversity; noncanonical relations of phages and bacteria; phage's migrations; phage genomes instability; pseudolysogeny and pseudovirulence; phage therapy as a kind of personalized medicine

1. Introduction

Gram-negative bacteria of species *Pseudomonas aeruginosa* may be found in different natural habitats, because they easily adapt to different conditions. The capability for quick adaptation is the main reason that identifies them as opportunistic pathogens. They cause infections in immune compromised patients or patients with cystic fibrosis—a frequently occurring hereditary disease in Caucasians. *P. aeruginosa* strains are common components in microbial communities of different origins. They have acquired the status of hospital pathogens, and may be isolated from clinical samples taken from the wounds, sputum, bladder, urethra, vagina, ears, eyes and respiratory tract. The emergence of resistance to the most powerful new antibiotics in such clinical *P. aeruginosa* strains, occurring even during treatment [1–9], makes the fight with *P. aeruginosa* hospital pathogens a great problem.

Genomes in most hospital strains of *P. aeruginosa* contain pathogenic islands, where genes, coding many factors of pathogenicity and virulence of this bacterial species such as phospholipase C elastase, protease, siderophore, DNAse, pyocyanin *etc.*, are located simultaneously with genes controlling multiple drug resistance. One of these large genomic islands, PAPI-1, can be transformed into an extrachromosomal circular form, plasmid, after precise excision from a bacterial chromosome [10]. Sometimes such plasmid acquires the ability to be transferred into other recipient *P. aeruginosa* strains by a conjugative mechanism, via a type IV pilus [11]. Such migration quickly disseminates an antibiotic resistance into new strains, thus making the use of antibiotics useless. This is the reason for a quite unexpected renaissance of phage therapy, the use of bacterial viruses in the treatment of bacterial infections, which was proposed by Felix D'Herelle in 1917, immediately after the finding of bacteriophages [12]. The bacteriophage treatment was applied in medical practice with varying success until the introduction of antibiotics; then, because of the great success and simplicity in use of antibiotics, phages were no longer regarded as a serious tool in anti-infective therapy. However, in Russia, Poland and Georgia the use of phage therapy has not ceased and continues to the present day. Given the frequent epidemics of food borne diseases and the increase of many enteric pathogens resistant to antibiotics [13–17], it is useful to remember that some specific phage compositions introduced by F. D'Herelle (but permanently updated) are used with success in the treatment and in the prevention of intestinal infections (see Figure 1).

Figure 1. Commercially produced mix of phages "Intesti" (ImBio Nizhny Novgorod, Russia).

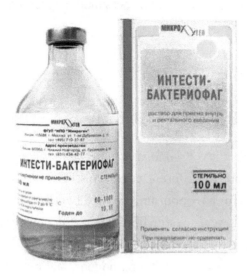

In Eastern Europe, bacteriophages are widely used in surgical wards [18–20]. Apparently, this trend in phage therapy of *P. aeruginosa* caused infections will be used further, considering the failure of the development of safe vaccines against *P. aeruginosa* [21–23].

Bacteriophages were an excellent model of genetic research. Many of the basic concepts of modern biology and the different elements of the methodology in biological and medical sciences have emerged as a result of studying the genetics of phages from the 1940s to 1970s. Being used in other areas, they have become a powerful boost to research of various pro- and eukaryotic systems. In the last two decades, interest in bacteriophages increased significantly again. One reason is the above mentioned occurrences of multiple antibiotic resistant bacteria and the hope that the use of live phages or their products—structured or molecular bacteriocins—antibacterial peptide phage origin—can help in the treatment of bacterial infections. In our opinion, there is another, no less significant reason for a detailed study of bacteriophages. As it has turned out, bacteriophages (both temperate and virulent) are actively involved in the evolution of bacteria, including pathogens, accomplished through different kinds of transduction (general and specialized) horizontal gene transfer (HGT). Studies of the structure of pathogenic islands of different bacterial species confirmed the presence in them of complete or fragmented genomes of temperate phages. Many of the new forms of infective diseases (so-called "emerging diseases") are caused by the appearance in pathogenic islands of previously known pathogens of new genes introduced by phages, plasmids, and transposons (which can be considered as largely related genetic elements that form a common pool of evolutionary active genes). The increased interest in the use of bacteriophages as a replacement for antibiotics in the West has led to an increase in the number of publications illustrating the possibility of actual use of phages in human and veterinary medicine, in finding and describing new phages. Here are the numerous review papers related somehow to bacteriophages that were published recently [24–30]. Extremely bold proposals have been made of the possible use of bacteriophages as antiseptic additives in ready to eat foods, which may potentially be infected with pathogenic bacteria. There are ideas for use of bacteriophages and their products to retard spoilage of foodstuffs during the time of storage or for the prevention of endemic infections and food poisoning associated with the consumption of potentially contaminated food (phage preparations of the «Intesti» type) (see above). The need to quickly find, some alternative to antibiotics is evident. However, in the case of phages there is an evident problem: how to eliminate the possible undesirable activity of phages with the emergence of new pathogens in the process of introducing phage therapy. Some other problems must also be solved if there is an intention to raise the level of phage therapy up to accepted standards in modern medicine. Among such problems are ensuring reproducible results of phage therapy, its long-term use and achieving the required standards for therapeutic phages and their mixtures. Can the introduction of phage therapy be a cause of the emergence of pathogens with new properties? There are also some questions about the prerequisite of high level cooperation, including a quick exchange with bacteriophages capable of stopping infectional outbreaks. We consider some of these issues later, but first we discuss the natural limitations in the use of bacteriophages that distinguish them from antibiotics. This assesses the real prospects of live phage therapy and also helps understand why phage therapy is not the main procedure of treatment for some diseases in the countries where its use was not interrupted after introduction of antibiotics.

2. The Goal, Objectives and Content of the Present Review

Many studies related in one way or another to bacteriophages have been published in the last 10–15 years. They concern, the numerous advantages of phage therapy and specific features of a large number of newly isolated phages have been discussed. The list of phages with sequenced and annotated genomes is in constant enlargement in NCBI.

Currently, there has been established a certain unofficial algorithm in the studies of newly isolated bacteriophages: the isolation of the phage itself, extraction of genomic nucleic acid (in most cases, DNA), electron microscopy of phage particles for classification purposes, genome sequencing and annotation, estimation of evolutionary relationships with other phages and prospects for practical use. Previously initiated as well as new companies are in action, producing and advertising their numerous commercial phage mixtures; newly organized specialized therapeutic bacteriophage treatment centers are operating. We have not provided links to these activities: all this can be found on the Internet.

The purpose of this review is to draw attention to the need for a close examination of the possible features in the behavior of different types of bacteriophages in their actual application, which can be found on the stage of laboratory studies [31], in order to understand the possible consequences and the necessary modifications of the bacteriophages and the methods of their use. The system of phage therapy is fundamentally different from the organization of antibiotic therapy. In supporting the use of bacteriophages in medical practice, we want first of all to justify the need for the introduction of such a system. We start by looking at the natural limitations in the use of phages as therapeutic agents.

3. Limitations in the Use of Live Bacteriophages in a Treatment

(1) Phage therapy with living phages is applicable only in cases where it is possible to provide direct contact with the virus and bacteria. However, the direct introduction of even highly purified

phage particles in the blood, such as in the case of septicemia (although there are described unique cases of this kind) [32] is very risky. The view that oral administration of phages may cause them to penetrate into the bloodstream, and then into the urine (the concept of "phages are absorbed into blood and cure"), is not justified by rigorous research (although such cases have also been described in the literature) [33]. Our own laboratory experiments did not confirm this. Requirements for direct contact limits the use of phage therapy with wound infections (other than pressure ulcers), urogenital infections, intestinal infections, eye infections and infections of the ear, nose and throat organs. Cystic fibrosis, where bacteria of the species P. aeruginosa are in contact with different phages cannot yet be attributed to the list of infections that are acceptable for phage therapy, although research in this area (in a mouse model) continues [34].

(2) Another reason for the limited use of phage therapy is the inability to provide long-term effectiveness of a particular mixture of therapeutic phages. In the manufacturing companies, producing commercial phage mixtures, it is generally accepted that included in such products are phages with the broadest spectrum of lytic activity for each bacterial species involved. However, regardless of the spectrum of lytic activity of phages in the mixtures, bacterial mutants of a species that reveal resistance to all phages in the mixture may arise even after a few days after the start of treatment. Thus, a single mutation in the genome of Gram-negative bacteria such as *P. aeruginosa*, leading to resistance to adsorption, may prevent the growth and lytic activity several unrelated phages [35]. It is known from personal contact with surgeons of departments of purulent infections, that phage-resistant clones often arise after short time use of polyvalent phage mixtures. Such resistant clones can then replace the previous hospital pathogens. We consider the properties of these pathogens and their relation to phages later. Usually renewal of phages in commercial mixtures is quite rare, a few times a year. Therefore, the rapid accumulation of phage-resistant mutants can lead very quickly to loss of activity of the applied phage mixture. The only possible solution to the problem is the replacement of such an ineffective phage mixture with the mixture from another manufacturer, in the hope that the phages in the new mixture have a different spectrum of lytic activity. This is a temporary albeit not very reliable solution.

(3) The obvious aim of each producer of therapeutic phages is to create a preparation with the highest range of final lytic activity, which can be used without replacement for as long as possible. Therefore, since the time of F. D'Herelle, in order to ensure maximal activity of therapeutic phages against pathogenic strains, phages for therapy have been produced with the use of pathogenic clinical isolates. Another important reason for such use of clinical (pathogenic) isolates is that it is not always possible to find appropriate non-pathogenic variants with sensitivity to a particular phage, which must be incorporated into the therapeutic mix. This is the positive side of the use in the batch phage production of clinical isolates as hosts. On the other hand, the real pathogen used as a production strain, even when non-lysogenic, obviously contains genes with harmful effects. As a result, phages obtained with the use of such a pathogen, when introduced into the therapeutic mixture, may in the course of treatment, with any kind of transduction, transfer such genes into clinical strains with the resulting emergence of bacteria that cause the disease with symptoms of "emerging diseases". The

only solution is to use only well studied bacterial hosts. In principal, it is necessary to create production strains that are sensitive to different phages, based on well-studied model strains that do not contain dangerous genes in their genomes.

(4) Some weaknesses in planning and use of phage therapy also hinder the successful introduction as a standard treatment or preventive procedure. There is a need to create a well thought out organization at all levels from the isolation of a pathogenic strain to the fast selection of the best phage for treatment. The creation of such an organization requires support on an inter-state level, because its development goes beyond the possibilities of individual researchers or research laboratories, and apparently requires coordination on a number of different levels. The existence of such a system could play a positive role in quick reactions to the occurrence of local epidemics, such as the emergence of the outbreak caused by an enterohemorrhagic (EHEC) strain of *E. coli* serotype O104: H4 in early May 2011 in northern Germany, or in the case of the outbreak of listeriosis in September and October 2011 in the U.S.A. In the U.S.A., for instance, there is a phage mixture active against *Listeria* (integrated drug ListShield™; [36]) which is permitted by the US Food and Drug Administration for use in «ready-to-eat» products. It is possible that the timely distribution of therapeutic phage mixtures amongst the population in areas threatened by epidemiological characteristics could reduce the number of subsequent cases. In the same hypothetical system of organization of phage therapy should be included laboratories involved in research of phages for therapy as well as international collections of bacteriophages [37]. An optimal situation is the case when only such phages active against local pathogens, but not polyvalent mixtures are used in clinics. It is good in the case when there is the possibility to choose such phages *in situ*, in the hospital, and then treatment must be accompanied by constant monitoring to check for arising phage-resistant variants. The choice of inappropriate phages and use of polyvalent mixtures without checking for the effectiveness of their component phages against particular wound pathogens will affect the efficiency of treatment and, as a consequence, will discredit the idea of phage therapy among doctors and patients.

Therapeutic phages, which are intended for personal use, should definitely go through the standard procedure of classification up to species. This significantly lowers the probability of HGT, revealing the presence in the genomes of phages containing any features that pose a potential danger. Perhaps, phage therapy at some future time could be included into the arsenal of personalized medicine.

4. Non-Canonical Interaction of Phages and Bacteria

The original idea of phage therapy is based on the assumption that interactions of phages and bacteria are similar with predator-prey relationships and the development of virulent phage proceeds within the framework of the lytic cycle. This ideal ("canonical") lytic cycle is as follows: adsorption of the phage to the surface cell structures (adsorption receptors specific for each type of phage); injection of the phage genome into the cell (using the natural pores or after local melting of the cell wall by a lytic enzyme incorporated into the structure of the tail of the phage particle); the implementation of phage development programs written in phage genome; the destruction of the cell envelope from inside by specific phage enzymes and release progeny phage particles; readiness to repeat the same cycle. The ability of cells to divide normally disappears soon after the injection of the phage genome.

In the case of infection cells with temperate phage there can be two outcomes: a typical lytic cycle or lysogenization of the infected cell with a phage genome in prophage condition; prophage can be integrated into the bacterial chromosome or be in a state of plasmid, whose division and distribution in each of the daughter bacterial cells are strictly coordinated with the division of bacteria and replication of bacterial chromosome. A stable state of bacterial cell with prophage is supported through the blockade of lytic genes expression in prophage with a special protein repressor. Sometimes the repressor undergoes inactivation, and the phage starts the lytic cycle of development. However, under certain conditions, there is a deviation from such a course of events that could significantly change the nature of the phage behavior and, as a consequence, the final result of phage therapy. These deviations can result in a pseudolysogenic state of infected bacterial cells. Developmental disorders in the case of temperate phages can prevent the establishment of a lysogenic state (pseudovirulent state).

A good example of the lytic cycle is the development of phage T4. As a model for temperate phage, the development of bacteriophage lambda can be accepted. Nevertheless, it is necessary to recognize that the properties, even of these models, are still far from being fully explored. For example, the functions of 20 proteins coded by the lambda genome were not possible to assign in comparison with the known database. Their functions were studied with specially designed experiments [38]. Authors predicted functions for 12 of the mentioned proteins. So even in the case of phage for which careful research has lasted several decades, the functions of many of the genes are still unknown.

5. Pseudolysogeny and Pseudovirulence

5.1. Pseudolysogenic Conditions

5.1.1. Some Examples of Reasons for Pseudolysogenic Conditions

Studies of bacteriophages in laboratories are usually accomplished under conditions which are optimal for bacterial host growth. However, interactions of bacteria and phages at the place of infection may occur in different situations. To check their influence on deviations in the canonical interaction of phage and bacteria before introduction of a phage into a therapeutic mixture, it is necessary to conduct special studies under different conditions. The importance of such studies is confirmed by the existence of particular types of phage development, which differ from the canonical lytic and temperate infections. We do not mention here specific bacterial viruses with chronic type infection as their use in therapy or occasional occurrence in phage therapeutic mixtures in our opinion may be very dangerous [37]. Let us now consider some different examples of pseudolysogeny:

> (a) Frequent reasons for establishment of pseudolysogeny and its maintenance are changes of cultural conditions in the background of specific bacterial and phage genotypes. Typical results were obtained in the study of *P. aeruginosa*, slowly growing in the presence of its

phage in a chemostat continuous culture model. The frequency of pseudolysogens increased as cells were starved. According to the authors, pseudolysogeny is specific phage strategy to survive in periods of starvation of their hosts [39].

(b) Transition of lysogenic cells from active multiplication into stationary phase can also stimulate high frequency of phage loss, as has been found in the study of the gram-positive bacteria *Propionibacterium acnes* with a group of closely related temperate inducible phages. Bacterial clones surviving such an event were susceptible to new infection by the same phages. Most likely, the phage genome in this case was not integrated into the bacterial chromosome [40].

(c) Bacteria may acquire a capability for lysogeny as result of mutation. Phage VHS1 produces clear plaques on its host *Vibrio harveyi* (strain VH1114). Bacterial clones have been isolated which could be lysogenized but being lysogens they produced cured cells after approaching the stationary growth phase. These bacteria had inherited differences proving their mutational origin [41].

(d) Sometimes pseudolysogeny is a result of several different genetical modifications in phage genome and/or in a bacterial genome. Neurotoxins C and D of *Clostridium botulinum* are encoded by genes of bacteriophages. The genome of converting phage c-st was sequenced and annotated. As it turned out, c-st prophage is present as a circular plasmid. Plasmids (and plasmid prophages) code special proteins, necessary to resolve plasmid multimer partition, and segregation, but it is possible that their activity is not sufficient to ensure the stable inheritance of c-st as plasmid. Besides, there is another important reason for the plasmid instability. A remarkable feature of the c-st genome is the abundance of IS elements (altogether 12 copies). The presence of such a number of IS is unexpected for a viable phage. The authors suggest that possible recombination between these multiple elements is the basic reason for the plasmid prophage instability. This case of pseudolysogeny may be considered as a typically weakened impaired lysogenic state [42].

Detailed studies of the regulation systems of the temperate phage lambda *E. coli* demonstrated the dependence of the stability of the lysogenic state of functioning of a group of genes that affect the expression of the gene encoding the repressor. However, apparently with new phages allocated from the natural environment there are many cases that cannot be explained by reference to known patterns.

5.1.2. Pseudolysogeny and Therapeutic Phages for *P. aeruginosa*: General Considerations

Our laboratory, for a number of years, has studied different bacteriophages, active in pseudomonades of different species. In the course of such work, we have met some situations which evidenced the "non-canonical" behavior of phages and phage-sensitive bacteria for *P. aeruginosa*. In relation to phage therapy, it may be of interest to consider some cases of non-canonical interactions for phages from existing commercial mixtures. Besides, it is necessary to take into account the special role of temperate phages in horizontal genetic transfer. Such observations can have certain practical significance and can be taken into consideration to enhance phage therapy of infected wounds or hospital infections. The studies of non-canonical interactions of phages and bacteria are intended primarily to assess the possibility and probability of occurrence during the treatment of hybrid phages capable of transposition, conversion, and lysogenization, *i.e.*, participating in the process of HGT. The need for such studies has been proved because even the sequencing and annotation of the genome cannot always predict the behavior of bacteriophages in the real world. Similarly, *in silico* experiments on the interaction of controlled phage proteins cannot confirm the safety of their functions or prove the significance of the function of these proteins for the viability or lytic activity of the phage. In relation to *P. aeruginosa* caused infections, there are detailed studies of genomes of several phage groups which may be considered as favorites for phage therapy, such as KMV-like, PB1-like, N4-like and phiKZ-like phages, which are be found in some commercial mixtures. Different phages of these four species are the frequent sources used to replenish commercial mixtures. It has been assumed that the representatives of individual phages belonging to certain species are not very different from each other. However, it is impossible to suggest the identity of the behavior of all phages of the same species without their detailed comparison. Certainly, there are some limitations in the attempts for immediate applications of results obtained in the laboratory ("Petri dish studies") applied to conditions existing in a real infected wound. Nevertheless the basic processes in relation of *P. aruginosa* and its bacteriophages in the case of surface infections reveal a good similarity in both cases (biofilm formation, adhesion, alginate production, basic features of phage infections as transpositions, plasmid and phage migrations, *etc.*). We consider here several different deviations from canonical interactions of phages and *P. aeruginosa*, which may manifest themselves in the infected wound. In addition, we discuss the possible use of temperate phages in therapy and the influence of wound bacterial *P. aeruginosa* strains on phage growth. Some *P. aeruginosa* temperate transducing phages ((B3, G101, F116) [43] reveal a natural ability to infect strains as *Burkholderia cepacia* which may be considered a proof of common origin of the species [44].

5.1.3. Pseudolysogeny in the Case of phiKZ-Like Phages

PhiKZ (access number NC_004629) is a very special giant phage, which has the specific structure of capsid with a spiral inner body [45,46]. The phiKZ-like phages active on *P. aeruginosa* were classified into three species (phiKZ, Lin68, EL) [47,48]. Different phiKZ-lke phages are permanent components of commercial phage therapeutic polivalent mixtures. One of the attractive features of this phage species when used in commercial preparations is its capability to produce high final yields. The phiKZ-like phages are a very common component of phage biota in water and soil, and may be isolated in different regions [49]. PhiKZ was found to contain several genes coding for orthologs of proteins of pathogenic prokaryotes, unrelated phages, and some eukaryotes [50]. It is not known up to now, whether these genes belong to the phage genome and code for proteins essential for phage development or whether they represent "genetic noise", captured by the phage during migration and potentially playing a detrimental role.

These issues are of interest for several reasons. First, commercial mixtures used in phage therapy sometimes show a regional specificity. However, phages of species phiKZ, isolated from geographically distant regions, have just small genetic differences from phage phiKZ. Each of the phages has a quite broad (more than 20%) lytic activity spectrum. Such phages are extremely useful to kill specific bacterial strains resistant to other phages. We have found that the drastic differences in interactions of phiKZ-like phages with their host are dependent on the multiplicity of infection (m.o.i.). In a one step growth cycle experiment (m.o.i. in a range of 1–5 particles), phiKZ behaves as a typical virulent phage. All infected cells were killed with liberation of a not very high number of phage particles [45]. However, there was a significant increase of m.o.i. of sensitive bacteria with phiKZ-like phages, which arises when they are grown in biofilm conditions. As a result, these phages produce huge amounts of progeny particles, which exceed phage concentrations many times in the case of low m.o.i. In the genomes of all phiKZ-like phages that have been sequenced up to now, no gene coding DNA-polymerases have been found as closely related with the ones described earlier [50]. Thus, at first it was suggested that phiKZ-like phages could use bacterial DNA polymerase for their replication. Recently, with the use of modern computational analysis, sequences have been found supposedly coding some proteins, which have a distant similarity with some DNA-polymerases domains, which were found in genomes of several different phiKZ-like phages [51,52].

It is assumed that the mechanism of replication of phiKZ-like phages requires two components, and probably the transition of multiply infected cells into the pseudolysogenic state may be associated with the change of conditions in DNA polymerase activity. For instance, this may be due to deficit of a bacterial component which is necessary for replication. Anyway, the bacterial cell recognizes multiple infection with phiKZ-like phage particles and stops servicing phage replication. This is one of the possible explanations. However, there is another opportunity to develop an adequate explanation for induction in a cell's pseudolysogenic state after multiple infection with phiKZ and EL-like phages. The genomes of these phages are coded proteins similar to repressors of some temperate phages [50,53]. It is possible that the activity of the gene's products expressed from a small number of genomes (in low m.o.i.) is not enough to stop or to slow down the lytic development of these phages. An increase in the multiplicity of infection leads to an increase in repressor concentration (more template DNA), which stimulates transfer to a lysogenic condition. In this model, the cell estimates the level of repressor activity and prevents phage development, not with a common mechanism (blocking of a specific site in the phage genome), but with turn-off of bacterial functions necessary for lytic development. In one way or in another, it is evident that in conditions that arise after the infection of cells with high multiplicity, intracellular development of virulent phage is turned off for a while.

The phiKZ-like wild type phages are not appropriate components for phage therapeutic mixtures. These phages can lead to HGT (because pseudolysogenic cells can support development of other phages, including temperate ones, transfer of plasmids, and transduction). We have isolated mutants of phiKZ and EL, unable to induce pseudolysogeny and, moreover, actually killing pseudolysogenic cells. The mutants have properties of classical virulent phages. Interestingly, phiKZ and EL behave like phages with repressors of different specificity (with phiKZ revealing dominance) [54], see Figure

2. Probably, the substitution of virulent mutants of phiKZ-like phages for wild type phages in therapeutic mixtures decreases the possibility of HGT [55,56].

Figure 2. Phylogenetically related bacteriophages phiKZ and EL have repressors of different specificity. Virulent mutant ELvir5 cannot lyse phiKZ wild type infected pseudolysogenic cells (plated on a lawn of bacteria *P. aeruginosa* PAO1).

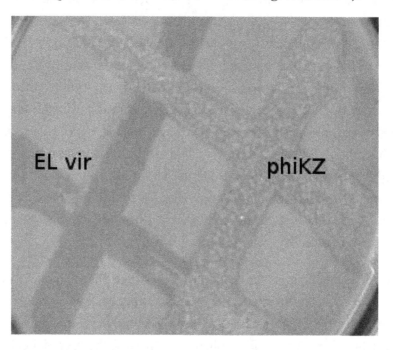

In Figure 3, the pseudolysogenic growth of *P. aeruginosa* PAO1 cells, infected by phiKZ with high m.o.i. in a Petri dish with nutrient media is shown. Besides the opalescent "bluish" growth of colonies with cells in pseudolysogenic conditions, there are some other additional features, which are of interest. First, material taken from the "bluish" (opalescent) initial growth (Figure 3a), after replanting, produces highly viscous colonies (Figure 3b). One of the possible explanations for such unexpected results may be the conversion of infected bacteria into an alginate producing state. Indeed, as it has been shown, the genes encoding the alginate biosynthetic enzymes are clustered in a single operon, which is under transcriptional control [57]. It may be very important for the activator of the alginate operon is AlgZ, a proposed ribbon-helix-helix DNA binding protein, which reveals some similarity with the repressors of phiKZ and phiEL phages (ORF 196 for phiKZ and ORF 163 for phiEL). Thus, overproduction of phage repressor proteins can be a reason for the activation of AlgZ operon with corresponding alginate overproduction. The second interesting conclusion that can be made from Figure 3c is that after additional prolongation of incubation for several days around the initial growth of opalescent colonies, a growth of phage sensitive colonies with a bluish border can be seen. This means that the sensitive bacteria and phage are able to move a certain distance from the place of initial planting. Phage particles are incapable of active movement. We suggest that phage particles were transported by bacteria resistant (at least for a time) to phage killing ability. One possible reason may be the blocking of intracellular phage development or interference with phage DNA injection (transfer of phages particles on the cell, as a "rider"). The most sensitive bacteria are killed in the process of moving to the border of the colony, as evidenced by a high concentration of phages in the intermediate area.

Figure 3. Properties of pseudolysogens which arise after infection of *P. aeruginosa* PAO1 cells with phage phiKZ. (**a**) the growth of phage phiKZ after two days of incubation, the appearance of opalescence in growth of pseudolysogens; (**b**) replanting of the material taken from growth seen on "a" on the surface of the cultural solid medium; single colonies formed with pseudolysogens after two days of incubation. The colonies produced a lot of mucous material; (**c**) segregation of phage sensitive bacteria and their "runaway" of pseudolysogenic colonies.

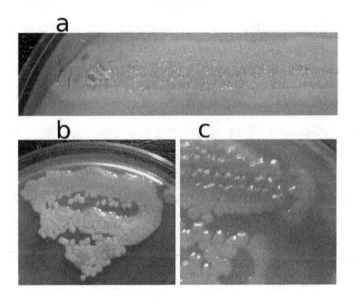

Pseudolysogenic condition for cells infected with phiKZ-like phages at high m.o.i. in natural conditions may have a biological sense of increasing final phage production. However, the use of these phages for phage therapy in infected wounds may increase viscosity by release of large amounts of DNA. This may obstruct the penetration of other phages or of antibiotics.

5.2. Pseudovirulence

The properties of infected bacteria have the ability to influence the development of temperate phages. In some cases temperate phages of P. aeruginosa cannot accomplish lysogenization. For instance Les-phenotype *P. aeruginosa* cells (mutations in control of the recombination genes) cannot be lysogenized with phage D3 [58]. This is a general effect and it can be found in the temperate phages of other bacterial species. For example, the temperate phage GIL01 *Bacillus thuringiensis* has no ordinary C1 repressor to maintain a lysogenic state. Presumably the LexA protein of host bacteria binds to specific 14-bp palindromic sequences within the promoter region of the phage. This prevents phage genes expression in lysogenic bacteria and provides the switch necessary to enter lytic development. Thus any damage of LexA protein creates the condition for lytic development of the phage [59].

5.2.1. Pseudovirulence of Phages with Mosaic Genomes and Migration of Non-Transposable
Temperate Phage

In the course of classification of phages growing on *P. aeruginosa* PAO1 based on the evaluation
levels of genome homology, all temperate phages were assigned to two groups [60]. One group
included four species of transposable phages and the other one six species of non-transposable phages.
The genomes of temperate non-transposable phages of different species showed a low level of DNA
homology in pairwise comparison. The level of homology varied when comparing different phages
from different species. Sequencing the genomes of several phages of this group confirmed the
presence of homology between them [61,62]. A mosaic structure of genomes was found in general,
fairly common in phages of different species of bacteria and usually found in phages active in related
species of bacteria [63]. However, there are some examples wherein genomic mosaicism arises
through the exchange of genes or blocks of genes between phage of unrelated bacterial species. This
suggests the possibility of migration of phage genomes or their fragments between distant species of
bacteria; although this transfer could happen before full speciation of the hosts.

We consider several such cases to support this point as well as to look from a different point of
view. In study [64], evidence was presented of significant relation at the level of genomes of two
temperate phages active on unrelated bacteria, namely *P. aeruginosa* phage phi CTX and *E. coli* phage
P2. There are not a large number of genome fragments with complete homology in phiCTX and P2
genomes but pronounced similarity at 28.9%–65.8% is observed in some ORFs. As shown by
comparison of the total structure of the genomes of these phages, a high level of similarity is also
exhibited. Besides, very importantly, phi CTX and P2 have some similar phenotypic traits, including
the similar capsid structure, the lack of response to the effect of inducing agents, and the important
feature to accomplish interspecies migration, which is the use of similar lipopolysaccharides for
adsorption. Authors of the study suggested that in this case the introduction of intestinal phage P2 in
pseudomonade has occurred and considered the case as a clear proof for the ability of phages to
overcome interspecific and intergeneric barriers in bacterial hosts. The phage phiCTX is one of the few
phage species that is related to the function CRISPR mechanism in *P. aeruginosa* (see Chapter
"Temperate *P. aeruginosa* phages and CRISPR effects"). Evidence supporting interspecies migration
appears as a result of the isolation and study of new bacteriophages. The phages of *P. aeruginosa*
described in [65] are the first representatives of a novel kind of *P. aeruginosa* phages having a
similarity in genome structure with N4-like viruses active in *E. coli*. The finding supports the notion of
interspecies migrations of bacteriophages.

Another example of a mosaic genome that arose due to exchange with fragments of the genomes of
phages, which belong to different species, including phages active on an unrelated bacterial species, is
the genome of phage phi297 (NCBI access number NC_016762) [66]. A significant part of the genome
phi297 exhibits a high level of homology to the genomes of phages F116 and D3 of *P. aeruginosa*. At
the same time, some special properties F116 and D3 are only partially manifested in phi297. The
bacteriophage D3 performs lysogenc conversion of PAO1 surface antigens (serotype O5 in serotype

O16) [67]. Such conversion requires the activity of three genes, located in a single fragment of the genome [68]. These ORFs encode three proteins: alpha-polymerase inhibitor; O-beta-polymerase and O-acetylase. Their sequential action leads to a change in the surface structure of lysogenic bacteria and in adsorption specificity. As it turned out, phi297 contains only gene O-beta-polymerase from this group of genes [66]. Apparently, this is not sufficient to ensure complete conversion, and as a result, the phage D3 can be adsorbed onto the surface of the cells of *P. aeruginosa* (phi297) and produce small plaques on its lawn. Further, the two regions homologous to F116, with the coordinates of 1–544 bp, and 47,119–47,438 bp in the genome phi297 match phage integrase gene F116 and its fragments [62]. At the same time, F116 prophage is not integrated into the bacterial chromosome and is present in the cell as a plasmid [69]. Usually, interaction between homologues of the bacterial partitioning proteins ParA and ParB, coded with a plasmid of low copy number, is necessary for stable maintenance and distribution of copies of plasmids in the division of bacterial cells. Phage phi297 genome contains a homologue of parA gene. However, parB is not found in the phi297 genome. Perhaps, lack of parB is the cause of an unstable lysogenization by phi297. Having said that, there is a view that the ParB function is not required for proper distribution of plasmids to daughter cells [70]. Usually, in the newly isolated phages, after sequencing the genome, it is commonly found that plenty of genes coding gene products have no similarities with functions of other gene products in databases. In genome of phi297 there were just eight of such genes. However, even in those cases where phi297 gene products show similarity with some of the gene products available in databases, their real functions are unknown. This is the common case. It can be assumed that in such cases, to determine a value of a gene product for phage viability, it would be useful to apply, along with other approaches, direct genetic research; for example, the selection of mutants of phage with subsequent allocation of the mutations to ORFs. The confirmation of phage genes flowing between these bacterial species has been found in the comparison of phiKZ-like bacteriophages active on different pseudomonades and in the recently discovered large bacteriophage SPN3US active on *Salmonella* [71]. Perhaps this indicates the presence of open-channel exchange between *Salmonella* and *Pseudomona*s. In the structure of phi297 genome there is also definite confirmation of the possibility of migration of large blocks of genes between *Salmonella* and *Pseudomonas*. It would be important to find out the mechanism and ways of such exchanges (participation of unique bacteria, specific bacteriophages, plasmids, migration of native phage genomes, transduction, *etc.*) (see Figure 4).

5.2.2. The Behavior of Wild-Type Phage phi297 on Different *P. aeruginosa* Pathogenic Strains of Different Origin

When comparing the growth of the various phages on a group of clinical isolates from a burn center, it was found that wild-type phage phi297 grows on the lawns of some clinical isolates in a manner of a virulent phage (see Figure 5). One such strain, Che1, shows resistance to most phages used for the treatment of burn wound infections. Properties of *P. aeruginosa* Che1 have similar properties to Les mutants of *P. aeruginosa* PAO1 (see above), as described in phage D3. However in the case when we accept that *P. aeruginosa* Che1 is indeed a variant of the Les-phenotype, it is necessary to recognize that its Les phenotype is repressor dependant because mutants of phi297 (as phi297ci, phi297vir) with diminished or absent lysogenization capability do not grow on this strain at all. The very existence of these bacterial isolates suggests that such generally accepted concepts, such as temperance and virulence of phages, may lose their ultimate significance in medical practice. Indeed, in the selection of phages needed to treat real infected wounds, it is better to abandon the dogmas that the use, in treatment, of temperate phages is impossible or undesirable because of the danger of HGT. If a temperate phage acts as a virulent in relation to a particular wound pathogen it is obvious that the probability of HGT in such a case will not be higher than for inherent genetic virulent phages.

Figure 4. Graphical qualitative layout of the comparative degree of relatedness among a group of phages which reveal some similarity with phiKZ, active on *Psedomonas aeruginosa* (phiKZ and EL), *Pseudomonas putida* (phage Lu11), *Pseudomonas fluorescens* (phage OBP), *Salmonella enterica* (phage SPN3US), *Pseudomonas chlororaphis* (phage 201φ2-1) and *Escherichia coli* (contigs PA3). The thickness and length of connection lines characterize levels of genome relatedness.

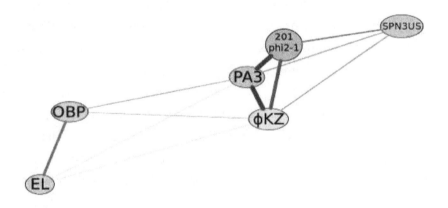

Figure 5. The growth of different phages on *P. aeruginosa* PAO1 and on clinical isolate *P. aeruginosa* Che1. Temperate phage phi 297w does not lysogenize cells of Che1 and behaves as a virulent phage.

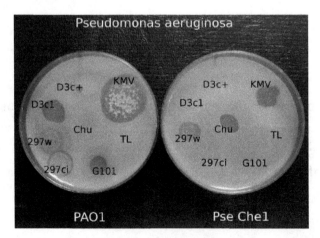

6. Growth of Phages in the Presence of Plasmids

The other examples of disturbances in canonical phage growth types of bacteriophages may be related to extremely diverse and specific effects of different plasmids in clinical isolates of *P. aeruginosa*. Conjugative plasmids can migrate between different strains of the same species and also between strains of unrelated species. The plasmids are active participants in the modification of pathogenic islands of bacterial genome transfer gene cassettes, controlling multiple antibiotic resistance, phage resistance or growth modification and genes influencing virulence and pathogenicity. We discuss here some plasmid-phage interactions and their significance for the choice of phages for therapy. The most differentiated activity in phage development inhibition is demonstrated by plasmids of IncP2 group of incompatibility. These plasmids, being specific to pseudomonades, are capable of recombination with IncP1 plasmids of a wide host range and thus can easily migrate to other species of gram-negative bacteria [72]. Such interplasmid recombination provides a powerful means for genetic diversity in plasmids. All IncP2-group plasmids inhibit phage growth, blocking intracellular phage development [73–75]. With the use of interspecies crosses between *P. aeruginosa* cells with IncP2 plasmids and *P. putida* it was possible to select plasmid variants with different inhibitive activity. As an example, plasmid pMG53 produced variants with at least six different types of phage growth inhibition. The other IncP2 plasmid, RpL11 also revealed high specificity in phage growth inhibition. It selectively reduces the frequency of lysogenization for temperate transducing phage G101 and transposable phages B39 and D3112 but not for transposable phage B3. It was found that phage D3112 functions, controlling establishment of lysogeny and the lytic cycle, were not expressed after infection of cells with RPL11. However, in the case when the cell carries both plasmid RPL11 and D3112 as prophage, there is no interference with repressor synthesis or with vegetative phage development after prophage induction. Thus, plasmid activity differentiates between the processes of primary integration after infection and that of reintegration of DNA after prophage induction [76]. Moreover, plasmids may recognize fine phenotypic differences among closely related phages with identical repressor immunity. In the presence in cells of plasmid Rms163 (IncP5), transposable phage B39 increase its lysogenization efficiency and as a result has efficiency of plating (e.o.p.) near 0.1, while mutants in c1 gene plate as clear plaques have e.o.p. of 1.0. Two other transposable phages with immB39 reveal a much more profound reaction on Rms163 and practically all infected cells become lysogenized. Mutations in their c1 genes of both phages restore lytic growth (e.o.p. 1.0). The site responsible for such differences in plasmid reaction for immB39 type repressor is located within the interval 1.1–3.9 kb of genome, being closely linked to gene cI [77]. Integration of *P. aeruginosa* transposable phages into plasmid RP4 induces different stable mutations [78,79]. Such hybrid plasmids can migrate among different gram-negative bacterial species, such as *Escherichia coli*, *Pseudomonas putida*, *Alcaligenes eutrophus* [80,81], they can be recipients for transposable phages of others species, such as Mu [82,83], and can influence the development of phages in other bacterial species [84]. Transfer of plasmid and phage genes capable of causing complicated genome rearrangements and dissemination among the strains as in the environment or as in the microbial wound community, may frequently be the cause of unpredictable results (production of filaments by *E. coli* (RP4::D3112) at 30 °C, stability and growth of *E. coli* (D3112) with a single phage genome copy in chromosome) [85–87].

In a general sense, it is possible to consider plasmids and phages as a common gene pool, organized in different structures, designed by nature for the fine regulation of evolutionary events. It is evident that plasmid effects can restrict the possibility of using some phages for the purposes of therapy. However, it is possible to obtain variants of bacteriophages of particular species (mutants or natural isolates), which overcome the inhibitory effect of plasmids. One such example is an isolation of a new phage phiPMG1 (NCBI access number NC_016765), capable to grow in the presence of pMG1, IncP2 plasmid. However, as it happens, phage phiPMG1 can also lyse bacteria with other IncP2 plasmids. It was found that the phage has some other unusual features. Being closely related with temperate phage D3, it cannot make lysogenic bacteria. Results of phiPMG1 genome sequencing [88] have shown (see Figure 6) that in the central part of its genome some very complicated restructuring events have occurred. In spite of its repressor gene being undamaged, the phage cannot accomplish stable lysogenization as D3 does, but only induce a temporary lysogenic state in infected cells.

Figure 6. Comparison of phage genomes phiPMG1 and D3 (with Dot-plot program GEPARD) shows the complicated rearrangements in phiPMG1 genome regulatory region. phiPMG1 produces repressor but there are no stable lysogens.

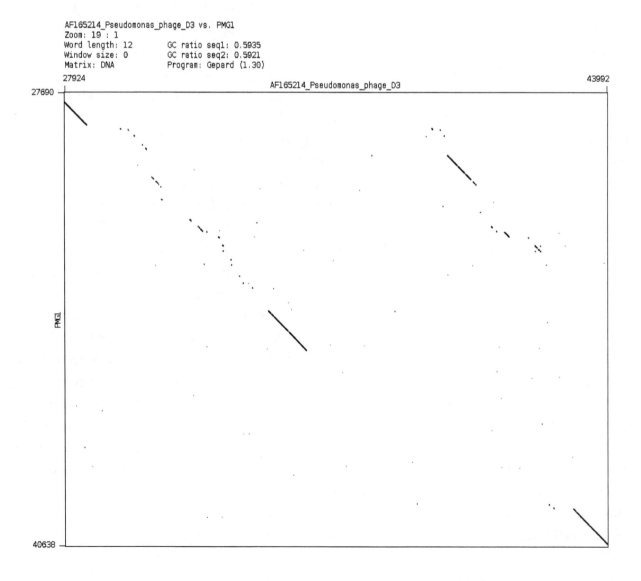

Such observed virtual virulence of the phage has arisen as a result of gene rearrangement in the regulatory region of the genome. Presence of plasmids can block lytic growth of recognized therapeutic phages. For example, plasmid RMS148 interferes with lytic development of phage phiKZ of *P. aeruginosa*. This fact has permitted the revelation of the transducing activity of phiKZ. It means that in the course of phage therapy that particular virulent giant phage may transduce in its own particle up to 5% of *P. aeruginosa* bacterial chromosome [89]. It is more than the largest of known pathogenic islands in genome of *P. aeruginosa* and more than the sizes of temperate prophages able to participate in construction or modification of pathogenic islands.

7. Possibilities for the Emergence of New Phages in the Course of Phage Therapy

It may be interesting to estimate the probability of the emergence of new recombinant phages as a result of the application of phage therapy to wounds infected with pathogens of different distantly related species. For example, *Pseudomonas aeruginosa* and species of *Burkholderia* complex often coexist in places of common infections (wounds, CF). For both bacterial groups similar temperate transducing phages are described. There are temperate phages specific for *Burkholderia sp.* growing on *Pseudomonas aeruginosa* [90] and there are strains of *Burkholderia*, which support the growth of temperate phages *P. aeruginosa* [44]. We see examples of interspecies migration for phages of several different types, including (1) transposable phages *P.aeruginosa*; (2) converting temperate phages related D3; (3) phage F116.

7.1. Transposable Phages

In our previous studies, we found plenty of different transposable phages of *P. aeruginosa* (all of them are temperate), which were distributed into groups of D3112-like phages (14 phages, one species) and B3-like phages (six phages, three species) [91–94].

D3112-like and B3-like phage genomes are composed in a similar pattern, but mostly from different non-homologous blocks. Phages belonging to different groups reveal only traces of DNA homology along most parts of their genomes. Some essential DNA homology can be found on the right ends of their genomes where genes responsible for specific adsorption to bacterial IV type pili are located [95]. Nevertheless, it has been possible to isolate several D3112/B3 hybrid phages with very strange genome structures. They have arisen through an even number of crossovers with low frequencies. The comparison of D3112 and B3 genomes by dot-plots has found just several of the 30–50 bp long DNA homology sites distributed along the genomes. We suggest that these small sites of homology are used in interspecies recombination. The compulsory even number of crossovers may be explained by the possible incompatibility of specific gene modules in D3112 and B3 [96,97]. D3112-like and B3-like phages have a difference in their transposition specificity. Although phage D3112 has multiplicity of the integration sites in *P. aeruginosa* chromosome and plasmids [79], nevertheless, it does not induce auxotrophic mutations in lysogenized bacterial cells. However, phage B3 and its relative PM105 induce such mutations with different nutritional requirements [98]. We have suggested that D3112 has a high transposition specificity and mostly uses sites that are located in the part of the *P. aeruginosa* chromosome controlling catabolism. Indeed, Rehmat and Shapiro [99] have localized different D3112-induced mutations in the amidase gene.

It was found that genome of D3112 may be expressed in different species of gram-negative bacterial species (*Escherichia coli*, *Pseudomonas putida*, *Alcaligenes eutrophus*) when delivered as a part of hybrid plasmid RP4::D3112 [80,81,85,86]. In the new hosts, the expression of D3112 can change some of the bacterial properties. Thus, the hybrid strain *E. coli* (RP4::D3112) grows well only at 42 °C but dies quickly at 30 °C (forming long filaments as a result of blockade of cell division).

The real reason is that D3112 in *E. coli* cannot establish stable lysogenic conditions but its transposase is active. However, at 42 °C D3112 genome transposition is blocked, and cells survive and form colonies [86]. It is possible to isolate quite stable *E. coli* strains where phage genome is inserted into bacterial chromosome with loss of RP4 plasmid. Such *E. coli* (D3112) clones produce a very low number of mature D3112 phage particles. It means that in such clones, a foreign phage becomes a component of *E. coli* genome and its stability is ensured by rare expression in single cells. It may be considered one of the examples of phage migrations. Such phage interspecies migrations (including migrations of plasmids with complete phage genomes or their fragments) may lead to emergence of new pathogens.

There are some other effects related to activities of transposable phages of *P. aeruginosa*. It has been shown that after thermal induction of *P. aeruginosa* cells carrying heat inducible prophage D3112 cts15, it is possible to observe plenty of different survivors producing highly mucous colonies [100]. Most of them quickly (after 1–2 replatings) lost their mucous phenotype but some of the survivors showed good stability.

Frequently among related transposable phages the effects of mutual influence on vegetative growth may be found. This is usually found in the interaction of phages of the same species, but with different specificities of repressors; for example, the suppression of the vegetative growth of phages with immB39 on the lawns of bacteria carrying prophage with immD3112. The suppression was found to be determined by the activity of locus *cip*. The *cip* is located in the genome of D3112 in the range of 1.3–2.45 kb and reveals its inhibitive effect only in lysogenic bacteria. In the case of multiple lysogeny the suppression increases substantially. There have been isolated mutants of phage B39 which can not be inhibited with Cip-activity. Mutations in gene cip which lost the ability to suppress B39 also can be selected [101,102].

As was found in the study of the genomes in different *P. aeruginosa* transposable phages, all of them have a pronounced modular structure that shows evidence of the possibility of free exchange with gene modules between the genomes of phages of the same species. Interestingly, in some individual genes, their composite nature can be detected. So, the number of phage repressor genes is composed of three sub-modules, which are supposed to control the various functions of repressors [103,104]. Apparently, the genomes of *P. aeruginosa* transposable phages, even those which now show no homology along their genomes genes (phages of D3122 and B3 groups), have the single pattern of a genome of an ancient predecessor. This is confirmed firstly by a similar sequence in groups of genes controlling similar functions; secondly, by the presence of a considerable number of genes with good homology which control the flexible structure of the tail and its ability to adsorb to bacterial pili. However, transposable phage BcepMu, active on *Burkholderia cenocepacia*, has a different origin, because it reveals some common features with phage Mu of *E. coli* (they have a contractible tail), although they have differences in specificity transposase compared with the phage Mu [105]. Likely divergences in the evolution of different types of phage-transposons for gram-negative bacteria occurred at the moment of selecting the main host. It is possible to suggest that most of the time the genome evolution of BcepMu like phages was in bacteria belonging to the family Enterobacteriaceae. In the genome BcepMu are present homologs of many genes of different strains of salmonella prophages.

We can assume that the differences in the genomes of modern *P. aeruginosa* transposable phages belonging to different groups arose as a result of their numerous migrations. This is confirmed by the results of genome annotation of transposable phages. Thus, in D3112 phage genome there were found genes which are orthologs and not only genes of pseudomonads, but of other bacteria or phages. For example, ORFs, 18, 19, 24 and 26 of D3112 encode proteins similar to protein defective prophages PNM1 PNM2 and two different strains of *Neisseria meningitidis* MC58 and Z2491, the next group ORFs 27, 28, 29, 32, 34 encodes proteins similar to proteins of the phage Mu, *etc.* [106]. In case of phage B3 it was found that its proteins have mostly another origin, revealing relatedness with proteins of phage Mu, BcepMu or *Salmonella typhimurium*. In accordance with results of phage B3 genome annotation, not less than 10 genes have their origin from genes of other phages of different bacterial species: five genes of salmonella phage P22, five genes of two BsepMu-like phages and one gene is a homologue of the gene in salmonella phage Sti3. Genes found in phage B3 genome reveal different levels of relatedness (on the level of controlled products) with such potential intermediate bacterial hosts as *Bordetella bronchiseptica*, *Burkholderia cenocepacia*, *Escherichia coli*, *Haemophilus ducreyi*, *Haemophilus influenzae*, *Neisseria meningitidis*, *Salmonella enterica*, *Vibrio cholerae*, soil species and plant pathogens of different families including non-pathogenic and pathogenic strains with various ecological niches [107]. In this respect, it should be interesting to continue the study of two other B3-like species of transposable phages, active on *P. aeruginosa*, but with a significant difference of B3 in the level of DNA homology [93,94]. It may be especially interesting considering the significance of *P. aeruginosa* transposable phages in CRISPR-effects (see later).

The origin of several phage genes in D3112 and B3 cannot be determined through annotation. To elucidate their functional significance in phage development it is necessary to accomplish detailed phenogenetical studies (directed mutagenesis of such genes, looking for mutants with a specific phenotype and study of their behavior under different conditions).

7.2. Converting Temperate Phages of P. aeruginosa

Temperate *P. aeruginosa* phage phiCTX carries a cytotoxin gene which is expressed in prophage state. The phage has a large genome homology with *E. coli* phage P2 as a result of highly probable interspecies migration. Both phages are identical in particle morphology and their genomes have similar structures. However, the fact that genome of phiCTX is similar in GC-content with *P. aeruginosa* suggests that the divergence of P2 and phiCTX occurred a long time ago [64].

D3 is the other converting temperate phage of *P. aeruginosa*, which in the prophage state changes the structure of the surface lipopolysaccharides, and prevents adsorption of several bacteriophages, including D3 itself [61,68]. Phage phi297, showing DNA homology to DNA of D3 and of F116 [61,108], also has converting activity. The distribution of DNA homology regions in genome of phi297 with genomes D3 and F116 is shown in Figure 7. The mosaic structure in the phi297 genome is evident. Phi297 inherited from phages D3 and F116 just some of the genes that give these phages unique features, including the ability of D3 to perform bacterial surface modification and, possibly, the ability of phage F116 to be plasmid in prophage conditions. As a result, phi297 carries only a partial modification of cell covers (as shown by only partial blockade of phage adsorption). At the same time, as it is proposed, inheriting only one of the F116 genes, necessary to support prophage as plasmid, phi297 phage does not form stable lysogens.

Figure 7. Genome of phage phi297 has homology regions of different sizes and locations with D3 and F116 genomes (yellow: phi297 own genes; green: fragments of homology with D3 DNA; blue: fragments of homology with F116 DNA).

We have found that phi297 is capable of recombination with the bacterial PAO1 strain chromosome, producing lytic non-reverting variants, incapable of accomplishing lysogenisation. The resulting hybrids exhibit lytic properties towards certain groups of clinical isolates and phage resistant mutants arising after use of commercial phage mixtures [66,88]. We consider that the addition of such lytic derivates of temperate phages into commercial phage mixtures for the cure of infected wounds may be useful for enlarging their spectrum of lytic activity. The lytic activity of the hybrid phi297vir will stimulate the search for other similar possibilities. It will require more detailed studies of temperate phages of different species and the isolation of new species of temperate phages.

7.3. Phage F116: Interspecies Wanderer?

Bacteriophage F116 is of special interest for the assessment of the possibility of experimental and inter-specific migration and study of its mechanisms. The fact that F116 genome fragments (or an intact genome of other phages with great similarity to the genome of F116) were found in an integrated state in *Neisseria* support the need for special attention on the further study of F116-like phages. Indeed, pathogenic *Neisseria gonorrhoeae* strains do not support growth of any of the presently known tailed phages. However, as a result of careful studies in bacterial genomes, several regions were found showing similarity to genes of different phage species. One of these regions, NgoPhi2 is unique in showing a very high level of similarity with the genome of the phage F116 [109]. Genes of prophage, which reveal relatedness with F116, are apparently intact, and exhibit multiple effects, including the inhibition of the growth of *E. coli* and the propagation of phage lambda. The repressor in the NgoPhi2 region was able to inhibit transcription genes of *N. gonorrhoeae* and of phage *Haemophilus influenzae* HP1, and the gene for choline replaces the function of the homologous gene of phage lambda. After induction of cells with mitomycin C and microscopy of the supernatant, complete phage particles were found. It can be assumed that the phage genome exhibiting affinity with F116 was introduced into *N. gonorrhoeae* by an act of HGT (in the course of direct infection with phage particles, or in the state of prophage plasmid in the conjugation process or further being integrated into a transmissible plasmid).

7.4. Evidence for Migrations of phiKZ-Like Phages

Bacteriophage phiKZ active to *P. aeruginosa* reveals such features as large particle size, large genome and a spiral formation, the "inner body", and a unique packaging of DNA as spools of thread [45,46]. Later two other species were described of giant phages of *P. aeruginosa*, EL and Lin68, which are not distinguishable from phiKZ in particle morphology [47,48]. Recently, structural proteomes of phiKZ and EL were studied in detail [110]. However, all new phages exhibit significant differences from phiKZ. Phage EL has no detectable genome homology with phages of the other two species, and differs from them in GC-composition and in size of the genome [53]. Phages of another species, Lin68, show DNA homology of a low level with phiKZ in one of the restrictional DNA fragments [47]. In addition, phages of species Lin68 are unable to grow at 42 °C, and their mature particle reveals more instability than phiKZ particles at 60 °C. The phages of this species exhibit weak lytic activity on the lawn of psychrophilic bacteria *P. fluorescens*. It is of interest to compare the genomes of phages in the species Lin68 with phiKZ-like phages of other psychrophiles. The phiKZ-like phages can persist in pseudolysogenic clinical isolates in wounds. Thus, phage Che, a new EL-like phage, has been found in a clinical isolate from the Chelyabinsk burn center. *P. aeruginosa*, at least in other bacteria; similar phages were not found for a long time. However, as has been shown, bacteriophages of similar morphotype, showing signs of kinship with phiKZ, can be found in other bacterial hosts [48,51,111–113]. It is possible too that phiKZ-like phages will be found among other large phages of distant bacterial species. A recently described giant bacteriophage specific to *Salmonella* [71] and another one for *Ralstonia solanocearum* [114] have similarities with phiKZ-like phages. Apparently, the ability of all phiKZ-like phages to establish a pseudolysogenic state in infected cells has evolved from an ancient temperate phage as a result of successive multiple migrations between different hosts. If it is possible to confirm specific DNA packaging for these phages as inner bodies, as in phiKZ, it will prove not only the existence of a new phage superfamily, but also the existence of the open gene exchange channel between *Pseudomonas*, *Salmonella* and soil types of bacteria for this phage superfamily.

8. Optimizing the Selection of Phages for Therapy

The detailed phenogenetical study of a phage is a compulsory condition for its acceptance for direct use in phage therapy or inclusion into phage therapeutic mixtures. Although in each case the approach for this study may be different, the first aim is to show that a selected phage is genuinely virulent. In case the phage virulence is derivative of a temperate phage, it is necessary to prove its incapability to revert into a temperate condition in different situations (as multiplicity of infection, temperature, presence of plasmids or temperate phages, *etc.*). For example, the giant phiKZ-like phages of *P. aeruginosa* are used in various commercial mixtures [108] due to their wide spectrum of lytic activity. These phages in standard single-stage growth cycle experiments behave as typical virulent phages by lysing all infected bacteria, but at a high multiplicity of infection, bacteria continue dividing over several days (see above). During this time, the infected cells can be recipients of HGT through the activity of plasmids or temperate phages, which often occurs in wound microbial communities.

The number of genuinely virulent phage species for *P. aeruginosa* is quite limited. There are only nine species of different families, and, moreover, some species may have limited usefulness. In the list below are the phage species (in parentheses are the minimum and maximum size of genomes of the species, where it is known). Family Podoviridae: (1) the species of phiKMV-like (42,954–43,548 bp); (2) species Luz24-like (45,503–45,625 bp); (3) species N4-like (72,544–74,901 bp); (4) species LUZ7/PEV2*; family Syphoviridae: species (5) M6-like (58,663–59,446 bp); family Myoviridae: (6) species PB-1-like (64,427–66,530 bp); (7) species phiKZ-like (280,334 bp); (8) species Lin68 (is not sequenced); species (9) EL-like (211,215 bp). Different manufactures can use one or the other species. However, it is evident that the number of species is not unlimited.

Of this list phiKZ-like phages (two species) can be considered only as conditionally virulent. Virulent phages of other species—Luz24 and TL—exhibit high relatedness with temperate phage PaP3 [115,116]. So far the most promising species for phage therapy of P. aeruginosa infections are phages of species PB1-like phages, which exhibit significant variability on the basis of the specificity of adsorption [117–119], and species of phiKMV-, N4-, LUZ7- and M6-like phages. Bacteriophages of species Lin68 can have only limited application, being sensitive to the development of human body temperatures. We can conclude that the phage potential of *P. aeruginosa* is limited. Variations in lytic activities for phages inside each of the virulent species are also limited. A sequenced set of phages and their number are altogether insufficient to maintain long-term phage therapy. Obviously, the need to expand efforts to find new species of virulent phages exists. Moreover, there must be some coordination of specialists working with phages and interest in the development of phage therapy to overcome the randomness in the selection of phages for sequencing and phenogenetical studies.

For each prospective therapeutic phage it is desirable to estimate the probabilities and diversity of phage-resistant mutants. According to our observations both evaluations vary greatly in dependance on phage species and (obviously) on specific hosts: clinical isolates and laboratory variants of the standard host.

9. At First Glance, the Phages with a Broad Spectrum of Lytic Activity Are Preferred for Therapeutic Blends, but only at First Glance

The use of sophisticated phage cocktails, including many strains of phages with a broad spectrum of lytic activity against different bacterial pathogens, has as its purpose the fast interruption of infection without prior isolation of the pathogen. This is justified only in cases where the infection and the loss of time may create a life threatening situation. However, it may be that the long-term presence of pathogenic bacteria in the wound are caused by the unique bacterial strains that are resistant to most of the phages in the therapeutic mix. Because there is no previous information about strains in the wound, before applying emergency treatment with a phage mix, in fact, the attempt at treatment iscarried out blindly and if the first application of phages does not give positive results, it means that time is lost. However, in the majority of cases of wound or nosocomial infections due to *P. aeruginosa*, there is usually no such extreme urgency.

Typically, the clinical laboratory identifies pathogenic bacteria bv determining their sensitivity to various antibiotics and only if antibiotic application is ineffective, will the decision be made on the use of bacteriophages. Thus in the case of wound and nosocomial infections in different locations due to isolated strains, more rational is the use of a minimal set of phages, even a single phage with a narrow spectrum of lytic activity capable to lyse strains isolated from the patient. The choice of a set of phages or a single phage against a particular pathogen is not a more labor-consuming procedure than a standard check of bacterial sensitivity to antibiotics with paper discs. Such a restriction in the number of phage species is intended to limit the accumulation of mutant pathogens that are resistant to the most valuable phages with a broad spectrum of lytic activity and save them for use in really extreme cases. Thus bacteriophages with a narrow spectrum of lytic activity can also be very useful in phage therapy. We consider the properties of several of these phages.

10. Description of the Properties of New Phage in the Possible Application of Phage Therapy

10.1. The Properties of a New Bacteriophage CHU Pseudomonas aeruginosa

Phage CHU is one of the newly discovered phages for *P. aeruginosa*. This is a relatively small (diameter of the head is near 50 nm) DNA containing tailed Syphoviridae phage. The unique feature of the phage is an unusual spectrum of lytic activity (trace growth on PAO1 strain and good growth on several strains producing alginate). There are two such strains, Pse163 and CF013A, isolated from CF patients (from the collection of Prof. M. Vaneechoutte, Laboratory Bacteriology & Virology University Hospital Ghent, 9000, Gent, Belgium). These strains differ in their phage sensitivity and other properties (Pse163 is lysogenic for a transposable phage related with D3112). Phage CHU grows on lawns of washed alginate Pse163. The phage has no alginase activity, as there is no visible destruction of slime around the spots of phage growth. There is, however, a difference in the character of phage CHU growth on lawn of strain CF013A. After a prolonged (three days) incubation around a lysis spot CHU on the lawn of CF013A was formed, (see Figure 8). The reason for this effect may be related to the difference in properties of alginate producing strains. It was confirmed in the study of four additional stable P. aeruginosa alginate producers exhibiting selective sensitivity to phage CHU, which were found among clinical isolates of the burn center of the Chelyabinsk Regional Hospital. All these four strains grow similarly, being plated on the surface of the agar medium. However, when plated in semisolid agar layers, two of these strains do not produce a visible layer of alginate but the other two strains continue to produce alginate, visible on the surface of the upper agar layer. Bacterial strains, which do not produce visible alginate, on being placed into semisolid agar of the upper layer, reveal alginate around phage CHU spots of lysis. However, the reasons for the difference in properties from two kinds of alginate producers, remains unknown. As the last four strains from the burn center were isolated at different times from different patients they may be regarded as hospital pathogens. It is possible the differences between them arose in the burn center.

Figure 8. The raised slimy ring on the CF013A lawn around the phage CHU lysis spot.

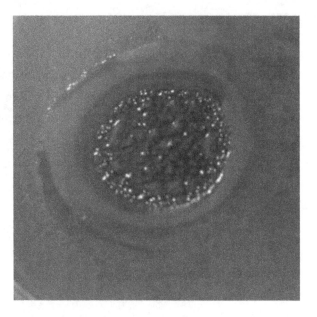

A feature of phage CHU is the high level of instability. The best differentiation of plaque types was achieved with use as lawns of strains *P. aeruginosa* 8–20 and 2–10 (obtained from Professor C. Pourcel, GPMS Institut de Genetique et Microbiologie Bat 400 Universite Paris-Sud 91405 Orsay cedex, France), see Figure 9. Up to now, despite repeated attempts, we could not isolate stable plaque derivatives of phage. In this relation CHU reveals some similarity with phage TL, described below.

Figure 9. Growth of CHU phage variants on the lawn *P. aeruginosa* 8–20.

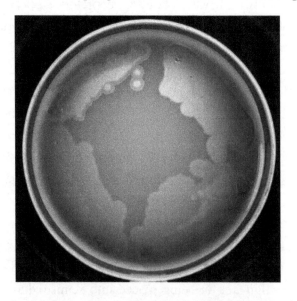

The use of this phage can be useful in open infected wounds against strains producing alginate.

10.2. Properties of a New Bacteriophage TL

Bacteriophage TL genome sequence (carried out in the laboratory of Professor L. Kulakov, (The School of Biological Sciences, The Queen's University of Belfast, Belfast BT9 7BL, Northern Ireland, United Kingdom), annotation made in our laboratory shows a high degree of relationship, with phage Luz24 [115,116]. TL reveals a relationship (as Luz 24) with the temperate phage PaP3. This leaves no doubt as to the common origin of the three phages. TL has poor growth on lawns of strain PAO1. A PAO1 mutant ELR2 has been isolated which is resistant to many phages, but supports good growth of TL. We suggest the loss or modification of surface adsorptional receptors led to the availability of receptor for phage TL [35,117–119]. Phage TL is produced on the lawn of ELR2 plaques with a different appearance, which cannot be stabilized, see Figure 10.

Figure 10. Growth of TL phage variants on the lawn *P. aeruginosa* ELR2.

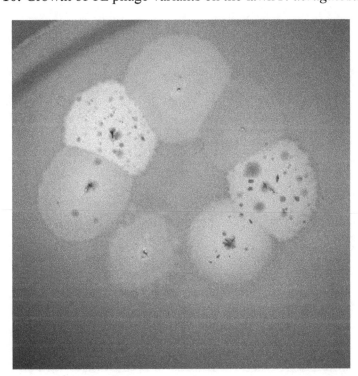

After the genome annotation, it was found that the TL genome contains gene coding for transposase gene in the region 5,000–6,000 bp. The gene was not detected in genomes of related phages Luz24 (AM910650), see Figure 11, and PaP3. The appearance of different plaque morphology variants and their instability is may be due to the activity of the phage transposase. As TL does not produce lysogens, like Luz24, it can be attributed to the lytic phages. TL is capable of lysing bacteria resistant to other phages used in anti-*P. aeruginosa* preparations. Therefore, the introduction of TL into any standard phage mixture against *P. aeruginosa* may be very useful. Indeed the phage will reveal its activity when the wound *P. aeruginosa* pathogens become resistant to other phages in the mixture. Although the features of this phage confirm the importance of the proper selection of phages in the preparation of the phage therapeutic mixture, nevertheless the use of Luz 24 related phages in phage therapy requires a more thorough study.

Figure 11. Genomes of phages TL and Luz24 have a high level of DNA homology (DotPlot on UGENE program parameters: 60 bp, 85% similarity) but TL genome contains transposase gene in the region of non-homology with Luz24.

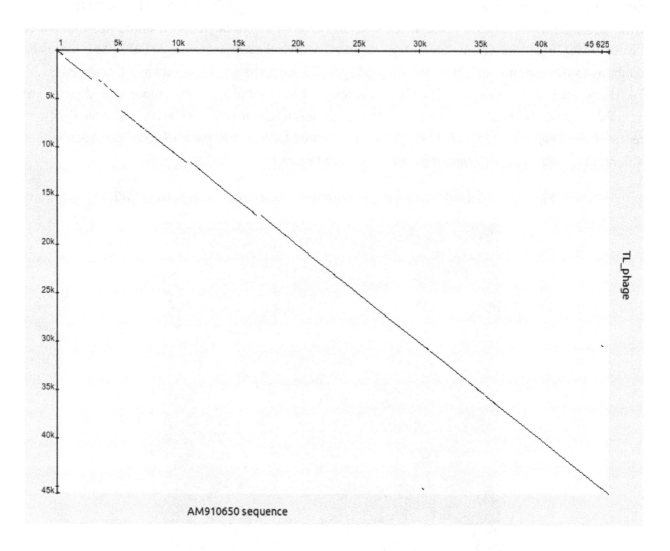

11. Significance of Specific Phage-Bacteria Interactions in Transfer to Personalized Phage Therapy

The mixed microbial community includes different types of bacteria (pathogenic and nonpathogenic) which may be in complicated mutual relations. For example, in the case of cystic fibrosis the frequent companion to *P. aeruginosa* is the bacterial species of the *Burkholderia* complex. One of the reasons for the persistence of *Burkholderia* is the presence of alginate [120]. There are described cases of acquisition, of the ability to grow at 37 °C, adhesiveness and cytotoxicity comparable with the corresponding features of *P. aeruginosa,* by psychrophilic soil pseudomonades as *P. fluorecens*, which were previously considered as harmless [121,122]. Furthermore, strains of *P. putida* with cassettes controlling antibiotic resistance in plasmid-borne integron from an *Achromobacter xylosoxidans* were isolated from several patients in the general intensive care unit, which suggests the possibility of evolutionary interactions between commensals in the clinical conditions [123]. It can be expected that the frequency of finding such new species from commensals will expand as will their significance in wound bacterial communities (carrying with them a set of new genetic elements—phages, plasmids, transposons—that can perform HGT). Thus, there is a need to continuously increase the number and diversity of bacteriophages that are acceptable for the treatment of infected wounds.

Accordingly, the number of truly virulent phage species is usually small in the case of different bacterial species, including *P. aeruginosa*, which obviously limits the possibilities of phage therapy. Moreover, it is clear that the continued use of the same phage will quickly lead to the emergence of multiple phage-resistant bacterial pathogens. It is impossible to predict the time of active use of the described phage species in the case of widespread introduction of phage therapy. It is not evident that we must expect to reveal a sufficient number of new virulent phage species. In this regard, there is a need to understand and accept decisions on the possibility of using specially modified temperate phages in phage therapy.

12. Let Us Discuss the Possibility to Use *P. aeruginosa* Temperate Phages in Phage Therapy

Temperate phages active on *P. aeruginosa* make up a significant part of the phage potential of this species. Refusal to use temperate phage of wild types capable to lysogenize infected bacteria is understandable and evident. However, as we have shown, among the recognized phage species acceptable for therapy, there are phiKZ-like species whose behavior (lytic against pseudolysogenic) may greatly depend on cotrolled or uncontrolled conditions in their application. In the case of these phages, special virulent mutants incapable of pseudolysogenization have been isolated.

We consider that it is possible in certain cases to use some temperate phages after their inheritable modifications or even to use phages showing temperance in infection of standard laboratory hosts, but having obvious virulence for real wound strain. Of course this can be done in conditions of individual selection of specific phage only under the supervision of physicians and clinical laboratories. Quite a lot of temperate *P. aeruginosa* phages have sequenced genomes. The most studied phages belong to the Syphoviridae family, including species: (1) transducing cytotoxic phiCTX-like: phiCTX (35,580–36,415 bp); (2) D3-related: D3 (56,425 bp), PAJU2 (46,872 bp), phi297 (NCBI HQ_711984), phiPMG1 (NCBI NC_016765); (3) F116-(65,195 bp); (4) Transposable phages: (a) D3112-like (13 different isolates): D3112 (37,611 bp), MP22 (36,409 bp); DMS3; (b) B3-like phages (six phages) B3 (38,439 bp), HW12, PM105 [93]. In addition, there is a group of phages of different families and species, which should be studied in more detail as Myoviridae, P2-like-MP29, MP38, F10; Podoviridae 119X, PaP2; Syphoviridae: 73, G101, 160 and unclassified phage PA11.

Phage phiPMG1 may be an example of an earlier temperate (D3-like) phage, which became acceptable (in our opinion) for use in therapy as a result of spontaneous genetical modifications [88]. Our attempts to isolate revertants of the phage similar to D3-like wild-type phage were unsuccessful. We tested the activity of this phage compared to the activity of two commercial mixtures of different manufacturers and showed that this phage successfully lyses spontaneous mutants of strain PAO1, resistant to phages in both commercial mixtures [108]. Phage phi297vir, which is a recombinant of phi297 wild type with bacterial chromosome also lyses resistant to commercial mixtures of bacteria, including bacteria in the pseudolysogenic state after multiple infections with phiKZ-like phages, such as bacteriophages.

The study of new species of temperate phages expands our understanding of their evolution [64]. For instance, the mosaic structure of phi297 genome occurred as a result of directly borrowing gene encoding products that are similar to the products of bacteria and phages of different taxonomic groups (including *Salmonella* and other enteric bacteria). This clearly supports the idea that genomes of different natural phages have arisen as a result of a combination of pre-engineered functionally related genes. Introduction to the temperate phage genome modifications, leading to irreversible virulence, can be used to expand the collection of therapeutic phages.

Of course, the likelihood of HGT using even non revertible virulent variants of temperate phages will be higher than in the case of truly virulent phages (because, for example, of the possibility of such a mutant recombining in infected wounds with related prophage in pathogenic bacteria). However, the immediate danger to the patient in the course of phage therapy will not be raised because the duration of phage therapy is usually not more than a few days. If during this time, the therapeutic effect is not achieved, it means that the mixture of phages in this particular case is not efficient and should be replaced [37]. Phage phi297vir behaves as lytical but it is not virulent, since it does not grow on bacteria which contain wild type prophage phi297w. Apparently phi297vir remains a functionally active operator.

13. Temperate *P. aeruginosa* Phages and CRISPR Effects

We have discussed the above possible migrations of intact phage genomes or their fragments between bacteria of different species. One of the obvious ways is the migration of phage genomes incorporated in the conjugative plasmids, as shown for transposable phages of *P. aeruginosa* (see above) and *E. coli* [123]. Another possible means for migration may be acceptable for phages by using for their adsorption type IV pili. Type IV pili are highly specialized structures on bacterial surfaces having immense significance in bacterial life in the expression of many phenotypes, such as motility, sensitivity to bacteriophages, natural genetic transformation, adherence and speciation for multicellular behavior. Modified bacterial pili of IV type may provide adaptation to specific tissues expanding the possibilities of the particular species of pathogenic bacteria and increasing their virulence [124–126]. The pili, being frequent appendages of different pathogens and adsorptional receptors for numerous phages (as F116, transposable phages of *P. aeruginosa*, possibly transposable phage specific for *Neisseria meningitidis*), may be used by such phages in interspecies migrations [127,128]. It is possible to assume that this interspecies migration path made the bacterial cells develop CRISPR as a specific mechanism of protection against expression of foreign genetic material, plasmids and phages. Loci CRISPR, control the synthesis of specific small RNA molecules that, under invasion by viruses or other genetic elements block their expression, binding the foreign DNA. The number and variety of such phage spicers evidences the number of previous contacts of bacteria with different foreign DNAs. At the same time, there is evidence that the CRISPR mechanism, a multifunctional system response, manifests itself not through the blockade of phage or plasmid penetration to save individual cells subjected to attack, but by modifying other essential features. Infection bacteria *P. aeruginosa* PA14 with phage DMS3 inhibits biofilm formation and swarming motility, both very important bacterial features associated with the activity of type IV pili. After introduction, mutations in the CRISPR region of both biofilm formation and swarming in DMS3 lysogenized strains recovered [129].

Reports of the suppression of the growth of temperate phages are rather contradictory. Some studies clearly stated the suppression of the growth group of temperate phages, which (according to the results of the comparison with database) are mainly related to the species transposable D3112-or B3-like phages and some other species as F116 and phiCTX. In other works, the inhibition of the growth of phages is denied. Thus, in two moderate phages, MP29 and MP42, whose genomes are similar to those of *P. aeruginosa* transposable temperate phages DMS3, there was no change of swarming [130]. Such discrepancy in results from different research groups may reflect subtle differences in the structure of phage genomes, as it is believed even point mutation in the genome of the phage can eliminate the effect of its CRISPR-suppression mechanism [131,132]. Strains in different laboratories are used under one name, which derived from the same parental strain, can rapidly accumulate point mutations, with unpredictable effects. Therefore, the hopes of using such a labile system for practical purposes (e.g., in the case of cystic fibrosis) may be considered premature. There is evidence that *P. aeruginosa* PA14, being capable of preventing replication of six newly isolated temperate phages can also acquire a new spacer content as a result of some lytic phage infection. The authors consider it as evidence of the adaptive nature of this CRISPR/Cas system [133]. CRISPR-bearing loci of the strains of *P. aeruginosa* are quite common among clinical isolates [134]. For example, among 122 clinical isolates sampled in different geographical areas 45 strains were found carrying CRISPR with 132 viral spacers matched to temperate bacteriophages/prophages capable of inserting into the host chromosome, but not to extrachromosomally replicating lytic *P. aeruginosa* bacteriophages. It is unclear exactly what the authors define as lytic phage. Because none of the tested temperate phages showed deterioration in growth or lysogenization in the presence of CRISPR, described in this paper, the authors actually reject the dogma for CRISPR protective effect.

The examination of the spacers (inserts) in CRISPR loci, found in all strains of *P. aeruginosa* studied to date that this system is aware of, only several definite temperate phages. At least for most of the phages, bacterial pili are compulsory receptors of adsorption. These pili are very important for the establishment of relationships in the microbial community. Bacterial conjugation is a form of such relations, carried by IV type pili in which DNA and proteins are transferred to the appropriate recipient cells [135].

A common feature of these phages is their high mobility. An existing relationship between the bacteria and most temperate phages are relations in a community where all interactions maintain a common gene pool and adapt to changing environmental conditions through natural selection (including humankind with its various activities as an integral part of the environment). There is an opinion that CRISPR activity is not designed to preserve the functional capabilities and integrity of the bacterial genome in a single bacterial cell. Moreover, it is protection from the expansion of destructive effects on all cells in a population, possibly in a state of biofilm. On the other hand, CRISPR can be viewed as a mechanism of active accumulation in the bacterial genomes of different phage genome fragments of different phages. It may create the possibility of re-infecting bacteriophages for further evolution through recombination. This may be of general interest, as there are many clinical isolates supporting the lytic cycle of temperate phages but not lysogenization. In the strict sense, it can also be seen as a way to protect the bacterial population from the invasion of new genetic elements into genomes.

Indeed, there is no described mechanism for protection of bacteria from virulent phages, similar to the mechanism of CRISPR. In this case, there may be situations where the death of one separate population of bacteria may be a more effective way to protect the entire population from the spread of genes that are incompatible with the survival of the bacteria. In this respect, it would be interesting to search among the known CRISPR strains for such options, which would respond to infection by certain temperate phages just as to infection by virulent lytic phages.

14. Future Therapeutic Phages: Different Directions

If phage therapy is destined to become an accepted and widely used medical procedure, its maintenance and development will depend entirely on the success of the constant expansion of the collection of therapeutic bacteriophages with lytic activities against newly arising phage resistance of bacterial populations. Apparently, there will be the need for permanent development and modification of genetically engineered phages with required properties. There are possible several different approaches to reach this purpose. For instance, such unified phages could be engineered on the basis of certain existing species with an attempt to create a unified assembly of phages having in their genomes individual blocks of genes which can improve the efficiency of infection and lysis of different bacterial species. It is known that in some cases of removing genes of phage, genome can improve the phage growth, as in the case of *P. aeruginosa* phage F116 [136]. There are other examples of loss of a substantial part of the genome *P. aeruginosa* phages without sacrificing growth efficiency.

Temperate phage SM is capable of growing in P. aeruginosa PAO1 cells with addition insertion in its intentionally deleted shortened genome up to 12 kb DNA insertion [137]. Even more striking is the difference in the amount of phage DNAs in phages phiKZ and EL (NCBI access number NC_007623). Capsids of these phages have the same size, but their genomes show a great difference (280,334 bp for phiKZ and 211,216 bp for EL). Thus, the redundant capacity of DNA to be packed in an EL capsid is not less than 69 kb (!).

One of the possible areas of work in the creation of modified phages would be affording different phage species with the ability to recognize the type IV pili of different bacterial species. In this regard, there is a promising group of species showing relations with phage F116, whose genome is present in the cell as plasmids and that use pili of IV type for adsorption.

In phage therapeutic mixtures, different phages of the same species are often used, with different activities against various clinical isolates of *P. aeruginosa* [108]. It would be interesting to investigate the possibility of combining their specific features within a single phage genome. This will also help to start the work on unification of therapeutic phages.

Another possible trend for creating unified phages may be related to a deeper investigation of the origin of similar genes in genomes of phages active against bacteria of different species as *Salmonella enterica* and *Pseudomonas aeruginosa*, and possible mutual channels for migration. In relation to the hypothetical work on construction of artificial bacteriophages, looking for the elaboration of their industrial hosts will be required.

We consider that some principal changes related to the further development and application of therapy with live bacteriophages should be accomplished in the near future, as long as society has a high level of confidence in their use.

Acknowledgments

This work was supported by the Russian Foundation for Basic Research (grant NO. 11-04-00270-a). Authors express their gratitude to E. Chesnokova (Laboratory for Bacteriophages Genetics. Mechnikov Research Institute of Vaccines and Sera, RAMS, Moscow, 105064, Russia) for the isolation of phage CHU and for assistance in preparation of manuscript. Authors are indebted to L. Kulakov (The School of Biological Sciences, The Queen's University of Belfast, Belfast BT9 7BL, Northern Ireland, United Kingdom) for results of the phage TL genome sequencing. Authors are grateful to C. Pourcel (GPMS Institut de Genetique et Microbiologie Bat 400 Universite Paris-Sud 91405 Orsay cedex, France), to M. Vaneechoutte (Laboratory Bacteriology & Virology University Hospital Ghent, 9000, Gent, Belgium), and to M. A. Popova (Clinical Laboratory of Chelyabinsk Regional Hospital, 454076, Russia) for the bacterial and phage strains used in our studies.

References

1.   Tejedor, C.; Foulds, J. Zasloff, M. Bacteriophages in sputum of patients with bronchopulmonary pseudomonas infections. *Infect. Immun.* **1982**, *36*, 440–441.

2.   Lombardi, G.; Luzzaro, F.; Docquier, J.D.; Riccio, M.L.; Perilli, M.; Coli, A.; Amicosante, G.; Rossolini, G.M.; Toniolo, A. Nosocomial infections caused by multidrug-resistant isolates of *Pseudomonas putida* producing VIM-1 metallo-Гџ-lactamase. *J. Clin. Microbiol.* **2002**, *40*, 4051–4055.

3.   Yomoda, S; Okubo, T; Takahashi, A; Murakami, M; Iyobe, S. Presence of *Pseudomonas putida* strains harboring plasmids bearing the Metallo-Гʊ-Lactamase gene blaIMP in a hospital in Japan. *J. Clin. Microbiol.* **2003**, *9*, 4246–4251.

4.   Finnan, S.; Morrissey, JP.; O'Gara, F.; Boyd, E.F. Genome diversity of *Pseudomonas aeruginosa* isolates from cystic fibrosis patients and the hospital environment. *J. Clin. Microbiol.* **2004**, *42*, 5783–5792.

5.   Ehrlich, N.; Jocz, J.; Kropp, L.; Wong, R.; Wadowsky, RM.; Slifkin, M.; Preston, R.A.; Erdos, G.; Post, J.C.; Ehrlich, G.D.; Hu, F.Z. Extensive genomic plasticity in *Pseudomonas aeruginosa* revealed by identification and distribution studies of novel genes among clinical isolates. *J. Infect. Immun.* **2006**, *74*, 5272–5283.

6.   Barbier, F.; Wolff, M. Multi-drug resistant *Pseudomonas aeruginosa*: Towards a therapeutic dead end? *Med. Sci.* **2010**, *26*, 960–968.

7.   Riou, M.; Carbonnelle, S.; Avrain, L.; Mesaros, N.; Pirnay, J.P.; Bilocq, F.; de Vos, D.; Simon, A; Pierard, D.; Jacobs, F.; *et al. In vivo* development of antimicrobial resistance in *Pseudomonas aeruginosa* strains isolated from the lower respiratory tract of Intensive Care Unit patients and receiving antipseudomonal therapy. *Int. J. Antimicrob. Agents* **2010**, *36*, 513–522.

8.   Oliver, A.; Mena, A. Bacterial hypermutation in cystic fibrosis, not only for antibiotic resistance. *Clin. Microbiol. Infect.* **2010**, *16*, 798–808.

9.   Madi, A.; Lakhdari, O.; Blottiere, H.M.; Guyard-Nicodeme M.; Le Roux, K.; Groboillot, A.; Svinareff, P.; Dore, J.; Orange, N.; Feuilloley, M.G.J.; *et al.* The clinical *Pseudomonas fluorescens* MFN1032 strain exerts a cytotoxic effect on epithelial intestinal cells and induces Interleukin-8 via the AP-1 signaling pathway. *BMC Microbiol.* **2010**, *10*, 215.

10. Qiu, X.; Gurkar, A.U.; Lory, S. Interstrain transfer of the large pathogenicity island (PAPI-1) of *Pseudomonas aeruginosa*. *Proc. Natl. Acad. Sci. U. S. A.* **2006**, *26*, 19830–19835.

11. Carter, M.Q.; Chen, J.; Lory, S. The *Pseudomonas aeruginosa* pathogenicity island PAPI-1 is transferred via a noveltype IV pilus. *J. Bacteriol.* **2010**, *192*, 3249–3258.

12. D'Herelle, F. *Le Bacteriophage: Son role dans l'immunite* (in French); Masson et cie: Paris, France, 1921. Russian translation, GIZ, 1926, p. 223.

13. Kudva, I.T.; Jelacic, S.; Tarr, P.I.; Youderian, P.; Hovde, C.J. Biocontrol of *Escherichia coli* O157 with O157-Specific Bacteriophages. *Appl. Environ. Microbiol.* **1999**, *9*, 3767–3773.

14. Buzby, J.C.; Roberts, T. The economics of enteric infections: human foodborne disease costs. *Gastroenterology* **2009**, *136*, 1851–1862.

15. Santos, S.B.; Fernandes, E.; Carvalho, C.M.; Sillankorva, S.; Krylov, V.N.; Pleteneva, E.A.; Shaburova, O.V.; Nicolau, A.; Ferreira, E.C.; Azeredo, J. Selection and characterization of a multivalent salmonella phage and its production in a non-pathogenic *E. coli* strain. *Appl. Environ. Microbiol.* **2010**, *21*, 7338–7342.

16. Patel, J.; Sharma, M.; Millner, P.; Calaway, T.; Singh, M. Inactivation of *Escherichia coli* O157:H7 attached to spinach harvester blade using bacteriophage. *Foodborne Pathog. Dis.* **2011**, *4*, 541–546.

17. Maura, D.; Debarbieux, L. Bacteriophages as twenty-first century antibacterial tools for food and medicine. *Appl. Microbiol. Biotechnol.* **2011**, *3*, 851–859.

18. Lakhno, V.M.; Bordunovski, V.N. Use of bacteriophage therapy in surgical practice. *Vestn. Khir. Im. II Grek.* **2001**, *160*, 122–125.

19. Lazareva, E.B.; Smirnov, S.V.; Khvatov, V.B.; Spiridonova, T.G.; Bitkova, E.E.; Darbeeva, O.S.; Mayskaia, L.M.; Parfeniuk, R.L.; Men'shikov, D.D. Efficacy of bacteriophages in compl treatment of patients with burn wounds. *Antibiot. Khimioter.* **2001**, *1*, 10–14.

20. Aslanov, B.I.; Yafaev, R.Ch.; Zueva, L.P. The ways of rational use of pyocianea bacteriophages in medical and anti-epidemic practice. *Zh. Mmikrobiol.* **2003**, *5*, 72–76.

21. Gilleland, H.E.; Gilleland, L.B.; Fowler, M.R. Vaccine efficacies of elastase, exotoxin A, and outer-membrane protein F in preventing chronic pulmonary infection by *Pseudomonas aeruginosa* in a rat model. *J. Med. Microbiol.* **1993**, *38*, 79–86.

22. Johansen, H.K.; Getzsche, P.C. Vaccines for preventing infection with *Pseudomonas aeruginosa* in cystic fibrosis. *Cochrane Database Syst. Rev.* **2008**, *8*, CD001399.

23. Kamei, A.; Coutinho-Sledge, Y.S.; Goldberg, J.B.; Priebe, G.P.; Pier, G.B. Mucosal vaccination with a multivalent, live-attenuated vaccine induces multifactorial immunity against *Pseudomonas aeruginosa* acute lung infection. *Infect. Immun.* **2011**, *79*, 1289–1299.

24. Chibani-Chennoufi, S.; Sidoti, J.; Bruttin, A.; Kutter, E.; Sarker, S.; Brussow, H. *In vitro* and *in vivo* bacteriolytic activities of *Escherichia coli* phages: Implications for phage therapy. *Antimicrob. Agents Chemother.* **2004**, *48*, 2558–2569.

25. Kropinski, A.M. Phage therapy — Everything old is new again. *Can. J. Infect. Dis Med. Microbiol.* **2006**, *17*, 297–306.

26. McVay, C.S.; Velasquez, M.; Fralick, J.A. Phage therapy of *Pseudomonas aeruginosa* infection in a mouse burn wound model. *Antimicrob. Agents Chemother.* **2007**, *6*, 1934–1938.

27. Denou, E.; Bruttin, A.; Barretto, C.; Ngom-Bru, C.; Brussow, H.; Zuber Brussoff, S. T4 phages against *Escherichia coli* diarrhea: potential and problems. *Virology* **2009**, *388*, 21–30.

28. Ahiwale, S.; Tamboli, N.; Thorat, K.; Kulkarni, R.; Ackermann, H.; Kapadnis, B. *In vitro* management of hospital *Pseudomonas aeruginosa* biofilm using indigenous T7-Like lytic phage. *Curr. Microbiol.* **2011**, *62*, 335–340.

29. Garbe, J.; Wesche, A.; Bunk, B.; Kazmierczak, M.; Selezska, K.; Rohde, C.; Sikorski, J.; Rohde M.; Jahn, D.; Schobert, M. Characterization of JG024, a *Pseudomonas aeruginosa* PB1-like broad host range phage under simulated infection condition. *BMC Microbiol.* **2010**, *10*, 301.

30. Connerton, P.L.; Timms, A.R.; Connerton, I.F. Campylobacter bacteriophages and bacteriophage therapy. *J. Appl. Microbiol.* **2011**, *111*, 255–265.

31. Hens, D.K.; Chatterjee, N.C.; Kumar, R. New temperate DNA phage BcP15 acts as a drug resistance vector. *Arch. Virol.* **2006**, *151*, 1345–1353.

32. Stone, R. Bacteriophage therapy: Stalin's forgotten cure. *Science* **2002**, *298*, 728–731.

33. Weber-Dabrowska, B.; Dabrowski, M.; Slopek, S. Studies on bacteriophage penetration in patients subjected to phage therapy. *Arch. Immunol. Ther. Exp. (Warsz)* **1987**, *5*, 563–568.

34. Morello, E.; Saussereau, E.; Maura, D.; Huerre, M.; Touqui, L.; Debarbieux, L. Pulmonary bacteriophage therapy on *Pseudomonas aeruginosa* cystic fibrosis strains: first steps towards treatment and prevention. *PLoS One* **2011**, *6*, e16963.

35. Pleteneva, E.A.; Shaburova, O.V.; Krylov, V.N. A formal scheme of adsorpbtional receptors in *Pseudomonas aeruginosa* and possibilities for its practical implementation. *Russ. J. Genet.* **2009**, *1*, 35–40.

36. Intralytix, Phages for Food Safety. Available online: http://www.intralytix.com/ (accessed on 19 January 2011).

37. Krylov, V.N. Use of live phages for therapy on a background of co-evolution of bacteria and phages. *Int. Res. J. Microbiol.* **2011**, *2*, 315–332.

38. Rajagopala, S.V.; Casjens, S.; Uetz, P. The protein interaction map of bacteriophage lambda. *BMC Microbiol.* **2011**, *26*, 213.

39. Ripp, S.; Miller, R.V. Dynamics of the pseudolysogenic response in slowly growing cells of *Pseudomonas aeruginosa*. *Microbiology* **1998**, *8*, 2225–2232.

40. Lood, R.; Collin, M. Characterization and genome sequencing of two Propionibacterium acnes phages displaying pseudolysogeny. *BMC Genomics* **2011**, *12*, 198.

41. Khemayan, K.; Pasharawipas, T.; Puiprom, O.; Sriurairatana, S.; Suthienkul, O.; Flegel, T.W. Unstable lysogeny and pseudolysogeny in Vibrio harveyi siphovirus-like phage 1. *Appl. Environ. Microbiol.* **2006**, *2*, 1355–1363.

42. Sakaguchi, Y.; Hayashi, T.; Kurokawa, K.; Nakayama, K.; Oshima, K.; Fujinaga, Y.; Ohnishi, M.; Ohtsubo, E.; Hattori, M.; Oguma, K. The genome sequence of *Clostridium botulinum* type C neurotoxin-converting phageand the molecular mechanisms of unstable lysogeny. *Proc. Natl. Acad. Sci. USA* **2005**, *29*, 17472–17477.

43. Holloway, B.W.; Egan, J.B.; Monk, M. Lysogeny in *Pseudomonas aeruginosa*. *Aust. J. Exp. Biol. Med. Sci.* **1960**, *38*, 321–329.

44. Nzula, S.; Vandamme, P.; Govan, J.R.; Sensitivity of the *Burkholderia cepacia* complex and *Pseudomonas aeruginosa* to transducing bacteriophages. *FEMS Immunol. Med. Microbiol.* **2000**, *28*, 307–312.

45. Krylov, V.N.; Zhazykov, I.Zh. Pseudomonas bacteriophage phiKZ — A possible model for studying the genetic control of morphogenesis. *Genetika* **1978**, *14*, 678–685.

46. Krylov, V.N.; Smirnova, T.A.; Minenkova, I.B.; Plotnikova, T.G.; Zhazikov, I.Z.; Khrenova, E.A. Pseudomonas bacteriophage phiKZ contains an inner body in its capsid. *Can. J. Microbiol.* **1984**, *30*, 758–762.

47. Bourkal'tseva, M.V.; Krylov, V.N.; Pleteneva, E.A.; Shaburova, O.V.; Krylov, S.V.; Volkart, G.; Sykilinda, N.N.; Kurochkina, L.P.; Mesianzhinov, V.V. Phenogenetic characterization of a group of giant Phi KZ-like bacteriophages of *Pseudomonas aeruginosa*. *Genetika* **2002**, *38*, 1470–1479.

48. Krylov, V.N.; Dela Cruz, D.M.; Hertveldt, K.; Ackermann, H.W. "phiKZ-like viruses", a proposed new genus of myovirus bacteriophages. *Arch. Virol.* **2007**, *152*, 1955–1959.

49. Krylov, V.N., Bourkal'tseva, M.V.; Sykilinda, N.N.; Pleteneva, E.A.; Shaburova, O.V.; Kadykov, V.A.; Miller, S.; Biebl, M. Comparison of genomes of new gigantic *Pseudomonas aeruginosa* phages from native populations from different regions *Russ. J. Genet.* **2004**, *40*, 363–368.

50. Mesyanzhinov, V.V.; Robben, J.; Grymonprez, B.; Kostyuchenko, V.A.; Bourkal'tseva, M.V.; Sykilinda, N.N.; Krylov, V.N.; Volckaert, G. The Genome of Bacteriophage phiKZ of *Pseudomonas aeruginosa. J. Mol. Biol.* **2002**, *317*, 1–19.

51. Cornelissen, A.; Hardies, S.C.; Shaburova, O.V.; Krylov, V.N.; Mattheus, W.; Kropinski, A.M.; Lavigne, R. Complete genome sequence of the giant virus OBP and comparative genome analysis of the diverse phiKZ-related phages. *J. Virol.* **2012**, *86*, 1844–1852.

52. Kazlauskas, D.; Venclovas, C. Computational analysis of DNA replicases in double-stranded DNA viruses: relationship with the genome size. *Nucleic Acids Res.* **2011**, *39*, 8291–305.

53. Hertveldt, K.; Lavigne, R.; Pleteneva, E.; Sernova, N.; Kurochkina, L.; Korchevskii, R.; Robben, J.; Mesyanzhinov, V.; Krylov, V.N.; Volckaert, G. Genome comparison of *Pseudomonas aeruginosa* large phages. *J. Mol. Biol.* **2005**, *2*, 536–545.

54. Pleteneva, E.A.; Krylov, S.V.; Shaburova, O.V.; Bourkal'tseva, M.V.; Miroshnikov, K.A.; Krylov, V.N. Pseudolysogeny of *Pseudomonas aeruginosa* bacteria infected with phiKZ-like bacteriophages. *Russ. J. Genet.* **2010**, *46*, 20–25.

55. Krylov, S.V.; Pleteneva, E.A.; Bourkal'tseva, M.V.; Shaburova, O.V.; Miroshnikov, K.A.; Lavigne, R.; Cornelissen, A.; Krylov, V.N. Genome instability of *Pseudomonas aeruginosa* phages of the EL species: examination of virulent mutants. *Russ. J. Genet.* **2011**, *2*, 162–167.

56. Krylov, V.N.; Miroshnikov, K.A.; Krylov, S.V.; Veyko, V.P.; Pleteneva, E.A.; Shaburova, O.V.; Bourkal'tseva, M.V. Interspecies migration and evolution of bacteriophages of the genus phiKZ: The purpose and criteria of the search for new phiKZ-like bacteriophages. *Russ. J. Genet.* **2010**, *46*, 138–145.

57. Ramsey, D.M.; Baynham, P.J.; Wozniak, D.J. Binding of *Pseudomonas aeruginosa* AlgZ to sites upstream of the algZ promoter leads to repression of transcription. *J. Bacteriol.* **2005**, *187*, 4430–4443.

58. Mondello F.J.; Miller, R.V. Identification of *Pseudomonas* plasmids able to suppress the lysogeny-establishment-deficiency (Les-) phenotype. *Plasmid* **1984**, *11*, 185–187.

59. Fornelos, N.; Bamford, J.K.; Mahillon, J. Phage-borne factors and host LexA regulate the lytic switch in phage GIL01. *J. Bacteriol.* **2011**, *193*, 6008–6019.

60. Krylov, V.N.; Tolmachova, T.O.; Akhverdian, V.Z. DNA homology in species of bacteriophages active on *Pseudomonas aeruginosa. Arch. Virol.* **1993**, *1–2*, 141–151.

61. Kropinski, A.M. Sequence of the genome of the temperate, serotype-converting, *Pseudomonas aeruginosa* bacteriophage D3. *J. Bacteriol.* **2000**, *182*, 6066–6074.

62. Byrne, M.; Kropinski, A.M. The genome of the *Pseudomonas aeruginosa* generalized transducing bacteriophage F116. *Gene* **2005**, *346*, 187–194.

63. Casjens, S.R.; Gilcrease, E.B.; Winn-Stapley, D.A.; Schicklmaier, P.; Schmieger, H.; Pedulla M.L.; Ford, M.E.; Houtz, J.M.; Hatfull, G.F.; Hendrix, R.W. The generalized transducing salmonella bacteriophage ES18: Complete genome sequence and DNA packaging strategy. *J. Bacteriol.* **2005**, *187*, 1091–1104.

64. Nakayama, K.; Kanaya, S.; Ohnishi, M.; Terawaki, Y.; Hayashi, T. The complete nucleotide sequence of, a cytotoxin-converting phage of *Pseudomonas aeruginosa*: Implications for phage evolution and horizontal gene transfer via bacteriophages. *Mol. Microbiol.* **1999**, *31*, 399–419.

65. Ceyssens, P.J.; Brabban, A.; Rogge, L.; Lewis, M.S.; Pickard, D.; Goulding, D.; Dougan, G.; Noben, J.P.; Kropinski, A.; Kutter, E.; Lavigne, R. Molecular and physiological analysis of three *Pseudomonas aeruginosa* phages belonging to the "N4-like viruses". *Virology* **2010**, *15*, 26–30.

66. Krylov, S.V.; Kropinski, A.; Shaburova, O.V.; Chesnokova, E.N.; Miroshnikov, K.A.; Krylov, V.N. Features of the genome structure of phi297—Temperate phage of new species active on *Pseudomonas aeruginosa*. *Russ. J. Genet.* **2013**, in press.

67. Holloway, B.W.; Cooper, G.N. Lysogenic conversion in *Pseudomonas aeruginosa*. *J. Bacteriol.* **1962**, *84*, 1321–1324.

68. Kuzio, J.; Kropinski, A.M. O-antigen conversion in *Pseudomonas aeruginosa* PAO1 by bacteriophage D3. *J. Bacteriol.* **1983**, *155*, 203–212.

69. Miller, R.V.; Pemberton, J.M.; Clark, A.J. Prophage F116: Evidence for extrachromosomal location in *Pseudomonas aeruginosa* strain PAO. *J. Virol.* **1977**, *22*, 844–847.

70. Rodionov, O.; Yarmolinsky, M. Plasmid partitioning and the spreading of P1 partition protein ParB. *Mol. Microbiol.* **2004**, *52*, 1215–23.

71. Lee, J.H.; Shin, H.; Kim, H.; Ryu, S. Complete genome sequence of salmonella bacteriophage SPN3US. *J. Virol.* **2011**, *85*, 13470–13471.

72. Jacoby, G.A.; Jacob, A.E.; Hedges, R.W. Recombination between plasmids of incompatibility groups P-1 and P-2. *J. Bacteriol.* **1976**, *127*, 1278–1285.

73. Freizon, E.V.; Kopylova, Iu.I.; Cheremukhina, L.V.; Krylov, V.N. The effect of the IncP-2-group plasmid on the growth of *Pseudomonas aeruginosa* bacteriophages. *Genetika* **1989**, *25*, 1168–1178.

74. Jacoby, G.A.; Sutton, L. Properties of plasmids responsible for production of extended-spectrum beta-lactamases. *Antimicrob. Agents Chemother.* **1991**, *35*, 164–169.

75. Philippon, A.M.; Paul, G.C.; Thabaut, A.P.; Jacoby, G.A. Properties of a novel carbenicillin-hydrolyzing beta-lactamase (CARB-4) specified by an IncP-2 plasmid from *Pseudomonas aeruginosa*. *Antimicrob. Agents Chemother.* **1986**, *29*, 519–520.

76. Krylov, V.N.; Eremenko, E.N.; Bogush, V.G.; Kirsanov, N.B. Interaction of *Pseudomonas aeruginosa* plasmids and mu-like phages: the suppression of the early stages of cell infection by phage D3112 in the presence of plasmid RPL11. *Genetika* **1982**, *18*, 743–752.

77. Gerasimov, V.A.; Eremenko, E.N.; Khrenova, E.A.; Gorbunova, S.A.; Krylov, V.N. Stimulation of the lysogenization of *Pseudomonas aeruginosa* cells by phage transposon B39 induced by plasmid Rms163. *Genetika* **1984**, *20*, 1080–1087.

78. Krylov, V.N.; Plotnikova, T.G.; Kulakov, L.A.; Fedorova, T.V.; Eremenko, E.N. Integration of the genome of the Mu-like *Pseudomonas aeruginosa* bacteriophage D3112 into plasmid RP4 and its hybrid plasmid transfer into *Pseudomonas putida* and *Escherichia coli* C600 bacteria. *Genetika* **1982**, *18*, 5–12.

79. Plotnikova, T.G.; Akhverdian, V.Z.; Reulets, M.A.; Gorbunova, S.A.; Krylov, V.N. Multiplicity of the integration sites of Mu-like bacteriophages into *Pseudomonas aeruginosa* chromosome and plasmids. *Genetika* **1983**, *19*, 1604–1610.

80. Gorbunova, S.A.; Ianenko, A.S.; Akhverdian, V.Z.; Reulets, M.A.; Krylov, V.N. Expression of *Pseudomonas aeruginosa* transposable phages in *Pseudomonas putida* cells. I. The establishment of a lysogenic state and the effectiveness of lytic development. *Genetika* **1985**, *9*, 1455–1463.

81. Krylov, V.; Merlin, C.; Toussaint, A. Introduction of *Pseudomonas aeruginosa* mutator phage D3112 into *Alcaligenes eutrophus* strain CH34. *Res. Microbiol.* **1995**, *146*, 245–250.

82. Kaplan, A.M.; Akhverdian, V.Z.; Reulets, M.A.; Krylov, V.N. Compatibility of transposable phages of *Escherichia coli* and *Pseudomonas aeruginosa*. I. Co-development of phages Mu and D3112 and integration of phage D3112 into RP4::Mu plasmid in *Pseudomonas aeruginosa* cells. *Genetika* **1988**, *4*, 634–640.

83. Kaplan, A.M.; Akhverdian, V.Z.; Krylov, V.N. Integration of phage Mu into the RP4::D3112A-plasmid in *Escherichia coli* cells. *Genetika* **1989**, *25*, 1384–1390.

84. Kopylova, I.I.; Gorbunova, S.A.; Krylov, V.N. Interaction of heterologous bacteriophages: Growth suppression of temperate PP56 phage of *Pseudomonas putida* by the transposable phage D3112 of *Pseudomonas aeruginosa*. *Genetika* **1988**, *5*, 803–807.

85. Plotnikova, T.G.; Kulakov, L.A.; Eremenko, E.N.; Fedorova, T.V.; Krylov, V.N. Expression of the genome of Mu-like phage D3112 specific for *Pseudomonas aeruginosa* in *Escherichia coli* and *Pseudomonas putida* cells. *Genetika* **1982**, *7*, 1075–1084.

86. Plotnikova, T.G.; Ianenko, A.S.; Kirsanov, N.B.; Krylov, V.N. Transposition of the phage D3112 genome in *Escherichia coli* cells. *Genetika* **1983**, *10*, 1611–1615.

87. Trenina, M.A.; Akhverdian, V.Z.; Kolibaba, L.G.; Rebentish, B.A.; Krylov, V.N. Features of genome expression of phage transposon D3112 of *Pseudomonas aeruginosa* in *Escherichia coli* bacteria: Dependence of bacterial phenotype on copy number of D3112 genome. *Genetika* **1991**, *27*, 1324–1335.

88. Krylov, S.V.; Kropinski, A.M.; Pleteneva, E.A.; Shaburova, O.V.; Bourklal'tseva, M.V.; Miroshnikov, K.A.; Krylov, V.N. Features of new D3-like phage phiPMG1 of *Pseudomonas aeruginosa*: Genome structure and prospects for use the phage in phage therapy. *Russ. J. Genet.* **2012**, *48*, 902–911.

89. Dzhusupova, A.B.; Plotnikova, T.G.; Krylov, V.N. Detection of transduction by virulent bacteriophage phiKZ of *Pseudomonas aeruginosa* chromosomal markers in the presence of plasmid RMS148. *Genetika* **1982**, *18*, 1799–1802.

90. Langley, R.; Kenna, D.T.; Vandamme, P.; Ure, R.; Govan, J.R. Lysogeny and bacteriophage host range within the *Burkholderia cepacia* complex. *J. Med. Microbiol.* **2003**, *52*, 483–490.

91. Krylov, V.N.; Bogush, V.G.; Shapiro, J. Bacteriophages of *Pseudomonas aeruginosa* with DNA similar in structure to that of phage Mu1. I. General description, localization of sites sensitive to endonucleases in DNA, and structure of homoduplexes of phage D3112. *Genetika* **1980**, *16*, 824–832.

92. Akhverdian, V.Z.; Khrenova, E.A.; Reulets, M.A.; Gerasimova, T.V.; Krylov, V.N. The properties of transposable phages of *Pseudomonas aeruginosa* belonging to two groups distinguished by DNA-DNA homology. *Genetika* **1985**, *21*, 735–747.

93. Akhverdian, V.Z.; Khrenova, E.A.; Lobanov, A.O.; Krylov, V.N. Role of divergence in evolution of group B3 *Pseudomonas aeruginosa* transposable phage evolution. *Russ. J. Genet.* **1998**, *34*, 699–701.

94. Akhverdian, V.Z.; Lobanov, A.O.; Khrenova, E.A.; Krylov, V.N. Recombinational origin of natural transposable phages of related species. *Russ. J. Genet.* **1998**, *34*, 697–700.

95. Bogush, V.G.; Plotnikova, T.G.; Kirsanov, N.B.; Rebentish, B.A.; Permogorov, V.I.; Krylov V.N. Bacteriophages of *Pseudomonas aeruginosa* with DNA similar in structure to that of phage Mu1. III. Isolation and analysis of hybrid phages D3112 and B39: Localization of the immunity region and some genetic factors. *Genetika* **1981**, *17*, 967–976.

96. Krylov, V.N.; Akhverdian, V.Z.; Khrenova, E.A.; Cheremukhina, L.V.; Tiaglov, B.V. Two types of molecular structure (composition) of a genome in one species of transposable bacteriophages of *Pseudomonas aeruginosa*. *Genetika* **1986**, *22*, 2637–2648.

97. Krylov, V.N.; Bogush, V.G.; Ianenko, A.S.; Kirsanov, N.B. Bacteriophages of *Pseudomonas aeruginosa* with DNA similar in structure to that of phage Mu1. II. Evidence of relatedness of bacteriophages D3112, B3 and B39: analysis of DNA digestion with restrictional endonucleases, isolation of recombinant of phages D3112 and B3. *Genetika* **1980**, *16*, 975–984.

98. Zemlyanaya, N.Y.; Kozma, A.R.; Krylov, V.N. Differences in mutator acrivity of phage-transposons of *Pseudomonas aeruginosa*. *Genetika* **1992**, *28*, 160–163.

99. Rehmat, S.; Shapiro, J.A. Insertion and replication of the *Pseudomonas aeruginosa* mutator phage D3112. *Mol. Gen. Genet.* **1983**, *192*, 416–423.

100. Krylov, V.N.; Solov'eva, T.I.; Bourkal'tseva, M.V. Mucoid clones of *Pseudomonas aeruginosa* PAO1, surviving after induction of prophage transposons. *Russ. J. Genet.* **1995**, *31*, 1170–1174.

101. Gerasimov, V.A.; Ianenko, A.S.; Akhverdian, V.Z.; Krylov, V.N. Interaction between *Pseudomonas aeruginosa* phages-transposons: Genetic analysis of the trait of inhibition by prophage D3112 of phage B39 development. *Russ. J. Genet.* **1996**, *32*, 1068–1073.

102. Gerasimov, V.A.; Ianenko, A.S.; Akhverdian, V.Z.; Krylov, V.N. Phage-transposon interaction: The cip locus of prophage D3112 responsible for the inhibition of integration and transposition of related phage B39 of *Pseudomonas aeruginosa*. *Genetika* **1985**, *21*, 1634–1642.

103. Krylov, V.N.; Akhverdian, V.Z.; Bogush, V.G.; Khrenova, E.A.; Reulets, M.A. Modular structure of the genes of phages-transposons of *Pseudomonas aeruginosa*. *Genetika* **1985**, *21*, 724–734.

104. Ianenko, A.S.; Bekkarevich, A.O.; Gerasimov, V.A.; Krylov, V.N. Genetic map of the transposable phage D3112 of *Pseudomonas aeruginosa*. *Genetika* **1988**, *24*, 2120–2126.

105. Summer, E.J.; Gonzalez, C.F.; Carlisle, T.; Mebane, L.M.; Cass, A.M.; Savva, C.G.; LiPuma, J.; Young, R. *Burkholderia cenocepacia* phage BcepMu and a family of Mu-like phages encoding potential pathogenesis factors. *J. Mol. Biol.* **2004**, *340*, 49–65.

106. Wang, P.W.; Chu, L.; Guttman, D.S. Complete sequence and evolutionary genomic analysis of the *Pseudomonas aeruginosa* transposable bacteriophage D3112. *J. Bacteriol.* **2004**, *186*, 400–410.

107. Braid, M.D.; Silhavy, J.L.; Kitts, C.L.; Cano, R.J.; Howe, M.M. Complete genomic sequence of bacteriophage B3, a Mu-like phage of *Pseudomonas aeruginosa*. *J. Bacteriol.* **2004**, *186*, 6560–6574.

108. Bourkaltseva, M.V.; Krylov, S.V.; Kropinski, A.M.; Pleteneva, E.A.; Shaburova, O.V.; Krylov, V.N. Bacteriophage phi297, a New Species of *Pseudomonas aeruginosa* Temperate phages with a mosaic genome: potential use in phage therapy. *Russ. J. Genet.* **2011**, *47*, 794–798.

109. Piekarowicz, A.; Kłyz, A.; Majchrzak, M.; Adamczyk-Popławska, M.; Maugel, T.K.; Stein, D.C. Characterization of the dsDNA prophage sequences in the genome of *Neisseria gonorrhoeae* and visualization of productive bacteriophage. *BMC Microbiol.* **2007**, *7*, 66.

110. Lecoutere, E.; Ceyssens, P.J.; Miroshnikov, K.A.; Mesyanzhinov, V.V.; Krylov, V.N.; Noben, J.P.; Robben, J.; Hertveldt, K.; Volckaert, G.; Lavigne, R. Identification and comparative analysis of the structural proteomes of phiKZ and EL, two giant *Pseudomonas aeruginosa* bacteriophages. *Proteomics* **2009**, *9*, 3215–3219.

111. Serwer, P.; Hayes, S.J.; Thomas, J.A.; Demeler, B.; Hardies, S.C. Isolation of novel large and aggregating bacteriophages. *Methods Mol. Biol.* **2009**, *501*, 55–66.

112. Adriaenssens, E.M.; Mattheus, W.; Cornelissen, A.; Shaburova, O.; Krylov, V.N.; Kropinski, A.M.; Lavigne, R. Complete genome sequence of the giant *Pseudomonas* phage Lu11. *J. Virol.* **2012**, *86*, 6369–6370.

113. Shaburova, O.V.; Hertveldt, K.; de la Cruz, D.M.A.; Krylov, S.V.; Pleteneva, E.A.; Bourkaltseva, M.V.; Lavigne, R.; Volckaert, G.; Krylov, V.N. Comparison of new giant bacteriophages OBP and Lu11 of soil pseudomonads with bacteriophages of the phiKZ-supergroup of *Pseudomonas aeruginosa*. *Russ. J. Genet.* **2006**, *42*, 877–885.

114. Yamada, T.; Satoh, S.; Ishikawa, H.; Fujiwara, A.; Kawasaki, T.; Fujie, M.; Ogata, H. A jumbo phage infecting the phytopathogen *Ralstonia solanacearum* defines a new lineage of the *Myoviridae* family. *Virology* **2010**, *398*, 135–147.

115. Ceyssens, P.J.; Hertveldt, K.; Ackermann, H.W.; Noben, J.P.; Demeke, M.; Volckaert, G.; Lavigne, R. The intron-containing genome of the lytic *Pseudomonas* phage LUZ24 resembles the temperate phage PaP3. *Virology* **2008**, *377*, 233–238.

116. Kulakov, L. The School of Biological Sciences, The Queen's University of Belfast, Northern Ireland, UK. The genome sequence of phage TL. Personal communication, 2012.

117. Pleteneva, E.A.; Shaburova, O.V.; Sykilinda, N.N.; Miroshnikov, K.A.; Krylov, S.V.; Mesianzhinov, V.V.; Krylov, V.N. Study of the diversity in a group of phages of *Pseudomonas aeruginosa* species PB1 (*Myoviridae*) and their behavior in adsorbtion-resistant bacterial mutants. *Genetika* **2008**, *44*, 185–194.

118. Ceyssens, P.J.; Miroshnikov, K.; Mattheus, W.; Krylov, V.; Robben, J.; Noben, J.P.; Vanderschraeghe, S.; Sykilinda, N.; Kropinski, A.M.; Volckaert, G.; *et al.* Comparative analysis of the widespread and conserved PB1-like viruses infecting *Pseudomonas aeruginosa*. *Environ. Microbiol.* **2009**, *11*, 2874–2883.

119. Pleteneva, E.A.; Bourkal'tseva, M.V.; Shaburova, O.V.; Krylov, S.V.; Pechnikova, E.V.; Sokolova, O.S.; Krylov, V.N. TL, the new bacteriophage of *Pseudomonas aeruginosa* and its application for the search of halo-producing bacteriophages. *Russ. J. Genet.* **2011**, *47*, 5–9.

120. Chattoraj, S.S.; Murthy, R.; Ganesan, S.; Goldberg, J.B.; Zhao, Y.; Hershenson, M.B.; Sajjan U.S. *Pseudomonas aeruginosa* alginate promotes *Burkholderia cenocepacia* persistence in cystic fibrosis transmembrane conductance regulator knockout mice. *Infect. Immun.* **2010**, *78*, 984–993.

121. Chapalain, A.; Rossignol, G.; Lesouhaitier, O.; Merieau, A.; Gruffaz, C.; Guerillon, J.; Meyer, J.M.; Orange, N.; Feuilloley, M.G. Comparative study of 7 fluorescent pseudomonad clinical isolates. *Can. J. Microbiol.* **2008**, *54*, 19–27.

122. Chapalain, A.; Rossignol, G.; Lesouhaitier, O.; Le Roux, K.;Groboillot, A.; Svinareff, P.; Dore, J.; Orange, N.; Feuilloley, M.G.J.; Connil, N. The clinical *Pseudomonas fluorescens* MFN1032 strain exerts a cytotoxic effect on epithelial intestinal cells and induces Interleukin-8 via the AP-1 signaling pathway. *BMC Microbiol.* **2010**, *10*, 215.

123. Groisman, E.A.; Casadaban, M.J. Cloning of genes from members of the family *Enterobacteriaceae* with mini-Mu bacteriophage containing plasmid replicons. *J. Bacteriol.* **1987**, *169*, 687–693.

124. Ishiwa, A.; Komano, T. Thin pilus PilV adhesins of plasmid R64 recognize specific structures of the lipopolysaccharide molecules of recipient cells. *J. Bacteriol.* **2003**, *185*, 5192–5199.

125. Rabel, C.; Grahn, A.M.; Lurz, R.; Lanka, E. The VirB4 family of proposed traffic nucleoside triphosphatases: Common motifs in plasmid RP4 TrbE are essential for conjugation and phage adsorption. *J. Bacteriol.* **2003**, *185*, 1045–1058.

126. Stewart, R.M.; Wiehlmann, L.; Ashelford, K.E.; Preston, S.J.; Frimmersdorf, E.; Campbell, B.J.; Neal, T.J.; Hall, N.; Tuft, S.; Kaye, S.B.; Winstanley, C. Genetic characterization indicates that a specific subpopulation of *Pseudomonas aeruginosa* is associated with keratitis infections. *J. Clin. Microbiol.* **2011**, *49*, 993–1003.

127. Masignani, V.; Giuliani, M.M.; Tettelin, H.; Comanducci, M.; Rappuoli, R.; Scarlato, V. Mu-like Prophage in serogroup B *Neisseria meningitidis* coding for surface-exposed antigens. *Infect. Immun.* **2001**, *4*, 2580–2588.

128. Winther-Larsen, H.C.; Wolfgang, M.C.; van Putten, J.P.; Roos, N.; Aas, F.E.; Egge-Jacobsen, W.M.; Maier, B.; Koomey, M. *Pseudomonas aeruginosa* Type IV pilus expression in *Neisseria gonorrhoeae*: Effects of pilin subunit composition on function and organelle dynamics. *J. Bacteriol.* **2007**, *189*, 6676–6685.

129. Zegans, M.E.; Wagner, J.C.; Cady, K.C.; Murphy, D.M.; Hammond, J.H.; O'Toole, G.A. Interaction between bacteriophage DMS3 and host CRISPR region inhibits group behaviors of *Pseudomonas aeruginosa*. *J. Bacteriol.* **2009**, *191*, 210–219.

130. Chung, I.Y.; Cho, Y.H. Complete genome sequences of two *Pseudomonas aeruginosa* temperate phages MP29 and MP42, which lack the phage-host CRISPR interaction. *J. Virol.* **2012**, *86*, 8336.

131. Barrangou, R.; Horvath, P. CRISPR: New horizons in phage resistance and strain identification. *Annu. Rev. Food Sci. Technol.* **2012**, *3*, 143–162.

132. Cady, K.C.; O'Toole, G.A. Non-identity-mediated CRISPR-bacteriophage interaction mediated via the Csy and Cas3 proteins. *J. Bacteriol.* **2011**, *193*, 3433–3445.

133. Cady, K.C.; Bondy-Denomy, J.; Heussler, G.E.; Davidson, A.R.; O'Toole, G.A. The CRISPR/Cas adaptive immune system of *Pseudomonas aeruginosa* mediates resistance to naturally occurring and engineered phages. *J. Bacteriol.* **2012**, doi:10.1128/JB.01184-12.

134. Cady, K.C.; White, A.S.; Hammond, J.H.; Abendroth, M.D.; Karthikeyan, R.S.; Lalitha, P.; Zegans, M.E.; O'Toole, G.A. Prevalence, conservation and functional analysis of *Yersinia* and *Escherichia* CRISPR regions in clinical *Pseudomonas aeruginosa* isolates. *Microbiology* **2011**, *157*, 430–437.

135. Lang, S.; Kirchberger, P.C.; Gruber, C.J.; Redzej, A.; Raffl, S.; Zellnig, G.; Zangger, K.; Zechner, E.L. An activation domain of plasmid R1 TraI protein delineates stages of gene transfer initiation. *Mol. Microbiol.* **2011**, *82*, 1071–1085.

136. Caruso, M.; Shapiro, J.A. Interactions of Tn7 and temperate phage F116L of *Pseudomonas aeruginosa*. *Mol. Gen. Genet.* **1982**, *188*, 292–298.

137. Kul'ba, A.M.; Gorelyshev, A.S.; Evtushenkov, A.N.; Fomichev, Iu.K. Temperate SM phage from *Pseudomonas aeruginosa* as a vector for cloning genetic information. *Mol. Gen. Mikrobiol. Virusol.* **1991**, *12*, 26–29.

8

Phage Lambda P Protein: Trans-Activation, Inhibition Phenotypes and their Suppression

Sidney Hayes *, Craig Erker, Monique A. Horbay, Kristen Marciniuk, Wen Wang and Connie Hayes

Department of Microbiology and Immunology, College of Medicine, University of Saskatchewan, Saskatoon, S7N 5E5 Canada; E-Mails: sidney.hayes@usask.ca (S.H.); craigaerker@gmail.com (C.E.); mah134@mail.usask.ca (M.A.H.); kdm449@mail.usask.ca (K.M.); wew153@mail.usask.ca (W.W.); clh127@mail.usask.ca (C.H.)

* Author to whom correspondence should be addressed; sidney.hayes@usask.ca

---

Abstract: The initiation of bacteriophage λ replication depends upon interactions between the *ori*λ DNA site, phage proteins O and P, and *E. coli* host replication proteins. P exhibits a high affinity for DnaB, the major replicative helicase for unwinding double stranded DNA. The concept of P-lethality relates to the hypothesis that P can sequester DnaB and in turn prevent cellular replication initiation from *oriC*. Alternatively, it was suggested that P-lethality does not involve an interaction between P and DnaB, but is targeted to DnaA. P-lethality is assessed by examining host cells for transformation by ColE1-type plasmids that can express P, and the absence of transformants is attributed to a lethal effect of P expression. The plasmid we employed enabled conditional expression of *P*, where under permissive conditions, cells were efficiently transformed. We observed that ColE1 replication and plasmid establishment upon transformation is extremely sensitive to P, and distinguish this effect from P-lethality directed to cells. We show that alleles of *dnaB* protect the variant cells from *P* expression. P-dependent cellular filamentation arose in *ΔrecA* or *lexA*[Ind⁻] cells, defective for SOS induction. Replication propagation and restart could represent additional targets for P interference of *E. coli* replication, beyond the *oriC*-dependent initiation step.

Keywords: *E. coli* DnaB replicative helicase; bacteriophage lambda (λ) replication initiation protein P; allelic alterations of *dnaB* and P; ColE1 plasmid curing and replication inhibition; cellular filamentation

---

1. Introduction

The mechanism for bi-directional initiation of bacteriophage λ DNA replication involves a complex interaction of λ proteins O and P with the *E. coli* host DNA replication machinery. The arrangement of genes *O* and *P* on the λ genome is shown in Figure 1. Their expression depends upon *pR-cro-cII-O-P* mRNA synthesis from promoter *pR* which is negatively regulated by the CI repressor, made from gene *cI*, binding to *oR* operator sequences overlapping *pR*. The gp*O* (= O) acts to bind *oriλ*, the λ origin of replication, situated midway within the *O* sequence [1]. The gp*P* = P protein facilitates replication initiation through recruitment of an *E. coli* host protein(s) to form a preprimosomal initiation complex at *oriλ*. In agreement with earlier genetic studies, P was found to physically interact [2] with DnaB the major replicative helicase for unwinding double-stranded (ds) DNA. DnaB promotes the advancement of a growing replication fork [3] by using energy provided by ATP hydrolysis; it functions as a "mobile promoter" in the general priming reaction [4,5] to aid DnaG primase in producing RNA primers for extension by DNA polymerase III; and it is able to promote progression of Holliday Junctions in the repair of DNA damage [6,7]. DnaB is a hexameric homomultimer with up to six bound ATP molecules [8,9]. Its intrinsic ssDNA binding activity is inhibited when the free form of DnaB forms a complex with the host replication initiation protein DnaC ([10] and included references). P can commandeer DnaB away from DnaC [11]. P:DnaB participates in loading DnaB onto DNA during formation of the DnaB:P:O:*oriλ* preprimosomal complex. The interaction between P and DnaB inactivates the helicase activity of DnaB. Restoration of DnaB activity, coupled with activation of *oriλ*-dependent replication initiation involves the dissociation of P bound to DnaB by *E. coli* heat shock proteins DnaK, DnaJ, and GrpE [12,13]. The ability of P to dissociate preformed cellular DnaB-DnaC complexes, coupled with a low cellular concentration of DnaB ($\sim$20 hexamers/cell) [14,15], suggests the hypothesis that λ P expression can inhibit *E. coli* DNA replication initiation from *oriC*, alluded to herein as P-lethality.

Figure 1. Relevant λ genes in expression plasmids and prophage. **(A)** Arrangement of synthetic plasmid pcIpR-[GOI]-timm. Plasmids employed include the precise coding sequence for GOI (gene of interest) genes *O, P,* or alleles of *P*, including P^{π}, *P*-SPA, $P^{\Delta 76}$, each inserted, in precisely the same orientation and position as gene *cro* in λ, directly downstream of *pR*. The coding sequence for the GOI is terminated by an ochre stop codon inserted just ahead of the powerful *timm* termination sequence [16,17], previously named *ti* [18,19]. The regulatory regions of pcIpR-(GOI)-timm were described (Figure 1 plus supplemental sequence file in [20]). The expression of the GOI from promoter *pR* is negatively regulated by the lambda CI[Ts] repressor binding to the wild type *oR* operator sites overlapping *pR*. These plasmids do not contain the *oL* operator sequences so that the tightest repression of transcription from the *pR* promoter, requiring CI-mediated DNA looping via CI dimers binding both operator sites [21], is not possible. Transcriptional read through beyond the GOI is prevented by *timm*, shown in its natural map position between the left operators, *oL* and the C-terminal end of *rexB* (see gene map drawn in Figure 1B). The synthetic sequence from the translational termination sequence for GOI through *timm* to downstream *Eco*RI-*Sal*I is: *TAATCGAT*cccgg*GG*tcagc*C*ccggg*ttttc*ttt*TGAATTCGTCGAC*, where the bases that were modified from the wild type lambda DNA sequence are in capitalized italics. Shifting cells with a pcIpR-[GOI]-timm plasmid that were grown at 25 or 30 °C, to above 39 °C induces expression of the GOI from the plasmid [20]. **(B)** Cryptic prophage in strain Y836. The λ phage genes *cro-cII-O-P-ren* are transcribed from promoter *pR* that is embedded within the rightward operator sequence, *oR*, between genes *cI* and *cro*. Gene *cI*, encoding a temperature sensitive repressor is transcribed from promoter *pM*; and *cro* is transcribed in the opposite direction from *pR*. C. At temperatures where CI remains active, *i.e.*, at or below 38 °C, λ replication initiation is prevented, and the λ fragment is replicated as part of the *E. coli* chromosome by forks arising from *oriC*. At about 39 °C, CI becomes fully denatured, *pR* transcription is induced, and λ undergoes a few replication initiation events from *ori*λ [22].

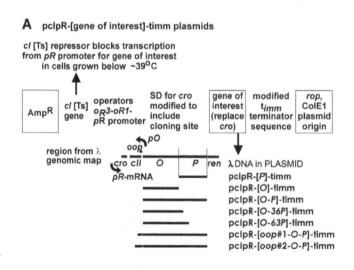

Figure 1. *Cont.*

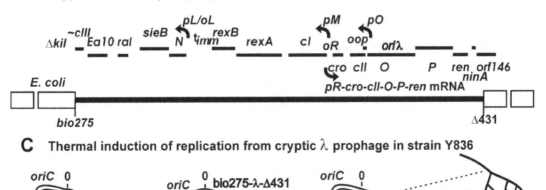

B Cryptic lambda (λ) prophage fragment in *E. coli* strain designated Y836

C Thermal induction of replication from cryptic λ prophage in strain Y836

Klinkert and Klein [23] examined the effect of cloning the λ fragment sequence bp's 39168–44972 (part of *O, P-ren*-and downstream *orf's 290-146-57-56-60-204-68-64*) into a plasmid expression vector regulated by the *lac* promoter. They showed that inducing *P* expression correlated with effects of shifting cells with temperature sensitive slow-stop mutations in replication initiation genes to the non-permissive temperature, *i.e.*, an eventual inhibition in *E. coli* replication was seen after about a 50 min induction. The effect paralleled the addition of chloramphenicol, and was opposite the effect of adding nalidixic acid, which produced an immediate (fast-stop) inhibition in chromosomal replication (their Figure 5). Maiti *et al.* [24] cloned λ genes into pBR322 and followed the survival of bacteria after transformation with a plasmid that constitutively expressed *pR-cro-O-P-ren-tR2*, measuring transformants (Amp^R colony forming units, cfu) per µg DNA. They observed that plasmids expressing active *P* did not yield survivor cfu, concluding that *P* expression is lethal to the host, *i.e.*, P-lethality. They found that three types of *groP* mutants of *E.coli*, which do not allow λ DNA replication, possibly due to the lack of interaction of P protein with the altered DnaB (*groP*A15, *groP*B558) or DnaJ protein (*groP*C259), were equally susceptible to killing by P protein. They concluded that P-lethality *does not involve interaction of* P with the host DnaB, DnaJ, or DnaK proteins, which are all essential for λ DNA replication [25-27] but is targeted to the *E. coli* DNA replication initiation protein DnaA, inhibiting its *oriC* DNA binding [28]. Datta *et al.* [29] identified mutations within the gene *dnaA* that confer cellular resistance to P-lethality. An alternative hypothesis for the absence of transformants formed by plasmids expressing *P* was not considered in prior studies, *i.e.*, that *P* expression from the transformed pBR322-derived plasmids employed interfered with / blocked the initiation of plasmid replication and copy increase within the newly transformed cell. Since the replication origin of pBR322 was derived from pMB1, closely related to ColE1 [30] we refer herein to the plasmids employed as ColE1-type plasmids.

In the present manuscript we examine if the interaction of P and DnaB provides an alternative mechanism for P-lethality. We assess the influence of P expression upon the maintenance of ColE1 plasmids, and we examine whether P expression, shown to stimulate cellular filamentation, depends upon triggering a cellular SOS response.

2. Result and Discussions

2.1. Transformation of Cells by Plasmids Expressing P, Complementation and Immunity Assays

Plasmid pHB30 (Table 10), encoding an intact lambda *P-ren* sequence whose expression is regulated by the CI[Ts] repressor, yielded similar transformation results as those reported by Maiti *et al.* [24], *i.e.* no transformants were observed at 42 °C for *dnaB*⁺ hosts, or for several host strains isolated as being defective for λ replication, *i.e.*, *groP*A15, *groP*558, *grpE*280, *grpC*2 [31]. However two mutants employed in this study, *grpA*80 and *grpD*55 [32] supported pHB30 transformation at 42 °C at about the same efficiency as at 30 °C [31]. The *grpD*55 mutation (initial reported *E. coli* linkage-map position ~71.5 min) was re-mapped as an allele of *dnaB* and was co-transduceable with malF3089::Tn*10* at ~91.5 min [33]. (We were unable to move the grpA80 mutation into another host.) These results were considered preliminary because pHB30 included several genes (a *cro-O* gene fusion, *P* and *ren*), and because the *grpA80* and *grpD55* mutations were not then confirmed to fall within *dnaB*. Table 1 shows that both *grpA*80 and *grpD*55 mutations each possess two missense changes within *dnaB*, one being in common, E426K.

Table 1. Sequencing analysis of alleles of *dnaB*.

dnaB allele	bp *E. coli* mutation	AA changed
grpD55	4,263,102 G to A	V256I
	4,263,612 G to A	E426K
grpA80	4,263,349 G to A	G338E
	4,263,612 G to A	E426K

The ability of the *dnaB-grpD55* allele within *E. coli* strain 594 to inhibit λ vegetative growth is shown in Table 2.

Table 2. Inhibition of λ*cI*72 vegetative growth on *dnaB-grpD55* host.

Host strains	EOP on host cells incubated at temperature [a]			
	30	37	39	42 °C
594	1.0	nd [b]	nd	0.94
594 *grpD55*	0.0147	4.4×10^{-7}	4.4×10^{-7}	4.6×10^{-7}

[a] Efficiency of plating (EOP) of λ*cI*72 was determined by dividing the average phage titer on 594 *dnaB-grpD55* host cells by the titer on 594 host at 30 °C (= 6.8×10^9 pfu/mL). [b] nd = not determined.

Plasmid pcIpR-*P*-timm (Figure 1A) includes λ gene *cI*[Ts]857, encoding a thermolabile repressor that is expressed from promoter *pM*, and a precise copy of wild type gene *P* DNA, situated downstream from λ promoter *pR* in exactly the same configuration as is occupied by gene *cro* in the λ genome. The expression of *P* from this plasmid is regulated by the CI[Ts] repressor binding to the *oR* sequences overlapping *pR*. The expression of *P* from pcIpR-*P*-timm was initially assumed similar to that shown for D-CAP or D-SPA gene fusions, as in pcIpR-D-CAP-timm (Figure 4 in [20]), where there was no detected expression from the non-induced gene fusion. The expression of genes *P* and *cI*[Ts]857 from pcIpR-*P*-timm was assessed by complementation, measuring the biological activity of P, and a CI-immunity assay to assess CI[Ts] repression of *immλ* phage plating. The complementation assay determined whether P expressed from the plasmid could support the plating of a phage with an amber mutation in *P*, *i.e.*, the heteroimmune infecting phage *λimm434Pam3*, which is insensitive to repression by CI$^\lambda$. The ability of cells with pcIpR-*P*-timm to produce CI[Ts] at a sufficient level to turn off *pR* transcription from both plasmid or an infecting phage was assessed by its inhibition of *λcI72* plating.

Table 3 shows the transformation of nonlysogenic 594 *dnaB*$^+$ cells (*dnaB* was sequenced [34]) and of 594(λnin5) lysogenic cells. The CI repressor made from the λnin5 prophage is not Ts. The lysogenic cells expressing CI$^+$ were transformable by pcIpR-*P*-timm at 37 °C. The nonlysogenic cells were transformed by pcIpR-*P*-timm only between 25-34 °C, where CI[Ts] retains sufficient repressor activity to shut off *P* expression in transformed cells. CI[Ts] is reported [35] marginally active at ≥37 °C and would not fully repress *P* expression, explaining the absence of transformants arising in nonlysogenic cells when the transformation temperature is raised above a temperature where CI[Ts] repressor retains sufficient DNA binding activity.

Table 3. Cellular transformation by pcIpR-*P*-timm plasmid.

Cells Transformed	Plating Temp. °C	Transformation Frequency [a]	Transformants per μg DNA
594	25	5.2×10^{-5}	1.5×10^5
	37	$<1.7 \times 10^{-8}$	0
594(λ wt nin5)	25	7.2×10^{-5}	4.7×10^4
	37	8.5×10^{-6}	5.5×10^3

[a] Transformation frequency represents the average cell titer on TB+Amp$_{50}$ plates per average cell titer on TB plates.

The 594[pcIpR-*P*-timm] cells complement for P, at 30 °C, and very marginally (in comparison to the Δ*P* plasmid) complement at 25 °C (Table 4), suggesting a low level of leaky *P* expression, even when CI[Ts] is active.

Table 4. Complementation of λ *Pam* mutation by plasmid cloned *P* gene.

Plating host	Incubation temperature	EOP [a] of *λimm434Pam3*
594[pcIpR-*P*-timm]	25 °C	0.0008
	30	0.82
	37	0.93
594[pcIpR-*P*$^{\Delta76}$-timm]	30	<0.00003
	37	2×10^{-7}

[a] EOP of 1.0 determined on host TC600 *supE* at 30 °C.

The noninduced P-SPA fusion complements for P in cells grown at 30 and 37 °C (data not shown). 594[pcIpR-$P^{\Delta76}$-timm] transformants were unable to complement for P, suggesting that the region of P removed by the in-frame deletion of 76 codons, localized to the N-teminal end of the 233 amino acid P protein (fusing part of codon 9 with 86) is essential for the role of P in λ replication. The immune state, *i.e.*, *imm*λ[Ts] phenotype, of the pcIpR-[ ]-timm plasmids was examined (Table 5, and [36]) by plating λ*cI*72 on hosts with plasmids encoding intact *P* or *O* genes.

Table 5. EOP [a] of λ*cI*72 on host strains with *imm*λ[TS] or hybrid *imm* plasmids [b].

Host strains	Plating temperature (°C)					
	25	**30**	**35**	**37**	**39**	**42**
594	0.5	0.9	0.9	1.0	1.0	0.4
594[pcIpR-*P*-timm]	$<4 \times 10^{-10}$	$<4 \times 10^{-10}$	0.2	0.5	0.7	1.0
594[pcIpR-*O*-timm]	$<4 \times 10^{-10}$	$<4 \times 10^{-10}$	$<4 \times 10^{-10}$	9×10^{-7} [c]	2×10^{-6} [c]	0.8
594[p434'pR-*O*-timm]	0.8	0.9	1.0	1.0	1.0	1.0

[a] EOP (efficiency of plating) of 1.0 determined for λ*imm*λ*cI*72 on host 594 at 37 °C. [b] The CI[Ts] repressor made from the pcIpR plasmids blocks the vegetative growth of an infecting *imm*λ phage as λ*cI*72, at or below ~39 °C, as well as the expression of genes *O* or *P*, inserted downstream of the *pR* promoter. Plasmid p434'pR-*O*-timm has a hybrid *cI* gene fusion made from λ and 434 *cI* genes and the resulting repressor does not repress transcription from *pR*, allowing constitutive *O* expression at all temperatures [36]. [c] Rare min plaques can only be seen under stereo microscope.

The plasmid control p434'pR-*O*-timm with a hybrid λ/434 CI repressor does not block *pR* transcription, expresses *O* constitutively at all temperatures, and showed no immunity to an infecting *imm*λ phage. Plasmid pcIpR-*O*-timm confers high *imm*λ immunity in cells grown between 25 and 39 °C, and loses it at 42 °C where the CI[Ts]857 repressor is fully inactive. Additional immune characterization of these plasmids was reported (Table 4 in [36]). Although pcIpR-*O*-timm only carries *oR* operators for repressor binding, CI immune repression is maintained through 39 °C, but lost at 42 °C where the CI repressor is denatured. Thus, the pcIpR-(GOI)-timm plasmids, even though they carry only *oR* site, can maintain CI immune repression, and presumably block the expression of GOI positioned downstream of *pR*, between 25-39 °C. However, the analogous plasmid pcIpR-*P*-timm exhibits *imm*λ immunity at 25-30 °C but the immune response is ineffective at and above 35 °C. Phage λ*cI*72 forms tiny to small plaques on 594[pcIpR-*P*-timm] cells at 35 and 36 °C where the CI[Ts]857 repressor is somewhat active [35], and larger plaques at 42 °C where the repressor is inactive (Figure 2).

Figure 2. Plaque type / formation by λc*I*72 on 594[pcIpR-*P*-timm] and 594 host cells. The ability of λc*I*72 to form plaques on 594[pcIpR-*P*-timm] host cells at 35-42 °C is attributed to loss, or considerable reduction, in plasmid copy number per cell. The *imm*λ interference phenotype (compare EOP, Table 5, 25 and 30 °C *vs.* 35 °C) is dependent upon CI[Ts] repressor expressed from the plasmid. The cellular loss of *imm*λ interference correlates with plasmid loss and the observed increase in plaque size between 35 and 42 °C; whereas, the plaque size was essentially constant between 30 to 42 °C when λc*I*72 was plated on 594 cells without the plasmid. The photos of the individual plaques shown in the top row were taken through the lens of a stereo microscope, from agar overlay plates (middle row) incubated at the indicated temperatures.

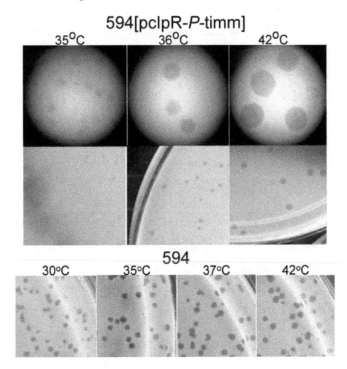

2.1.1. "Trans-activation" of P

The gene *P* function of an *imm*λ prophage was previously shown to be *trans*-activatable [37,38], *i.e.*, cells with a repressed λ prophage could fully complement and support the vegetative growth of an infecting *imm*434 phage defective for *P* (by a factor of 10^3), even though the expression of the prophage copy of *P* was repressed by the *imm*λ CI repressor blocking *pR* transcription in prophage with both *oL* and *oR* sites. Thomas [38] argued that "a small but significant amount of P product was synthesized by the prophage", which he rationalized as being explained by the existence of a minor constitutive promoter mapping between *pR* and *P* on the prophage genome. We saw *trans*-activation of *P* from repressed pcIpR-*P*-timm (slightly at 25 °C and nearly full complementation at 30 °C), which is explained by leaky transcription from *pR*, since there is no room for an unaccounted promoter upstream of *P* as only 18 bases, including the ribosomal binding site, separates the start site for *pR* mRNA transcription and the AUG for *P*. We conclude that trace, certainly not full, *P* expression is sufficient for *ori*λ dependent replication initiation.

2.2. Influence of P-Expression on Plasmid Retention.

The *in*ability of plasmids expressing *P* to form stable transformants is used as a measure of "P-lethality", based on an assumption that P made from the transforming plasmid kills the transformed cells. The influence of *P* expression from pcIpR-*P*-timm in 594 *dnaB*⁺ cells was examined for plasmid replication / maintenance / stability (Figure 3). Plasmid loss was observed in cultures of 594[pcIpR-*P*-timm] grown at 34 or 35 °C (in comparison to cultures grown at 30 °C), but curing was extensive in cell cultures grown at or above 36 °C (Figure 3B). Essentially identical results were seen for 594 *clpP*[pcIpR-*P*-timm] cultures (Figure 3B) showing that inactivation of the *clpP* protease, which participates along with ClpX in degrading λ O protein [39,40], did not significantly influence P-dependent plasmid loss relative to the *clpP*⁺ strain 594; the ClpP protease defect did not amplify plasmid loss, which is expected if its absence increased the P concentration per cell. Secondary cultures prepared from 594 *grpD55*[pcIpR-*P*-timm], or 594[pcIpR-*P*$^{\Delta76}$-timm] cells retained their plasmids, even when grown at 42 °C for at least seven doublings while constitutively expressing *P* (Figure 3C). The *grpD55* allele blocks λ vegetative growth (Table 2) and permits cell survival (at some level) in the presence of constitutive *P* expression at 42 °C, suggesting that allelic variations in DnaB *can* protect cells from *P* expression and its consequent negative influence on cell metabolism. This result is in disagreement with the hypothesis made by Maiti *et al.* [24], which in essence states that allelic variants of *dnaB* will not protect cells from P-lethality.

Figure 3. *P*-induced plasmid loss. (**A**) Cultures of 594[pcIpR-*P*-timm], 594 *clpP*::kan[pcIpR-*P*-timm], 594 *grpD55*[pcIpR-*P*-timm], and 594[pcIpR-*P*Δ76-timm] were grown to stationary phase in TB plus 50 μg/ml ampicillin for 48 hr at 25 °C. Cell aliquots were diluted into fresh TB medium (no ampicillin) as shown in the outline (**A**) and incubated for about 20 hr in shaking bath between 30 to 42 °C (refer to Experimental Section *3.8*). (**B,C**) Plasmid retention by culture cells (described in (**A**)) grown between 30 to 42 °C.

A.

Inoculate broth to 0.008-0.01 A575nm from 25°C culture → Grow at indicated temp. to saturation → Dilute, spread for single colonies, incubate 25°C

↓ Extract plasmid DNA   ↓ Assay colonies, 25°C (Amp^R cfu / total cfu)

B. pcIpR-*P*-timm retention in culture cells

Culture growth temp. (°C)	% cells retaining plasmid (Amp^R cfu / total cfu)	
	594[pcIpR-*P*-timm]	594 *clpP*[pcIpR-*P*-timm]
30	100 (375/375)	100 (126/126)
34	71.4 (30/42)	77.5 (45/48)
35	77.9 (282/362)	51.4 (37/72)
36	0 (0/50)	0 (0/58)
37	0.3 (1/362)	0 (0/72)
42	0 (0/56)	nd

Figure 3. *Cont.*

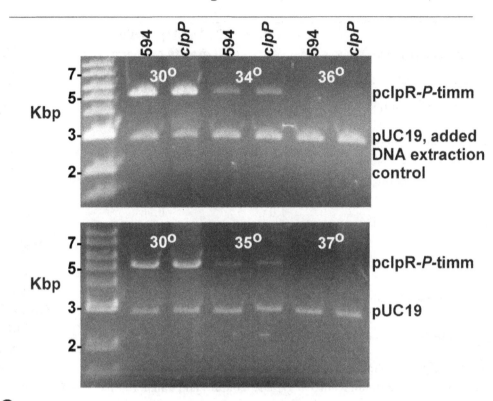

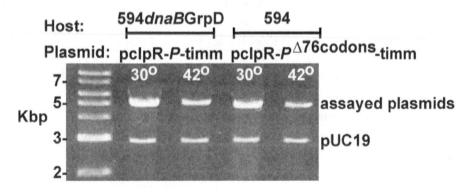

C. Plasmid retention in culture cells

2.3. P-Lethality Suppression

The results in Figure 3 do not account for P-lethality, only plasmid loss, and so their interconnection was examined (Table 6). Whereas cells within a starting culture that become cured of a plasmid expressing P can grow and eventually fully populate a culture, isolating individual cells by spreading dilutions on an agar plate will permit determination of cell viability at the incubation temperature. Less than 1% viability was observed for spread 594[pcIpR-*P*-timm] cells incubated at or above 37 °C, and any surviving cfu had lost the plasmid. *P* expression was not lethal in spread 594 *grpD55*[pcIpR-*P*-timm] cells incubated between 37 and 42 °C, where the cfu arising at or above 37 °C retained the plasmid. We examined whether the plasmids extracted from these survivor cfu retained P-lethality and plasmid-loss properties, when moved back into 594 *dnaB*⁺ cells. The reclaimed plasmids retained both the P-lethality and plasmid loss phenotypes (bottom lines, Table 6).

Table 6. Influence of modifying P or $dnaB$ on P lethality and plasmid loss.

Host cells and plasmids	Cell viability [a] and (plasmid retention/cfu; %) [b] at growth temperature (°C)		
	25	**Ave 37 & 39**	**42**
594[pcIpR-P-timm]	1.0 (195/196; ~100)	0.0049 (0/57; 0)	0.001 (0/28; 0)
594[pcIpR-P^{SPA}-timm]	1.0 (14/14; 100)	0.006 (0/150; 0)	0.004 (0/39; 0)
594[pcIpR-P^{π}-timm]	1.0 (103/103; 100)	0.635 (789/793; ~100)	0.003 (0/294; 0)
594[pcIpR-$P^{\Delta 76}$-timm]	1.0 (166/168; 99)	0.96 (334/336; 99)	0.12 (166/168; 99)
594 $grpD55$ [pcIpR-P-timm]	1.0 (35/35; 100)	0.99 (87/88; 99)	1.0 (14/14; 100)
594 $grpD55$ [pcIpR-P^{SPA}-timm]	1.0 (42/42; 100)	0.75 (88/94; 94)	0.57 (18/19; 95)
594 $grpD55$ [pcIpR-P^{π}-timm]	1.0 (54/54; 100)	0.08 (434/435; ~100)	0.01 (5/108; 5)
594 $grpD55$ [pcIpR-$P^{\Delta 76}$-timm]	1.0 (nd)	0.45 (nd)	0.001 (nd)
Re-claim pcIpR-P-timm from 594 $grpD55$ cultures and transform into 594 $dnaB^+$ cells			
Re-claim from 25°C cultures [c]	1.0 (54/56; 96)	0.08 (0/119; 0)	0.01 (0/91; 0)
Re-claim from 42°C cultures [d]	1.0 (130/130; 96)	0.059 (1/230; 0.4)	0.026 (0/201; 0)

[a] The cell viability shown in each column entry, in top line, was determined by dividing the cell titer obtained at each given incubation temperature by the cell titer at 25 °C.) Refer to Experimental Section *3.5*. [b] The values in parentheses in each column entry show the number of Amp[R] cfu / number of survivor cfu assayed per indicated temperature; the value following represents the percentage of Amp[R] cfu with plasmids, and was rounded up. All data show the results for two or more independently transformed single colonies. [c] Exp.'s for sc1, sc2. [d] Exp.'s for sc's 3,4,5, 6.

A stark contrast was noted between the O and P versions of pcIpR-timm plasmids regarding their ability to prevent the plating of $imm\lambda$ phage $\lambda cI72$. The O-version retained the $imm\lambda$ phenotype between 25 and 39 °C, losing it when the CI[Ts] repressor became denatured; the P-version lost it between 30 and 35 °C and supported plaque formation on the cell lawns at 35 and 36 °C. The simplest explanation is that the P-version plasmid was lost from the cells grown above 30 °C, and coordinately, cI repressor gene expression was lost. Gel analysis showed that 100% of cfu from cultures grown at 30 °C retained the P-version plasmid, but cultures grown at 34 and 35 °C were reduced in plasmid copies, and cultures grown at or above 36 °C were fully cured. Since cultures grown at 30 °C could complement for P, plasmid maintenance was very sensitive to trace P expression, which apparently increased with an increase in culture temperature from 30 to 36 °C. In support of this claim, we obtained 1.5×10^5 transformants per μg of pcIpR-P-timm DNA at 25 °C and none at 37 °C. The partial to full loss between 34–37 °C (Figure 3) of *P*-encoded plasmids, argues for trace levels of P causing ColE1 plasmid curing. This result suggests a new dimension to the concept of P-lethality. It appears that ColE1 plasmid replication is extremely sensitive to P$^\lambda$ protein and that P-inhibition of plasmid replication prevents plasmid establishment upon transformation. This provides a better explanation for the lack of transformants (found by us and other authors) by plasmids capable of expressing P than does cell killing by P (*trans* P lethality).

P$^\pi$ proteins exhibit lower affinity for DnaB, compared to P [41], which may explain why elevated P expression at 42 °C was needed to evoke the normal P phenotypes in the *grpD55* strain in order for it to compete for and sequester DnaB away from DnaC [11], effectively creating the possibility for a slow-stop type inhibition in *E. coli* replication initiation. A cautionary note here is that the dissociation of DnaB-DnaC complexes by P was only demonstrated *in vitro*.

As well as interacting with λ P [42], DnaB interacts with many cellular proteins, e., g., with DnaA [43], DnaC [44], DnaG [45], SSB [46], the Tau subunit of DNA polymerase III [47], and RNA polymerase [48], and as noted the functional DnaB hexamer is present in limiting amounts, *i.e.*, about 20 hexamers per cell [14,15]. Introducing two missense mutations into *dnaB* will permit the cells to survive constitutive *P* expression (P-lethality), and can suppress P-inhibition of ColE1 replication, yet these mutations do not overtly impede cellular growth/replication. A single missense mutation in *P* can suppress P-lethality and plasmid loss at 37 to 39 °C. Therefore, we suggest that an interaction between P and DnaB is an important component of both P-lethality and P-inhibition of ColE1 replication.

Datta *et al.*, [28,29] could transform *dnaA rpl* mutants (resistance to *P* lethality) of the same strain we employed, namely 594, with plasmids expressing P, concluding that an interaction between P and DnaA, and not with DnaB, was important for suppressing P-lethality. Their *dnaA*-rpl8 [29] mutant did not inhibit λ plating [49], and thus has a different phenotype from *E. coli* mutants arising in selections where λ development is inhibited, such as the one used by Saito and Uchida [32]. Sequence analysis revealed that our 594 strain is wild type for *dnaA* [34] and therefore, the *grpD55* mutation in *dnaB* appears sufficient for cellular resistance to P lethality; however, we have not compared the two 594 strains by whole genome sequence analysis. We suggest that these possibly contradictory results are explained by the participation of DnaA in mechanistic events shared with DnaB, allowing the *dnaA-rpl* allele(s) to modulate DnaB:P dependent P-lethality and P-inhibition. Alternatively, the reported effect of P protein on DnaA [28] is not at the level of DNA replication, but is in the context of transcription of genes regulated by DnaA.

2.3.1. Model for Plasmid Loss

How can P evoke plasmid loss? The expression and accumulation of lambda P protein, even in seemingly trace levels, interferes with ColE1-type plasmid maintenance replication (Figure 4).

Figure 4. The expression and accumulation of lambda P protein, even in trace levels, interferes with ColE1-type plasmid maintenance replication.

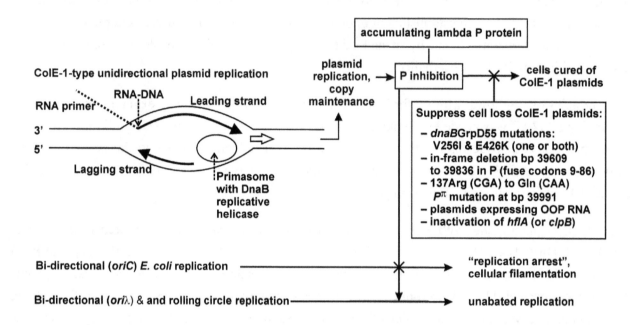

Although ColE1 is considered a theta-type replicating plasmid [50], at an early stage, leading-strand synthesis proceeds in the absence of lagging-strand synthesis, yielding unidirectional replication [51,52], and requires DNA polymerase I (Pol I) to initiate continuous leading-strand replication [53–55] (shown by open arrow, Figure 4). Replication initiation requires primer RNA II for leading strand synthesis [52,56], stable hybridization of RNA II to DNA [57,58], and processing by RNase H to generate the primer for leading strand synthesis [57,59–61]. Pol I extension of the primer unwinds the DNA exposing a primasome assembly signal (n' pas) or single-strand initiation A sequence (*ssiA*) that allows assembly of the primosome through recruitment and activation of the PriA protein [62]. *This represents a distinct form of DNA replication initiation associated with DNA repair*, in contrast to DnaA-dependent replication initiation at *oriC* [63,64]. Following PriA-primosome assembly, DnaB helicase and DnaG primase are loaded and work coordinately to initiate the discontinuous priming of lagging-strand [58,65], which opens a sequence for termination of lagging-strand synthesis, *terH*, effectively establishing unidirectional replication [53,66]. There is a *dnaA* box that is close to n' pas, and potentially serves as a DnaA-dependent DnaB-DnaC assembly site [67,68]. The simplest model for ColE1 plasmid replication inhibition and P-dependent plasmid curing is that formation of a cellular P:DnaB complex limits the availability, or perturbs the activity of DnaB to participate in ColE1 replication initiation, perhaps in a step very similar to replication restart of a stalled replication fork, which can involve DnaA. Replication restart proteins form multiple pathways to restart repaired replication forks (Figure 1 in [69]) and they function equivalently to DnaA in forming protein-DNA complexes so that DnaB helicase is loaded onto the DNA [70,71]. Of note, using *in vitro* ΦX174 DNA synthesis assays to monitor DnaB participation, Wickner [2] observed that P inhibited DnaB protein activity in ΦX replication, making a similar proposal to ours for P inhibition of ColE1 replication.

2.4. Influence of P-alteration, or O and OOP RNA Co-Expression on P-lethality and Plasmid Retention

Linking P with a 66 amino acid SPA tag [72] at its C-terminal end did not suppress P lethality, nor plasmid loss, but the lethality of P^{SPA} strongly decreased and plasmid retention sharply increased in the *grpD55* host at 37–42 °C (Table 6). The 76 codon in frame N-terminal deletion, $P^{\Delta76}$, suppressed both plasmid loss and P-lethality at 37 or 39 °C. Thus, amino acids 9–86 of P include a critical domain. Its removal knocks out P initiation of replication (our complementation data), the P-lethality phenotype, and P-induced ColE1 plasmid loss.

The allele $P^{\pi39991}$ was used to make pcIpR-$P^{\pi39991}$-timm. The mutation was identified in λ*cI72* mutants that were able to form plaques on the 594 *grpD55* host, and is identical to the πA7 mutation previously sequenced [73]. This R137Q single base change partially suppressed P-lethality and fully suppressed plasmid loss at 37-39 °C (but not at 42 °C where P^{π} was fully expressed) in the *dnaB*$^+$ host. The lethality of P^{π} significantly increased at 37-39 °C in the *grpD55* host even though plasmid retention remained near 100%. Cells with pcIpR-$P^{\pi39991}$-timm (or pcIpR-P^{SPA}-timm) could complement the amber mutation in λ*imm434Pam3*.

We asked if co-expression of *O-P*, or if the introduction into the plasmid of the coding sequence for the CII antisense micro-RNA termed OOP, was able to suppress P-lethality and plasmid loss (Table 7).

Table 7. Influence of *O*, *P* and OOP-RNA expression on cell killing and plasmid loss [a].

Plasmids in 594 host cells	Cell viability and (plasmid retention/cfu; %) at growth temperature (°C)				
	30	35	37	39	42
pcIpR-*P*-timm	1.0 (35/35; 100)	0.32 (33/33; 100)	0.01 (0/35; 0)	0.008 (0/35; 0)	0.07 (0/35; 0)
pcIpR-*O-P*-timm	1.0 (62/70; 89)	0.345 (28/70; 40)	0.12 (0/70; 0)	0.12 (0/70; 0)	Nd
pcIpR-*O-36P*-timm	1.0 (30/30; 100)	0.825 (30/30; 100)	0.793 (30/30; 100)	0.012 (1/30; 3)	Nd
pcIpR-*O-63P*-timm	1.0 30/30; 100)	0.895 (30/30; 100)	0.895 (30/30; 100)	0.055 (14/40; 35)	Nd
pcIpR-*O*-timm	1.0 (30/30; 100)	1.0 (30/30; 100)	1.0 (30/30;100)	1.0 (30/30; 100)	0.608 (29/30; 97)
p434'pR-*O*-timm [b]	1.0 (26/30; 87)	1.0 (29/30; 97)	1.0 (28/30; 93)	1.0 (26/30; 87)	0.615 (26/30; 87)
pcIpR-*oop#1-O-P*-timm [c]	1.0 (120/120; 100)	0.938 (120/120; 100)	0.20 (98/101; 97)	0.005 (115/120; 96)	Nd
pcIpR-*oop#2 O-P*-timm [c]	1.0 (117/120; 98)	0.988 (60/60; 100)	0.055 (76/154; 49)	0.048 (62/120; 52)	Nd

[a] As described for Table 6, except stationary phase cultures were grown up at 30 and not 25 °C. [b] Plasmid has constitutive expression of *O*. [c] The viability results represent the average for four independent plasmid isolates, each single experiments. The results in parentheses, for plasmid retention by survivor cfu, sums results for all the cfu's assayed from the four isolates. The cell viabilities at 30 and 35 °C were at or very near unity. pcIpR-*oop#1-O-P*-timm: 37°C (0.34, 0.49, 0.006, 0.008), 39°C (0.01, 0.006, 0.0006, 0.0015); pcIpR-*oop#2-O-P*-timm: 37°C (0.01, 0.01, 0.12, 0.08), and 39°C (0.01, 0.01, 0.09, 0.08).

The co-expression of *O* would provide an interactive partner for P, possibly lowering the cellular level of unbound P. The 77 nucleotide antisense OOP RNA, which is transcribed from promoter *pO* in the opposite orientation to the *pR-O-P* mRNA (Figure 1A), can hybridize between *pR* and *O*, forming an OOP-RNA: *pR-O-P*-mRNA hybrid that can provide a target for *P* mRNA degradation downstream from the OOP-binding site. The co-expression of *O-P* increased cell viability at 37–39 °C, but not plasmid retention. The constitutive or inducible expression of *O* without *P* did not influence cell viability or plasmid loss between 30–39 °C. Early studies [74] attempting to determine the coding sequence for *O*, reported a longer protein (O') made by read-through of the normal stop codon, terminating downstream somewhere within the N-terminal sequence of *P*. Plasmids were made that included the wild type λ sequence between the start of *O* through the UGA stop codon [75], plus 35 or 62 in-frame codons downstream into the *P* DNA sequence, so that if the stop codon for *O* was designated "1", the next in-frame stop codons would occur at positions 36 or 63 in frame with *O*, even though the downstream reading frame for *P* differs from *O*. While the plasmids expressing the intact *O* sequence were not toxic, the plasmids with O-36P or O-63P, when expressed at 39 °C, were toxic and caused plasmid loss. It is unclear if their toxicity depends on O-read-through downstream of its stop codon, or to polypeptides made from the N-terminal end of *P*. Plasmids pcIpR-*oop#1-O-P*-timm and pcIpR-*oop#2-O-P*-timm (Figure 1A, Table 10), when compared to versions expressing only *P* or *O-P* significantly suppressed plasmid loss at 37–39 °C, and partially suppressed P-lethality at 37 °C, with the *oop#1* plasmid showing stronger suppressing effects than the slightly larger #2 version. This data supports a hypothesis that *P* expression from the λ genome is reduced by OOP antisense micro RNA, possibly via a OOP:*pR*-mRNA complex serving as a target for mRNA degradation.

We observed that the co-expression of *O* and *P*, with the potential for formation of an O:P complex, did not suppress the P phenotypes. In contrast, placing the natural *oop-O-P* sequence orientation downstream from *pR* in the pcIpR-[ ]-timm plasmids markedly suppressed P-inhibition of ColE1 replication as monitored by assaying for plasmid curing. This is explained, possibly, by OOP micro-RNA serving in some manner as a competitor to RNA I made by the plasmid, resulting in the loss of plasmid copy control and higher plasmid copies per cell. We view this explanation as unlikely. Alternatively, OOP RNA may limit the adverse effects of *P* expression. It can bind to *pR*-promoted *cII* mRNA and target its degradation in an RNaseIII-dependent reaction [76,77], suggested to limit downstream *O-P* gene expression.

What is the possibility that inefficient termination of *pR*-GOI transcription and consequent run-on transcription could cause problems with plasmid maintenance and explain the loss of pcIpR-*P*-timm from cells, rather than a direct effect of P on plasmid maintenance? We have observed that the wild type *timm* terminator prevents read through from the *cI-rex* transcript into the *oL* operator region or through into *N* when CII stimulated transcription from *pE* occurs at 30 to 100-fold the level of *cI-rex* transcription from the *cI* maintence promoter *pM* (see [17] and contained references). This was the rationale for incorporating *timm* into the design of pcIpR-(GOI)-timm plasmids. Table 7 shows the influence of expression of lambda gene *O* from plasmids pcIpR-*O*-timm and p434'pR-*O*-timm on plasmid loss at 30, 35, 37, 39 and 42 °C. The immunity properties of these two plasmids was shown in Table 4 of [36]. The expression of gene O is constitutive at all temperatures from plasmid p434'pR-*O*-timm. It can be seen that 100% cellular maintenance / retention occurred for plasmid pcIpR-*O*-timm between 30–39 °C and 97% at 42 °C. Even with constitutive expression of *O* from p434'pR-*O*-timm over the range of cell growth temperatures, between 87%–97% of the cells retained the plasmid. A similar result is also seen in Table 6 for plasmid pcIpR-*P*$^{\Delta76}$-timm, where 99% of the cells grown between 25 and 42 °C retained the plasmid. These results strongly support an argument that *timm* in pcIpR-(GOI)-timm is likely a powerful terminator that prevents inefficient termination and consequent run-on transcription, which might otherwise cause problems with plasmid maintenance.

2.5. Replicative (cis) killing; P (trans) Lethality / Inhibition

We compared the kinetics of cell death resulting from *P* expression in cells with pcIpR-*P*-timm, *i.e.*, P-lethality (Table 8A) with Replicative Killing, resulting from initiating *ori*λ replication from the Y836 chromosome (Figure 1C) or in cells where this fragment was moved by transduction (Table 8B). Shifting these cell cultures to 42 °C denatures the reversibly-denaturable CI[Ts]857 repressor [17,78] resulting in transcription from promoter *pR*, and *P* expression from the plasmid, or *cro-cII-O-P-ren* expression from the chromosomally inserted λ fragment. The latter event results in the initiation of bidirectional replication from *ori*λ. After the period of gene de-repression the cells were swirled in an ice bath and then plated for survivors at 30 °C, which serves to shut off transcription from *pR* via the renaturation of CI, which, in turn stimulates *cI* expression.

Table 8. Contrasting *trans* P-lethality/inhibition and *cis* Replicative Killing.

A.	Cells tolerate short-term exposure to P (short-term interference, not lethality) [a]		
Strain with plasmid	Incubation at 37 °C	Cell viability (Amp^R cfu / total cfu)	Outcome of P expression from plasmid
594[pcIpR-*P*-timm]	1 h	0.87 (232/242)	most cells recover
	2 h	0.49 (258/281)	many cells recover
	6 h	0.13 (2/173)	high plasmid loss
B.	Inducing replication from a trapped cryptic prophage causes Replicative Killing [a]		
Strains with cryptic prophages	Prophage Induction time	Cell viability	Outcome of prophage Induction
Y836[~*cIII-cI857-O-P-ren*]	5 min	0.33	rapid cell killing
	20 min	0.13	rapid cell killing
	3 h	0.00018	extensive cell killing
	5 h	0.00008	extensive cell killing
594[~*cIII-cI857-O-P-ren*]	5 min	0.55	rapid cell killing
	20 min	0.25	rapid cell killing
	3 h	0.0022	extensive cell killing
	5 h	0.0022	extensive cell killing
Y836[~*cIII-cI857-O-P::kan-ren*]	3 h	5.1 [b]	cell growth
	5 h	6.1 [b]	cell growth

[a] Refer to Experimental Section *3.6*. [b] The increase in viability by for example 5.1 indicates somewhat more than two and less than three cell doublings.

The λ fragment is not excised from the chromosome of strain Y836 or from the (*cIII-ren*)λ-transduced 594 variant since each inserted λ fragment lacks the genes for *int-xis-kil* and is deleted for both *attL* and *attR* sites. The *ori*λ initiation event from these integrated λ fragments exerts a rapid *cis*-active Replicative Killing effect, likely causing interference with *E. coli* replication fork progression. In *cis*-killing 75%–87% of the cells die (*i.e.*, are not rescued upon spreading / incubating the cells at 30 °C) by 20 min after prophage induction (Table 8B). But the *trans*-P lethality/inhibition effect is reversible, to a point, and much slower as only 14% of the cells are killed after 1 h (55% after 2 h) of *P* expression, even though >90% of the cells retained the plasmid (Table 8A).

2.5. 1. Contrasting *trans* P-lethality / inhibition, and *cis* Replicative Killing

We showed that shifting culture cells transformed with pcIpR-*P*-timm and grown at 25 °C up to 37 °C for two hrs reduced cell viability to about half, and thus cells are capable of tolerating or metabolizing some level of P, with a reported half-life of up to an hr [79,80]. However, initiating irreversible *oriλ* replication from a trapped cryptic prophage, Figure 1C, reduced cell viability between 4 and 10-fold after only 20 min of induction. The Replicative Killing (RK$^+$ phenotype) survivor frequency of 10^{-6} to 10^{-8} is distinguished by rare RK$^-$ mutants in host or prophage (not merely in *P*) that block some aspect of λ replication initiation [78,81–83]. We suggest the terms *cis* Replicative Killing and *trans* P lethality to distinguish mechanistically these the two ideas. The RK$^+$ phenotype seems uniquely dependent upon multiple, non-repairable λ replication forks arising from *oriλ* (drawn in Figure 1C), but there are likely multiple possibilities for P-lethality. Several experiments suggest that the encoded gene *ren* downstream of *P* is not responsible for P-lethality. A modified version of pHB30 (*i.e.*, pHB31) encoding *P-ren,* but deleted for bases 39609–39836 within *P* (equivalent to $P^{\Delta 76}$ used herein) was fully capable of transforming 594 cells at 42 °C, suggesting that P, not Ren blocked transformation at 42 °C [31]. RK$^-$ mutants of Y836 with insertions or deletions (recombinant or natural) within/inactivating *O* (ilr208b, ilr223a, ilr541c, ilr200b, ilr203b, ilr207d, ilr201b), or *P* (ilr566a, Bib11t), yet which sequenced to be *ren$^+$*, lost the Replicative Killing competence phenotype when shifted from 30 to 42 °C, suggesting that replication initiation from *oriλ*, and not *ren* expression, is responsible for Replicative Killing; transduction of the λ fragments into 594 did not alter the phenotype observed for Y836 [RK$^+$], or the RK$^-$ mutants. Ren is not required for P-lethality phenotype, nor for Replicative Killing, *i.e.*, in the absence of *oriλ* replication initiation, *ren* expression will not produce the RK$^+$ phenotype, however, our experiments do not rule out some ancillary role for Ren.

Several examples reported suggest that P-lethality is a separate mechanism from P-inhibition of ColE1 replication: i) *grpD55* cells with pcIpR-P^π-timm were killed at 37–39 °C, yet 100% of the surviving cells retained the plasmid; ii) most of the cells with pcIpR-*oop-O-P*-timm plasmids retained the plasmids at 39 °C while suffering high P-lethality. Additionally (see Figure 4), cells defective in host protease genes *clpB* or *hflA* were sensitive to P lethality at 37–39 °C, yet the majority of the survivors retained the pcIpR-*P*-timm plasmid (data not shown; available from authors).

Is P-inhibition due only to a decrease in available DnaB, or does it influence / perturb ongoing replication, or replication restart? Our results do not indicate if P fully sequesters DnaB or if a single P monomer bound to DnaB is sufficient. Moreover, the genetic evidence does not show whether this interaction is persistent or transient. There seems little doubt that competing DnaB away from DnaC is an important function of P; but the longer term consequence is that the interaction of P and DnaB yields a P-DnaB-ATP dead-end ternary complex [10], and the only known role for P is in early *oriλ*-dependent initiation events, of which only a few are required. We propose that the full sequestration of DnaB by P is not necessary in order for P to interfere with ongoing plasmid replication, because of the sensitivity of ColE1 replication / copy maintenance to leaky (sub-induction) levels of P (arising from a plasmid encoding *P*). The simple explanation for plasmid loss is that while the cells continue to replicate and divide, plasmid replication is differentially inhibited. For example, *grpD55* cells very poorly support vegetative growth of λ at 30 °C, and not at all between 37–42 °C, whereas cellular growth is not noticeably perturbed. A low level of P, tolerated by cells, could fully inhibit ColE1 replication, in turn reducing plasmid copy number with each cell division, which could explain plasmid curing for cells grown between 34–36 °C. Less clear is how cells with pcIpR-*P*-timm grown at 30 °C are cured when shifted to 37–42 °C, where division is inhibited and cells elongate forming filaments (Sect 2.8.).

2.6. Does P Expression from pcIpR-P-timm perturb λ Vegetative Growth?

As previously noted, although present in very limited amounts in the cell, DnaB is a multifunctional protein involved in replication fork movement [3], serving as a mobile promoter in priming reactions [4,5], in the progression of Holliday junctions [6,84], and in replication restart reactions [70,71], some of which are likely involved in λ replication beyond the *oriλ*-O-P-dependent initiation step. Since the interaction of P and DnaB yields a P-DnaB-ATP dead-end ternary complex [10], we wondered if the expression of P from fully induced pcIpR-*P*-timm in cells shifted to 42 °C could serve to limit λ replication, as monitored by phage burst. Lysogenic cultures with a λ*cI*[Ts]857 *S*am7 prophage defective for natural cell lysis (because of the nonsense mutation in *S*) were synchronously induced by shifting the cells from 30 to 42 °C. One of the parallel cultures contained pcIpR-*P*-timm. Following induction, the cells were artificially lysed. Relative phage burst from each of the cultures was determined by dividing the released phage titer by the cell titer pre-induction. The cells with pcIpR-*P*-timm showed an increased relative burst by 60 min post induction over parallel cells without the plasmid (*i.e.*, bursts of 248 (cells with pcIpR-*P*-timm) and 42). Clearly, λ replication/maturation is not curtailed (and appears enhanced) by combined expression of *P* from both induced prophage and plasmid, which is dramatically opposite to the effect of *P* expression on ColE1 plasmid replication/maintenance.

2.6.1. λ Replication and Phage Maturation

Our results suggest that P is not inhibitory to λ replication, which, considering its influence on *E. coli* and ColE1 replication, raises some interesting questions. Doesn't λ need to deal with replication restart, likely one of several possible targets for P-DnaB interaction, or does it have an alternative mechanism? Does P or a P-complex have an unrecognized DNA helicase activity? While some *P* analogues of lambdoid-type phages encode their own DNA helicase activity (see suppl. Figure S2 in [36], and [85] for comparison of 20 *P*-like genes in lambdoid phages), P is not recognized as having this property and those P-like proteins with putative helicase activities are larger than λ *P*.

2.7. P-Induced Cellular Filamentation

We induced *P* expression from plasmids pHB30 and pcIpR-*P*-timm and followed filament formation (Table 9) in SOS⁺ strain 594, and in SOS-defective variants [86] of 594 made by transducing in the alleles of *lexA*3[Ind⁻], where the alanine-glycine protease cleavage site [87] is changed to alanine-aspartate [88] (Figure 5, Table 9), or Δ*recA* (Table 9)

Figure 5. Cellular filamentation resulting from induced P-expression. The photos are representative of data shown for one of the three sets of photos per strain per assay condition that were used for cell measurements in Table 9.

SOS-independent P-induced filamentation

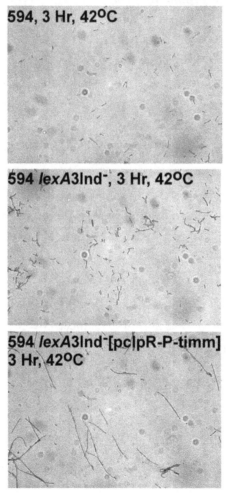

The results in Table 9 suggest that *P* expression negatively impacts *E. coli* cell division. The cellular filamentation observed upon shifting 594 *lexA*3[pcIpR-*P*-timm] cells to 37 °C (Table 9) supports the concept (e.g., from Figure 3) that some leaky *P* expression arises from this plasmid at 37 °C in sufficient level to prevent ColE1 plasmid replication maintenance.

Table 9. SOS-independent P-induced cellular filamentation. [a]

Strain [plasmid]	Time (Temp)	Relative cell length										Sum cells
		1X	2X	3X	4X	5X	6X	7X	8X	9X	≥10X	
594	0 (25)	26	19	0	0	0	0	0	0	0	0	45
"	1 (42)	15	11	11	8	0	0	0	0	0	0	45
"	3 (42)	28	15	2	0	0	0	0	0	0	0	45
"	5 (42)	29	14	2	0	0	0	0	0	0	0	45
594 *lexA3*	0 (25)	40	32	2	0	0	0	0	0	0	0	74
"	1 (42)	23	16	6	0	0	0	0	0	0	0	45
"	3 (42)	31	13	1	0	0	0	0	0	0	0	45
"	5 (42)	42	3	0	0	0	0	0	0	0	0	45
594 *lexA3*[pcIpR-*P*-timm]	0 (25)	71	19	0	0	0	0	0	0	0	0	90
"	1 (37)	21	23	1	0	0	0	0	0	0	0	45
"	3 (37)	7	9	9	6	4	3	2	1	2	2	45
"	5 (37)	6	20	0	0	2	0	3	3	3	8	45
594 *lexA3*[pcIpR-*P*-timm]	1 (42)	9	19	13	3	1	0	0	0	0	0	45
"	3 (42)	11	1	3	4	4	9	8	2	2	1	45
"	5 (42)	8	6	7	1	1	0	2	1	2	14	42
594 [pHB30]	0 (30)	6	20	4	0	0	0	0	0	0	0	30
"	1 (42)	0	2	10	9	3	4	1	1	0	0	30
"	3 (42)	2	4	8	3	2	5	2	1	2	1	30
"	5 (42)	2	8	10	2	2	2	3	0	0	1	30
594 *lexA* [pHB30]	0 (30)	5	17	8	0	0	0	0	0	0	0	30
"	1 (42)	0	6	6	8	4	1	4	0	1	0	30
"	3 (42)	3	5	2	6	4	0	5	3	1	1	30
"	5 (42)	1	7	5	6	5	2	1	1	0	2	30
594 Δ*recA*	0 (30)	16	12	2	0	0	0	0	0	0	0	30
"	1 (42)	4	12	7	4	1	2	0	0	0	0	30
"	3 (42)	9	16	2	4	2	0	0	0	0	0	30
"	5 (42)	12	11	4	2	0	0	1	0	0	0	30
594 Δ*recA* [pHB30]	0 (30)	8	17	4	1	0	0	0	0	0	0	30
"	1 (42)	2	5	8	10	3	1	1	0	0	0	30
"	3 (42)	5	4	5	3	6	4	2	0	0	1	30
"	5 (42)	3	5	4	8	0	4	2	0	1	3	30

[a] Refer to Experimental Section *3.7.*

2.7. 1. P-Dependent Cellular Filamentation

We previously observed extensive cellular filamentation upon the de-repression of λ*cI*[Ts]857*O*am8 prophage, but not for a λ*cI*[Ts]857*P*am3 prophage in lysogenic cells [31]. Klinkert and Klein [23] demonstrated that expression of *P* from a plasmid can cause cellular filamentation, which they attributed to an ability of P to impair bacterial DNA synthesis. Agents that block cellular DNA synthesis can stimulate a cellular SOS response [86], which in turn elevates expression of one of the SOS response genes, *sulA*, whose product prevents the action of FtsZ at the site of septum formation, causing a cell to stop dividing [89,90], and allows time for DNA repair (see [91]). Cells inhibited for FtsZ activity form filaments that are longer than dividing cells. Induction defective mutations in *lexA* encoding the repressor for the SOS response, or certain *recA* mutations prevent the induction of a cellular SOS response [86].

The hypothesis that P-lethality does not involve an interaction between P and DnaB, but is instead targeted to DnaA, implies that the P:DnaA interaction can prevent *E. coli* replication initiation at *oriC* by removing available DnaA. Previous studies have shown that a *dnaA*46[Ts] mutant exhibits a slow-stop effect on DNA synthesis when the cells are shifted to 42 °C, blocking re-initiation at *oriC*; however, these cells continue to divide, forming anucleate cells, not filaments [92,93]. Other cellular interactions beyond those between DnaA or DnaB and P are important. For example, as noted, P interacts with DnaK and DnaJ. The inactivation of *dnaK*, and apparently *dnaJ*, results in multiple cellular defects, including formation of cell filaments, abnormally segregated chromosomes, and loss of plasmid maintenance [94].

Does P function in replication restart? It was reported [69] that neither *lexA* nor *sulA* deletions abolished filamentation in a *priA sulB* mutant, suggesting models whereby repair/restart of chromosomal replication is essential for completing a round of replication, and that replicative forms are resolved to monomers before cell division can take place. Since DnaB is reloaded onto DNA for replication restart, P accumulation could serve to limit the availability of DnaB to replication restart pathways, or it may directly interact with the elongating replisome, neither previously suggested. The need for replication restart is influenced by collisions between transcription and replication forks. RNA polymerase mutations in *rpoB* and *rpoC* [95,96] facilitate replication progression. Indeed, CH has shown that two of 22 mutants obtained in screens for rifampicin-resistant *E. coli* cfu had significant resistance to P-lethality (remainder did not; Hayes lab, unpublished results). However, among the group of Rif[R] isolates, 30 were localized by DNA sequence analysis to two regions within *rpoB* (because *rpoB* is 4029 bp, we did not sequence the whole gene), but the site of mutation in the two resistant isolates remains undetermined, and the existence of additional suppressor mutations arising in these two mutants remains a possibility. We predict that mutations in *E. coli* which decrease conflicts between transcription and replication would make cells more resistant to P, and mutations which increase conflicts would make cells more sensitive.

3. Experimental Section

3.1. Strains Employed

The bacteria, plasmids (Figure 1A) and phages employed are listed in Table 10.

Table 10. Bacteria, plasmids and phages employed.

Bacterial strains	Characteristics or genotype	Source/Ref.'; Hayes lab #[a]
594	F⁻ lac-3350 galK2 galT22 rpsL179 IN(rrnD-rrnE)1; see [97]; called R594	[97], SH lab; B10
TC600	supE, Pm⁺	SH lab, B8
Ymel	supF, Pm⁺	SH lab, B71
DE407	lexA3[Ind⁻] malB::Tn9 Tet^R sulA211 sfiA11, UV^S	D. Ennis; B142
FC40 (=SMR624)	Δ(srlR-recA)306::Tn10 Tet^R UV^S	SM Rosenberg [98]; Y921
AB2834 aroE	grpD55, thi tsx^R λ^R at 42 °C from K552	H. Uchida [32,33]; NB83
W3874 malB5	dnaB grpA80 lac⁻ Str^R λ^R at 42 °C	[32], NB81
W3350 dnaB- grpD55	grpD55 malF3089::Tn10 Tet^R λ^R at 42°C, λrepP22^S	[33], NB15
594 dnaB-grpD55	grpD55 allele malF3089::Tn10; Tet^R, λ^R at 42°C, λrepP22^S	[34], NB295
594 lexA3[Ind⁻] malB::Tn9	LexA repressor induction defective	CE, NB293
594 Δ(srlR-recA)306::Tn10	deletion of recA Tet^R UV^S	CE, B318
W3350	F⁻ lac⁻3350 galK2 galT22 rpsL179 IN(rrnD-rrnE)1	SH lab, B12
Y836	SA500(λbio275cI[Ts]857 Δ431) his⁻	[78,82], NY1049
594::nadA::Tn10 [~cIII-ren]^λ	Tn10 [zbh29 at 16.8 min] bio⁺ transductant = 594 bio275 (λcIII-cI[Ts]857-O-P-ren) Δ431	A. Chu, SH lab, NY1057
Y836 P::kan (Bib11t)	SA500 (λbio275 cI[Ts]857 O⁺P::kan Δ431) his⁻ Kan^R	SH, NY1153
594(λcI857Sam7)	λ lysogen defective for cell lysis	C. Marek, SH lab, Y1163
594(λcI857Sam7)[pcIpR-P-timm]	as above with transformed plasmid	SH lab, P509
594 clpP::kan	clpP⁻, Kan^R from SG22159	S. Gottesman; [99], NB276
Plasmids	**Transformed into strain 594**	**Source/Ref.'; Hayes lab #[a]**
pUC19	Wild type Amp^R (New England Biolabs)	NP188
pcIpR-P-timm	BamHI-ClaI PCR fragment from λcI857, replacing D-CAP in P459 with λ bp's 39582-40280	CH, P466
pcIpR-P::kan-timm	PCR BamHI-ClaI fragment from Y836 P::kan (Bib11t) strain NY1153	KM, P510
PcIpR-P-SPA-timm [b]	Replace D in P462 between BamHI and AscI sites with BamHI-P(λ bp's 39582-40280)-AscI PCR fragment	KM, P467
pcIpR-P^Δ76-timm	In-frame deletion76 codons: λbp 39609-39836 in pcIpR-P-timm with HpaI, ligate	KM, P515
pcIpR-P^π-timm	BamHI-ClaI PCR fragment from λcI72π Lysate #3a, replacing D-CAP in P459 with λ bp's 39582-40280	KM, P505

Table 10. *Cont.*

Plasmids	Transformed into strain 594	Source/Ref.'; Hayes lab #[a]
pcIpR-*O*-timm	*Bam*HI-*Cla*I PCR fragment from λcI857, replacing D-CAP in P459 with λ bp's 38686-39582	[36], CH, P465
P434'pR-*O*-timm	Constitutive *O* expression; *Bam*HI-*Cla*I PCR fragment from λcI857, replacing D-CAP in P459 with λ bp's 38686-39582	[36], CH, P494
pcIpR-*O-P*-timm	*Bam*HI-*Cla*I PCR fragment from λcI857, replacing D-CAP in P459 with λ bp's 38686-40280	CH, P569
pcIpR-*O*-36*P*-timm	*Bam*HI-*Cla*I PCR fragment from λcI857, replacing D-CAP in P459 with λ bp's 38686-39687	CH, P565
pcIpR-*O*-63*P*-timm	*Bam*HI-*Cla*I PCR fragment from λcI857, replacing D-CAP in P459 with λ bp's 38686-39768	CH, P566
pcIpR-*oop*#1-*O-P*-timm	*Bam*HI-*Cla*I PCR fragment from λcI857, replacing D-CAP in P459 with λ bp's 38559-40280	CH, P567
pcIpR-*oop*#2-*O-P*-timm	*Bam*HI-*Cla*I PCR fragment from λcI857, replacing D-CAP in P459 with λ bp's 38546-40280	CH, P568
pHB30	λ bases 34499-34696, 36965-38103, 38814-40806 (see Section 3.2.)	[31,34], SH lab, P8
Bacteriophage	**Genotype**	**Hayes lab lysate #**
λ wild type (wt)	λpapa	[78], 944,1001
λ*cI*72	*cI*	[78], 951, 999
λnin5	made from λ wt	[78], CH, 698
λvir	λ*v2v1v3*	[78], 260
λ*cI*857	*cI*[Ts]857	[100], 1002
λ*cI*857Sam7	defective for cell lysis	[101], 963
λ*imm*434*P*am3	*imm*434, sequenced *P*am3 mutation C to T, λ base 39786 (CAG to TAG)	[83], SH lab, 518, 664
λ*imm*434nin5	*imm*434, Δnin5 region, forms very turbid plaques at 37°C	[22], CH, 963

[a] The strain numbers are from the Hayes laboratory collections. All gene inserts within the pcIpR-[ ]-timm plasmids were sequenced to confirm the genetic integrity of the inserted fragment. [b] Plasmid pcIpR-D-SPA-timm [20] (strain P462) was prepared from pcIpR-D-CAP-timm (strain P459), replacing 318 bp CAP from P459 by digestion with *Asc*I and *Cla*I and replacing with 239 bp SPA tag from pMZS3F [72] (from J. Greenblatt) isolated via PCR with primers L-Asc-CBP & R-ClaI-FLAG. SPA is a 66 amino acid tag with 3X FLAG sequences.

3.2. Construction of Expression Vectors for λ Genes, Gel Analysis of Plasmids, Insertion Localization

The plasmids were extracted from cells using Qiagen plasmid mini preps. They were separated by agarose electrophoresis on gels made and run using 1X TBE buffer (10X = 1M Tris, 1M Boric acid, 0.02 M Na$_2$EDTA, pH 8). The precursor plasmid pcIpR was made by degrading pBR322 with *Eco*RI and *Bam*HI and purifying the large *Eco*RI-*bla*-rep/rop-*Bam*HI fragment. This was ligated with an 833 bp PCR fragment derived from λ*cI*857 DNA amplified using primers that added MfeI and *Bam*HI tags to the ends of lambda bases 37203 and 38036 to produce plasmid pcIpR, where the ligation of the MfeI end to the *Eco*RI site removed both sites in the resulting construct. pcIpR was digested with *Bam*HI and SalI and the large fragment of 4817 bp was separated from the 276 bp region between the *Bam*HI and SalI sites. A synthetic DNA sequence including λ oR/pR region, a *Bam*HI site, gene fusion D-CAP-*Cla*I-timm-*Eco*RI-SalI sequence provided by IDT, Coralville IA, was digested with *Bam*HI and SalI and the 710 bp fragment was ligated to the 4817 bp fragment to produce pcIpR-D-CAP-timm [20]. This permitted positioning the AUG for the D-CAP fusion protein immediately to the right of the *Bam*HI sequence, and blended into the consensus Shine-Delgarno sequence, so that the D-fusion orf was positioned 18 bp downstream from the mRNA start site for the λ *pR* promoter (see Figure 1, [20]). The D-CAP orf was removed from pcIpR-D-CAP-timm λ by digestion with *Bam*HI and *Cla*I and the resulting large fragment was ligated with gene *P*, produced by generating a *Bam*HI-*P*-*Cla*I PCR fragment, including the precise sequence of *P* (λ bp's 39582 – 40280) from λ*cI*857 DNA to yield pcIpR-*P*-timm, where a TAA stop codon was added that immediately followed the *P* insertion. Plasmid pcIpR-$P^{\pi-39991}$-timm was generated by inserting a *Bam*HI-P^{π}-*Cla*I PCR fragment made from λ*cI*72-$P^{\pi-39991}$ phage. The plasmid pcIpR-$P^{\Delta76}$-timm was constructed by using restriction endonuclease *Hpa*I to delete λ bp 39609–39836, *i.e.*, bp 28 through 255 within the N-terminal end of *P*, then ligating to fuse codon 9 with codon 86. The plasmid pcIpR-*P*-SPA-timm was constructed by removing the *Bam*HI to AscI fragment from the 5155 bp plasmid pcIpR-D-SPA-timm [20], and ligating with the remaining 4807 bp fragment a *Bam*HI-*P*-AscI PCR fragment encoding lambda bases 39582 to 40280 to produce the 5524 bp plasmid pcIpR-*P*-SPA-timm. SPA is a 66 amino acid tag sequence [72] with both calmodulin and 3XFLAG binding recognition sequences. Plasmids pcIpR-*O*-timm and p434'pR-*O*-timm have the precise *O* sequence (ATG=38686-39582) plus TAA stop codon. In p434'pR-*O*-timm, the SD differed by one bp compared to SD in pcIpR-*O*-timm because of the slightly different sequence ahead of *cro* in *imm*434 DNA [102]. Plasmids pcIpR-*O*-36P-timm and pcIpR-*O*-63P-timm, respectively, have λ DNA sequences 38686–39687, or 38686–39768, each including an intact *O* sequence plus an extension comprising the N-terminal portion of *P*, followed by TAA stop at the end of inserted partial *P* sequence. Plasmids pcIpR-*O*-*P*-timm (with precise *O*-*P* sequence), pcIpR-*oop*#1-*O*-*P*-timm and pcIpR-*oop*#2-*O*-*P*-timm (each with DNA from within *cII* through *P*) have inserted sequences, respectively: 38686-40280, 38559-40280, and 38546-40280, followed by TAA stop codon. All GOI inserts within the pcIpR-[GOI]-timm plasmids were sequenced to confirm the genetic integrity of the inserted fragment. pHB30 [31] contains λ genes *cI*[Ts]857, a *cro*-*O* in frame fusion,

P-ren; *i.e.*, the pBR322 bases from 375–4286 and λ bases (*Bam*HI)34499-34696(*Cla*I)-(*Cla*I)36965-38103 (BglII)-(BglII)38814-40806(AatII) and was re-characterized and sequenced [34].

3.3. Plasmid Transformation; Phage and Culture Assays

Cells from a single colony of the *E. coli* strain being transformed with a plasmid were inoculated into 20 ml fresh LB (5 g NaCl, 10g Bacto Tryptone, 10 g Bacto Yeast Extract per liter), grown overnight, subcultured into fresh LB medium and grown at 30 °C to A_{575} = 0.4, which equals about 4×10^8 cfu (colony forming units) per ml. The cells (1 ml) were centrifuged in 1.5ml microtubes for 1 min at 12×10^3 rpm in an Eppendorf 5424 microcentrifuge. The supernatant was decanted and cell pellet washed with 750ul 0.01M NaCl. The cells were again pelleted, suspended in 750 μl ice cold 0.03 M $CaCl_2$, incubated on ice for 30 min, pelleted, and resuspended in 150 ul 0.03 M $CaCl_2$ = competent cells. 200 ng DNA of the plasmid in TE* buffer (0.01 M Tris, 0.001 M Na_2EDTA, pH 7.6) was combined with the competent cells and mixed gently. The mixture was held on ice for 60 min, the tubes were heat shocked at 42 °C for 90 seconds in a heating block and plunged on ice for 2 min. 850 ul of room temperature LB was added to each sample tube and these were incubated with gentle shaking in a 25 °C water bath for 90 min. At the end of incubation time the cell samples were diluted in ø80 buffer (1.2 g Tris, 5.8 g NaCl per liter, pH 7.6). Then aliquots (0.1 ml) were spread on LB or TB (10 g Bacto tryptone, 5 g NaCl per liter) agar plates. The TB or LB medium used for the solid support agar plates included the addition of 11 g Bacto agar per liter prior to autoclaving. The agar plates used for screening Amp^R cfu were supplemented with 50 μg/ml of ampicillin (=Amp50) added after the autoclaved agar medium had cooled, prior to pouring the plates. Molten TB top agar (10 g Bacto tryptone, 5 g NaCl and 6.5 g Bacto agar per liter) was used for plating phage.

3.4. Sequence Analysis of Alleles of dnaB

The *dnaB* genes were amplified with primers DnaB-1 and DnaB-6. The PCR fragments were sequenced with overlapping primer pairs DnaB-1 and DnaB-2, DnaB-3 and DnaB-4, and DnaB-5 and DnaB-6 [34]. The *grpA80* and *grpD55* GenBank accession nos. are DQ324464 and DQ324465.

3.5. Assessing Influence of Modifying P or dnaB on P Lethality and Plasmid Loss

The cultures were grown up to stationary phase in TB culture medium plus 50μg/ml ampicillin for 48 hr at 25 °C, diluted, spread on TB agar plates that were incubated at 25, 30, 37, 39, or 42 °C for 48 h and cfu per ml was determined. Survivor cfu arising on the plates were stabbed to TB and TB+Amp50 plates to estimate the proportion of cfu retaining the plasmid. We tried to assay all cfu per dilution plate sector(s) to avoid colony size discrimination. Individual cfu arising from 594 *grpD55* [pcIpR-*P*-timm] from the 25 or 42 °C plates (single colonies – sc1 through sc6) were inoculated into TB+Amp50 plates, incubated 48 hrs at 25 °C, and plasmid DNA was extracted. The extracted plasmid preparations from the six individual cultures were each transformed into 594 culture cells and Amp^R cfu were selected on agar plates incubated at 25 °C. Individual single colonies (sc's 1-6) from these plates were inoculated into TB with 50 µg/ml ampicillin and grown 48 h at 25 °C, then diluted and spread on TB agar plates that were incubated at 25, 37, 39, or 42 °C for 48 hr and cfu per ml was determined. Survivor cfu arising on the plates were stabbed to TB and TB+Amp50 plates to estimate the proportion of cfu retaining the plasmid. The experiment with the *re-claimed* plasmids was undertaken to determine if the survivor plasmids extracted from the 594 *grpD55* [pcIpR-*P*-timm] cultures grown at 42 °C retained the P-lethality phenotype.

3.6. Contrasting trans P-Lethality / Inhibition and cis Replicative Killing

The influence of transient *P* expression (*trans* P lethality) from pcIpR-*P*-timm on cell viability and plasmid loss was assessed by incubating cells diluted and spread on pre-heated agar plates, that were held at 37 °C for 1, 2, or 6 h, and then incubated at 25 °C for about 72 hrs for survivor cfu. The survivors were stabbed to TB+Amp50 agar plates to assess for plasmid loss. Replicative Killing, or *cis* killing, involves the irreversible effect of inducing gene expression from a trapped (nonexcisable) cryptic fragment of λ [~*cIII-ren* genes in Y836] inserted within bacterial chromosome. The result is the formation of an onion-skin replication bubble at *ori*λ site in bacterial chromosome as shown in Figure 1C. The λ fragment in strain Y836 was transduced into 594 in two steps by P1 transduction: a) moving *nadA*::Tn*10* into Y836, and b) moving the *nadA*::Tn*10* [~*cIII-cI857-O-P-ren*]$^\lambda$ fragment into 594. Gene *P* expression in strain Y836 was inactivated by recombineering involving use of primers L-P-stop-kan [5'-gaccgtgagcagatgcgtcggatcgccaacaacatgactaactagctctgatgttacattgcacaag] and R-kan-stop-P [5'-ggtcgattctgccgacgggctacgcgcattcctgcgctagttagtcagtcagcgtaatgctctgcca] and the insertion of Kan^R (same as in pBR322::kan) into *P* to make strain Y836 [~*cIII-cI857-O-P*::kan-*ren*] (isolate Bib11t), where Kan substituted the bases 39651-39838 of *P*.

3.7. SOS-Independent P-Induced Cellular Filamentation

Overnight cultures of each strain were prepared in TB (plus 50 µg/ml ampicillin for strains with plasmids) and grown at 30 °C for the controls and strains with pHB30, and at 25 °C for strains with pcIpR-*P*-timm. Subcultures were made into 50 ml TB or TB-Amp50 and grown at 25 or 30 °C, as shown for 0-time, to an A575nm of 0.01 to 0.15. An aliquot was removed for the 0-time assay and the flasks were transferred to 37 or 42 °C shaking water baths. For each assay point culture aliquots were removed (5 µl, and 1 µl then diluted with 4 µl φ80 buffer) and placed on glass slides. The samples were allowed to air dry and gently fixed over a flame. The slides were prepared by Gram stain, and then examined by light microscopy, taking three pictures per slide of areas with lower density cells. The digital pictures were projected onto a large screen and the length per cell was manually measured, 15 cells per slide. Representative culture absorbance 0 time, 1, 3 and 5 hrs at inducing temperature: 594 (0.09, 0.22, 0.74, 0.95); 594 *lexA3* (0.01, 0.25, 0.68, 0.95); 594 *lexA3*[pcIpR-*P*-timm] induced to 37 °C (0.10, 0.26, 0.54, 0.76); and 594 *lexA3*[pcIpR-*P*-timm] induced to 42 °C (0.10, 0.25, 0.50, 0.67). The value for 1X represented the average of ten smallest cell measurements for 594 cells, representing

35.6 mm. All values were rounded up into next category, e.g., cells with length >53.4 mm were presented as 2X average length.

3.8. P-Induced Plasmid Loss

The volume of cells utilized for plasmid extraction from each of the cultures was normalized to achieve a final A_{575} of 1.0. Aliquots were removed to extract plasmid DNA and isolate cfu on TB agar plates. The survivor cfu arising on the TB plates were stabbed to TB and TB+Amp50 plates to estimate the proportion of cfu retaining the Amp^R plasmid. Plasmid DNA was extracted from aliquots (5 ml) of the cultures and mixed with 0.5 ml of a stationary phase culture of 594[pUC19] (serving as an internal plasmid extraction / gel loading control). DNA was extracted from the cell pellet(s) using QIAgen spin mimiprep kits and suspended in 0.05 ml elution buffer. Aliquots from all DNA preparations were digested to completion with restriction endonuclease *Eco*RI, which cuts each plasmid once. The digests were run on 0.8% agarose gels in TBE buffer at 90 volts for 90 min and then stained with ethidium bromide for 10 min. DNA band sizes were estimated using a 1Kb DNA ladder (left gel lanes in B and C). Plasmid pcIpR-*P*-timm is 5292bp, pcIpR-$P^{\Delta76}$-timm is 5064bp, and the high copy pUC19 is 2686bp.

4. Conclusions

Our complementation results suggest that only trace levels of *P* expression are needed to catalyze the initiation of λ replication, providing an explanation for early observations about "trans-activation." It appears that ColE1 replication is extremely sensitive to P, and cells with repressed *P*-encoded plasmids (*i.e.*, with sub-induction levels of *P* expression) can lead to ColE1 plasmid curing. Both P-lethality to cells, and the observed P-dependent cell loss of ColE1 plasmids, were fully suppressed by dual missense mutations altering *dnaB,* or an in frame deletion near the N-terminal end of P. The P-inhibitory phenotypes were partially suppressed by a π missense mutation in *P*, and plasmids expressing the λ encoded OOP antisense micro-RNA. P-dependent cellular filamentation was observed in *ΔrecA* or *lexA*[Ind⁻] cells, considered defective for SOS induction. These studies suggest the hypothesis that cellular levels of P can directly interfere not only with *E. coli* replication initiation, but subsequent steps involving DNA propagation and replication restart.

Acknowledgments

This work was supported by NSERC Canada Discovery grant to SH.

References

1.   Tsurimoto, T.; Matsubara, K. Purified bacteriophage lambda O protein binds to four repeating sequences at the lambda replication origin. *Nucleic Acids Res.* **1981**, *9*, 1789–1799.

2.   Wickner, S.H. DNA replication proteins of *Escherichia coli* and phage lambda. *Cold Spring Harbor Symp. Quant. Biol.* **1979**, *43 Pt 1*, 303–310.

3.   LeBowitz, J.H.; McMacken, R. The *Escherichia coli* DnaB replication protein is a DNA helicase. *J. Biol. Chem.* **1986**, *261*, 4738–4748.

4.   McMacken, R.; Ueda, K.; Kornberg, A. Migration of *Escherichia coli* DnaB protein on the template DNA strand as a mechanism in initiating DNA replication. *Proc. Natl. Acad. Sci. USA* **1977**, *74*, 4190–4194.

5.   Zyskind, J.W.; Smith, D.W. Novel *Escherichia coli dnaB* mutant: Direct involvement of the *dnaB*252 gene product in the synthesis of an origin–ribonucleic acid species during initiaion of a round of deoxyribonucleic acid replication. *J. Bacteriol.* **1977**, *129*, 1476–1486.

6.   Bujalowski, W. Expanding the physiological role of the hexameric DnaB helicase. *Trends Biochem. Sci.* **2003**, *28*, 116–118.

7.   Kaplan, D.L.; O'Donnell, M. DnaB drives DNA branch migration and dislodges proteins while encircling two DNA strands. *Mol. Cell* **2002**, *10*, 647–657.

8.   Arai, K.; Kornberg, A. Mechanism of dnab protein action. Ii. ATP hydrolysis by DnaB protein dependent on single– or double–stranded DNA. *J. Biol. Chem.* **1981**, *256*, 5253–5259.

9.   Reha–Krantz, L.J.; Hurwitz, J. The *dnaB* gene product of *Escherichia coli*. I. Purification, homogeneity, and physical properties. *J. Biol. Chem.* **1978**, *253*, 4043–4050.

10. Biswas, S.B.; Biswas, E.E. Regulation of *dnaB* function in DNA replication in *Escherichia coli* by *dnaC* and lambda P gene products. *J. Biol. Chem.* **1987**, *262*, 7831–7838.

11. Mallory, J.B.; Alfano, C.; McMacken, R. Host virus interactions in the initiation of bacteriophage lambda DNA replication. Recruitment of *Escherichia coli* DnaB helicase by lambda P replication protein. *J. Biol. Chem.* **1990**, *265*, 13297–13307.

12. Alfano, C.; McMacken, R. Ordered assembly of nucleoprotein structures at the bacteriophage lambda replication origin during the initiation of DNA replication. *J. Biol. Chem.* **1989**, *264*, 10699–10708.

13. Zylicz, M.; Ang, D.; Liberek, K.; Georgopoulos, C. Initiation of lambda DNA replication with purified host– and bacteriophage–encoded proteins: The role of the DnaK, DnaJ and GrpE heat shock proteins. *EMBO J.* **1989**, *8*, 1601–1608.

14. Nakayama, N.; Arai, N.; Bond, M.W.; Kaziro, Y.; Arai, K. Nucleotide sequence of *dnaB* and the primary structure of the DnaB protein from *Escherichia coli*. *J. Biol. Chem.* **1984**, *259*, 97–101.

15. Ueda, K.; McMacken, R.; Kornberg, A. DnaB protein of *Escherichia coli*. Purification and role in the replication of phix174 DNA. *J. Biol. Chem.* **1978**, *253*, 261–269.

16. Hayes, S.; Bull, H.J.; Tulloch, J. The Rex phenotype of altruistic cell death following infection of a lambda lysogen by T4rII mutants is suppressed by plasmids expressing OOP RNA. *Gene* **1997**, *189*, 35–42.

17. Hayes, S.; Slavcev, R.A. Polarity within *pM* and *pE* promoted phage lambda *cI–rexA–rexB* transcription and its suppression. *Can. J. Microbiol.* **2005** *51*, 37 49.

18. Hayes, S.; Szybalski, W. Control of short leftward transcripts from the immunity and *ori* regions in induced coliphage lambda. *Mol. Gen. Gen.* **1973**, *126*, 275–290.

19. Landsmann, J.; Kroger, M.; Hobom, G. The *rex* region of bacteriophage lambda: Two genes under three–way control. *Gene* **1982**, *20*, 11–24.

20. Hayes, S.; Gamage, L.N.; Hayes, C. Dual expression system for assembling phage lambda display particle (LDP) vaccine to porcine circovirus 2 (PCV2). *Vaccine* **2010**, *28*, 6789–6799.

21. Revet, B.; von Wilcken–Bergmann, B.; Bessert, H.; Barker, A.; Muller–Hill, B. Four dimers of lambda repressor bound to two suitably spaced pairs of lambda operators form octamers and DNA loops over large distances. *Curr. Biol.* **1999**, *9*, 151–154.

22. Hayes, S.; Asai, K.; Chu, A.M.; Hayes, C. NinR– and Red–mediated phage–prophage marker rescue recombination in *Escherichia coli*: Recovery of a nonhomologous *imm*lambda DNA segment by infecting lambda*imm*434 phages. *Genetics* **2005**, *170*, 1485–1499.

23. Klinkert, J.; Klein, A. Cloning of the replication gene P of bacteriophage lambda: Effects of increased P–protein synthesis on cellular and phage DNA replication. *Mol. Gen. Gen.* **1979**, *171*, 219–227.

24. Maiti, S.; Mukhopadhyay, M.; Mandal, N.C. Bacteriophage lambda P gene shows host killing which is not dependent on lambda DNA replication. *Virology* **1991**, *182*, 324–335.

25. Georgopoulos, C.P.; Herskowitz, I. *Escherichia coli* mutants blocked in lambda DNA synthesis. In *The bacteriophage lambda*, Hershey, A.D., Ed. Cold Spring Harbor Laboratory: Cold Spring Harbor, NY, USA, 1971; pp. 553–564.

26. Georgopoulos, C.P. A new bacterial gene (*groPC*) which affects lambda DNA replication. *Mol. Gen. Gen.* **1977**, *151*, 35–39.

27. Sunshine, M.; Feiss, M.; Stuart, J.; Yochem, J. A new host gene (*groPC*) necessary for lambda DNA replication. *Mol. Gen. Gen.* **1977**, *151*, 27–34.

28. Datta, I.; Sau, S.; Sil, A.K.; Mandal, N.C. The bacteriophage lambda DNA replication protein P inhibits the *oriC* DNA– and ATP–binding functions of the DNA replication initiator protein DnaA of *Escherichia coli*. *J. Biochem. Mol. Biol.* **2005**, *38*, 97–103.

29. Datta, I.; Banik–Maiti, S.; Adhikari, L.; San, S.; Das, N.; Mandal, N.C. The mutation that makes *Escherichia coli* resistant to lambda P gene–mediated host lethality is located within the DNA initiator gene *dnaA* of the bacterium. *J. Biochem. Mol. Biol.* **2005**, *38*, 89–96.

30. Bolivar, F.; Betlach, M.C.; Heyneker, H.L.; Shine, J.; Rodriguez, R.L.; Boyer, H.W. Origin of replication of pBR345 plasmid DNA. *Proc. Natl. Acad. Sci. USA* **1977**, *74*, 5265–5269.

31. Bull, H.J. Bacteriophage lambda replication–coupled processes: Genetic elements and regulatory choices. University of Saskatchewan, Saskatoon, SK, Canada, 1995.

32. Saito, H.; Uchida, H. Initiation of the DNA replication of bacteriophage lambda in *Escherichia coli* K12. *J. Mol. Biol.* **1977**, *113*, 1–25.

33. Bull, H.J.; Hayes, S. The *grpD*55 locus of *Escherichia coli* appears to be an allele of *dnaB*. *Mol. Gen. Gen.* **1996**, *252*, 755–760.

34. Horbay, M.A. Inhibition phenotype specific for *ori*–lambda replication dependent phage growth, and a reappraisal of the influence of lambda *P* expression on *Escherichia coli* cell metabolism: P–interference phenotype. University of Saskatchewan, Saskatoon, SK, Canada, 2005.

35. Mandal, N.C.; Lieb, M. Heat–sensitive DNA–binding activity of the cI product of bacteriophage lambda. *Mol. Gen. Gen.* **1976**, *146*, 299–302.

36. Hayes, S.; Horbay, M.A.; Hayes, C. A cI–independent form of replicative inhibition: Turn off of early replication of bacteriophage lambda. *PLoS One* **2012**, *7*, e36498.

37. Herskowitz, I.; Signer, E. Control of transcription from the *r* strand of bacteriophage lambda. *Cold Spring Harbor Symp. Quant. Biol.* **1970**, *35*.

38. Thomas, R. Control of development in temperate bacteriophages. 3. Which prophage genes are and which are not *trans*–activable in the presence of immunity? *J. Mol. Biol.* **1970**, *49*, 393–404.

39. Bejarano, I.; Klemes, Y.; Schoulaker–Schwarz, R.; Engelberg–Kulka, H. Energy–dependent degradation of lambda O protein in *Escherichia coli. J. Bacteriol.* **1993**, *175*, 7720–7723.

40. Wegrzyn, A.; Czyz, A.; Gabig, M.; Wegrzyn, G. ClpP/clpX–mediated degradation of the bacteriophage lambda O protein and regulation of lambda phage and lambda plasmid replication. *Arch. Microbiol.* **2000**, *174*, 89–96.

41. Konieczny, I.; Marszalek, J. The requirement for molecular chaperones in lambda DNA replication is reduced by the mutation pi in lambda P gene, which weakens the interaction between lambda P protein and DnaB helicase. *J. Biol. Chem.* **1995**, *270*, 9792–9799.

42. Klein, A.; Lanka, E.; Schuster, H. Isolation of a complex between the P protein of phage lambda and the *dnaB* protein of *Escherichia coli. Eur. J. Biochem.* **1980**, *105*, 1–6.

43. Sutton, M.D.; Carr, K.M.; Vicente, N.; Kaguni, J.M. *Escherichia coli* DnaA protein – the N–terminal domain and loading of DnaB helicase at the *E. coli* chromosomal origin. *J. Biol. Chem.* **1998**, *273*, 34255–34262.

44. Wickner, S.; Hurwitz, J. Interaction of Escherichia coli *dnaB* and *dnaC(D)* gene products *in vitro. Proc. Natl. Acad. Sci. U.S.A.* **1975**, *72*, 921–925.

45. Lu, Y.B.; Ratnakar, P.; Mohanty, B.K.; Bastia, D. Direct physical interaction between DnaG primase and DnaB helicase of *Escherichia coli* is necessary for optimal synthesis of primer RNA. *Proc. Natl. Acad. Sci. USA* **1996**, *93*, 12902–12907.

46. Biswas, E.E.; Chen, P.H.; Biswas, S.B. Modulation of enzymatic activities of *Escherichia coli dnaB* helicase by single–stranded DNA–binding proteins. *Nucleic Acids Res.* **2002**, *30*, 2809–2816.

47. Gao, D.X.; McHenry, C.S. Tau binds and organizes *Escherichia coli* replication proteins through distinct domains – domain IV, located within the unique C terminus of Tau, binds the replication fork helicase, DnaB. *J. Biol. Chem.* **2001**, *276*, 4441–4446.

48. McKinney, M.D.; Wechsler, J.A. RNA polymerase interaction with DnaB protein and lambda P protein during lambda replication. *J. Virol.* **1983**, *48*, 551–554.

49. Maiti, S.; Das, B.; Mandal, N.C. Isolation and preliminary characterization of *Escherichia coli* mutants resistant to lethal action of the bacteriophage lambda P gene. *Virology* **1991**, *182*, 351–352.

50. Del Solar, G.; Giraldo, R.; Ruiz–Echevarria, M.J.; Espinosa, M.; Diaz–Orejas, R. Replication and control of circular bacterial plasmids. *Microbio. Mol. Biol. Rev.* **1998**, *62*, 434–464.

51. Tomizawa, J.; Sakakibara, Y.; Kakefuda, T. Replication of colicin E1 plasmid DNA in cell extracts. Origin and direction of replication. *Proc. Natl. Acad. Sci. U.S.A.* **1974**, *71*, 2260–2264.

52. Tomizawa, J.I.; Sakakibara, Y.; Kakefuda, T. Replication of colicin E1 plasmid DNA added to cell extracts. *Proc. Natl. Acad. Sci. USA* **1975**, *72*, 1050–1054.

53. Dasgupta, S.; Masukata, H.; Tomizawa, J. Multiple mechanisms for initiation of ColE1 DNA replication: DNA synthesis in the presence and absence of ribonuclease H. *Cell* **1987**, *51*, 1113–1122.

54. Kogoma, T. Absence of Rnase H allows replication of pBR322 in *Escherichia coli* mutants lacking DNA polymerase I. *Proc. Natl. Acad. Sci. U.S.A.* **1984**, *81*, 7845–7849.

55. Naito, S.; Uchida, H. Rnase H and replication of ColE1 DNA in *Escherichia coli*. *J. Bacteriol.* **1986**, *166*, 143–147.

56. Tomizawa, J. Two distinct mechanisms of synthesis of DNA fragments on colicin E1 plasmid DNA. *Nature* **1975**, *257*, 253–254.

57. Itoh, T.; Tomizawa, J. Formation of an RNA primer for initiation of replication of ColE1 DNA by ribonuclease H. *Proc. Natl. Acad. Sci. U.S.A.* **1980**, *77*, 2450–2454.

58. Masukata, H.; Tomizawa, J. A mechanism of formation of a persistent hybrid between elongating RNA and template DNA. *Cell* **1990**, *62*, 331–338.

59. Hillenbrand, G.; Staudenbauer, W.L. Discriminatory function of ribonuclease H in the selective initiation of plasmid DNA replication. *Nucleic Acids Res.* **1982**, *10*, 833–853.

60. Bird, R.E.; Tomizawa, J. Ribonucleotide–deoxyribonucleotide linkages at the origin of DNA replication of colicin E1 plasmid. *J. Mol. Biol.* **1978**, *120*, 137–143.

61. Tomizawa, J.I.; Ohmori, H.; Bird, R.E. Origin of replication of colicin E1 plasmid DNA. *Proc. Natl. Acad. Sci. USA* **1977**, *74*, 1865–1869.

62. Masai, H.; Nomura, N.; Kubota, Y.; Arai, K. Roles of phi x174 type primosome– and G4 type primase–dependent primings in initiation of lagging and leading strand syntheses of DNA replication. *J. Biol. Chem.* **1990**, *265*, 15124–15133.

63. Allen, J.M.; Simcha, D.M.; Ericson, N.G.; Alexander, D.L.; Marquette, J.T.; van Biber, B.P.; Troll, C.J.; Karchin, R.; Bielas, J.H.; Loeb, L.A.; *et al*. Roles of DNA polymerase I in leading and lagging–strand replication defined by a high–resolution mutation footprint of ColE1 plasmid replication. *Nucleic Acids Res.* **2011**, *39*, 7020–7033.

64. Masai, H.; Arai, K. Dnaa– and pria–dependent primosomes: Two distinct replication complexes for replication of *Escherichia coli* chromosome. *Front. Bios.* **1996**, *1*, d48–58.

65. Nomura, N.; Ray, D.S. Expression of a DNA strand initiation sequence of ColE1 plasmid in a single–stranded DNA phage. *Proc. Natl. Acad. Sci. U.S.A.* **1980**, *77*, 6566–6570.

66. Minden, J.S.; Marians, K.J. Replication of pBR322 DNA *in vitro* with purified proteins. Requirement for topoisomerase I in the maintenance of template specificity. *J. Biol. Chem.* **1985**, *260*, 9316–9325.

67. Seufert, W.; Dobrinski, B.; Lurz, R.; Messer, W. Functionality of the *dnaA* protein binding site in DNA replication is orientation–dependent. *J. Biol. Chem.* **1988**, *263*, 2719–2723.

68. Seufert, W.; Messer, W. DnaA protein binding to the plasmid origin region can substitute for primosome assembly during replication of pBR322 *in vitro Cell* **1987** *48*, 73–78.

69. McCool, J.D.; Sandler, S.J. Effects of mutations involving cell division, recombination, and chromosome dimer resolution on a *priA2*::Kan mutant. *Proc. Natl. Acad. Sci. USA* **2001**, *98*, 8203–8210.

70. Bramhill, D.; Kornberg, A. Duplex opening by DnaA protein at novel sequences in initiation of replication at the origin of the *E. coli* chromosome. *Cell* **1988**, *52*, 743–755.

71. Marians, K.J. Pria–directed replication fork restart in *Escherichia coli. Trends Biochem. Sci.* **2000**, *25*, 185–189.

72. Zeghouf, M.; Li, J.; Butland, G.; Borkowska, A.; Canadien, V.; Richards, D.; Beattie, B.; Emili, A.; Greenblatt, J.F. Sequential peptide affinity (SPA) system for the identification of mammalian and bacterial protein complexes. *J. Proteome. Res.* **2004**, *3*, 463–468.

73. Reiser, W.; Leibrecht, I.; Klein, A. Structure and function of mutants in the *P* gene of bacteriophage lambda leading to the pi phenotype. *Mol. Gen. Gen.* **1983**, *192*, 430–435.

74. Yates, J.L.; Gette, W.R.; Furth, M.E.; Nomura, M. Effects of ribosomal mutations on the read–through of a chain termination signal: Studies on the synthesis of bacteriophage lambda *O* gene protein *in vitro. Proc. Natl. Acad. Sci. U.S.A.* **1977**, *74*, 689–693.

75. Tsurimoto, T.; Hase, T.; Matsubara, H.; Matsubara, K. Bacteriophage lambda initiators: Preparation from a strain that overproduces the O and P proteins. *Mol. Gen. Gen.* **1982**, *187*, 79–86.

76. Krinke, L.; Wulff, D.L. Rnase iii–dependent hydrolysis of lambda *cII–O* gene mRNA mediated by lambda OOP antisense RNA. *Genes Dev.* **1990**, *4*, 2223–2233.

77. Krinke, L.; Wulff, D.L. The cleavage specificity of Rnase III. *Nucleic Acids Res.* **1990**, *18*, 4809–4815.

78. Hayes, S.; Hayes, C. Spontaneous lambda *oR* mutations suppress inhibition of bacteriophage growth by nonimmune exclusion phenotype of defective lambda prophage. *J. Virol.* **1986**, *58*, 835–842.

79. Miwa, T.; Akaboshi, E.; Matsubara, K. Instability of bacteriophage lambda initiator O and P proteins in DNA replication. *J. Biochem.* **1983**, *94*, 331–338.

80. Wyatt, W.M.; Inokuchi, H. Stability of lambda O and P replication functions. *Virology* **1974**, *58*, 313–315.

81. Hayes, S. Mutations suppressing loss of replication control. Genetic analysis of bacteriophage lambda–dependent replicative killing, replication initiation, and mechanisms of mutagenesis. In *DNA replication and mutagenesis*, Moses, R.E., Summers, W.C., Ed. American Society for Microbiology: Washington, D.C., 1988; pp 367–377.

82. Hayes, S.; Duncan, D.; Hayes, C. Alcohol treatment of defective lambda lysogens is deletionogenic. *Mol. Gen. Gen.* **1990**, *222*, 17–24.

83. Hayes, S.; Hayes, C.; Bull, H.J.; Pelcher, L.A.; Slavcev, R.A. Acquired mutations in phage lambda genes *O* or *P* that enable constitutive expression of a cryptic lambda–N^+cI[Ts]cro^- prophage in *E. coli* cells shifted from 30 degreesC to 42 degreesC, accompanied by loss of *imm*lambda and Rex$^+$ phenotypes and emergence of a non–immune exclusion–state. *Gene* **1998**, *223*, 115–128.

84. Kaplan, D.L.; O'Donnell, M. DnaB drives DNA branch migration and dislodges proteins while encircling two DNA strands. *Mol. Cell* **2002**, *10*, 647–657.

85. Horbay, M.A.; McCrea, R.P.E.; Hayes, S. OOP RNA: A regulatory pivot in temperate lambdoid phage development. In *Modern bacteriophage biology and biotechnology*, Wegrzyn, G., Ed. Research Signpost: Kerala, India, 2006; pp 37–57.

86. Walker, G.C. Mutagenesis and inducible responses to deoxyribonucleic acid damage in *Escherichia coli*. *Microbiol. Rev.***1984**, *48*, 60–93.

87. Horii, T.; Ogawa, T.; Nakatani, T.; Hase, T.; Matsubara, H.; Ogawa, H. Regulation of SOS functions: Purification of *E. coli* LexA protein and determination of its specific site cleaved by the RecA protein. *Cell* **1981**, *27*, 515–522.

88. Markham, B.E.; Little, J.W.; Mount, D.W. Nucleotide sequence of the *lex*A gene of *Escherichia coli* K–12. *Nucleic Acids Res.* **1981**, *9*, 4149–4161.

89. Gottesman, S.; Halpern, E.; Trisler, P. Role of SulA and SulB in filamentation by *lon* mutants of *Escherichia coli* K–12. *J. Bacteriol.* **1981**, *148*, 265–273.

90. Schoemaker, J.M.; Gayda, R.C.; Markovitz, A. Regulation of cell division in *Escherichia coli*: SOS induction and cellular location of the SulA protein, a key to Lon–associated filamentation and death. *J. Bacteriol.* **1984**, *158*, 551–561.

91. Hill, T.M.; Sharma, B.; Valjavec–Gratian, M.; Smith, J. Sfi–independent filamentation in *Escherichia coli* is LexA dependent and requires DNA damage for induction. *J. Bacteriol.* **1997**, *179*, 1931–1939.

92. Hirota, Y.; Jacob, F.; Ryter, A.; Buttin, G.; Nakai, T. On the process of cellular division in *Escherichia coli*. I. Asymmetrical cell division and production of deoxyribonucleic acid–less bacteria. *J. Mol. Biol.* **1968**, *35*, 175–192.

93. Mulder, E.; Woldringh, C.L. Actively replicating nucleoids influence positioning of division sites in *Escherichia coli* filaments forming cells lacking DNA. *J. Bacteriol.* **1989**, *171*, 4303–4314.

94. Bukau, B.; Walker, G.C. Delta *dnaK*52 mutants of *Escherichia coli* have defects in chromosome segregation and plasmid maintenance at normal growth temperatures. *J. Bacteriol.* **1989**, *171*, 6030–6038.

95. Baharoglu, Z.; Lestini, R.; Duigou, S.; Michel, B. RNA polymerase mutations that facilitate replication progression in the *rep uvrD recF* mutant lacking two accessory replicative helicases. *Mol. Microbiol.* **2010**, *77*, 324–336.

96. Trautinger, B.W.; Lloyd, R.G. Modulation of DNA repair by mutations flanking the DNA channel through RNA polymerase. *EMBO J.* **2002**, *21*, 6944–6953.

97. Bachmann, B.J. Derivations and genotypes of some mutant derivatives of *Escherichia coli* K–12. In *Escherichia coli and Salmonella typhimurium: Cellular and molecular biology*, Neidhardt, F.C., Ingraham, J.I., Low, K.B., Magasanik, B., Schaechter, M., Umbargr, H.E., Ed. American Society for Microbiology: Washington, D.C., 1987; Vol. 2, pp 1192–1219.

98. Harris, R.S.; Longerich, S.; Rosenberg, S.M. Recombination in adaptive mutation. *Science* **1994**, *264*, 258–260.

99.   Slavcev, R.A.; Hayes, S. Blocking the T4 lysis inhibition phenotype. *Gene* **2003**, *321*, 163–171.

100. Hayes, S.; Hayes, C. Control of lambda repressor prophage and establishment transcription by the product of gene *tof. Mol. Gen. Gen.* **1978**, *164*, 63–76.

101. Hayes, S. Control of the initiation of lambda replication, oop, lit and repressor establishment RNA synthesis. In *DNA synthesis, present and future*, Molineux, I.; Kohiyama, M., Eds. Plenum Press: New York, 1978; pp 127–142.

102. Grosschedl, R.; Schwarz, E. Nucleotide sequence of the *cro–cII–oop* region of bacteriophage 434 DNA. *Nucleic Acids Res.* **1979**, *6*, 867–881.

9

Spatial Vulnerability: Bacterial Arrangements, Microcolonies, and Biofilms as Responses to Low Rather than High Phage Densities

Stephen T. Abedon

Department of Microbiology, The Ohio State University, 1680 University Dr., Mansfield, OH 44906, USA; E-Mail: abedon.1@osu.edu

Abstract: The ability of bacteria to survive and propagate can be dramatically reduced upon exposure to lytic bacteriophages. Study of this impact, from a bacterium's perspective, tends to focus on phage-bacterial interactions that are governed by mass action, such as can be observed within continuous flow or similarly planktonic ecosystems. Alternatively, bacterial molecular properties can be examined, such as specific phage-resistance adaptations. In this study I address instead how limitations on bacterial movement, resulting in the formation of cellular arrangements, microcolonies, or biofilms, could *increase* the vulnerability of bacteria to phages. Principally: (1) Physically associated clonal groupings of bacteria can represent larger targets for phage adsorption than individual bacteria; and (2), due to a combination of proximity and similar phage susceptibility, individual bacteria should be especially vulnerable to phages infecting within the same clonal, bacterial grouping. Consistent with particle transport theory—the physics of movement within fluids—these considerations are suggestive that formation into arrangements, microcolonies, or biofilms could be either less profitable to bacteria when phage predation pressure is high or require more effective phage-resistance mechanisms than seen among bacteria not living within clonal clusters. I consider these ideas of bacterial 'spatial vulnerability' in part within a phage therapy context.

Keywords: adsorption; bacteriophage; biofilms; cellular arrangements; ecology; microcolonies; particle transport; phages; phage therapy

1. Introduction

Environments can be distinguished in terms of the degree of spatial structure that they exhibit, where spatial structure is a description of the extent to which diffusion, motility, and environmental mixing are constrained. Important spatially structured bacterial habitats include soils; sediments; surface tissues of plants, animals, and fungi; and bacterial biofilms in general. The latter are found both as suspended aggregates and on most submerged surfaces. It is an oft-repeated assertion that the majority of bacteria, or at least a large fraction, may be found within biofilms rather than as planktonic organisms [1,2]. Naturally occurring bacteria thus exist to a great extent as spatially structured populations or communities. Furthermore, and pertinent to fields as diverse as medicine [3] and civil engineering [4], pathogenic or nuisance bacteria found in the biofilm state can be resistant to both antibiotics and disinfectants—a resistance that can be more a function of the phenotypic plasticity of bacteria, *i.e.*, varying metabolic states, rather than due to either genetically acquired resistance or diffusion barriers to chemical penetration into biofilms. Development of alternative methods of biofilm removal therefore is desirable [5].

Bacteriophages, the viruses of bacteria, are a possible alternative to antibiotics, or disinfectants, as antibacterial agents. Such phage therapy or phage-mediated bacterial biocontrol [6,7] has shown promise against bacterial biofilms [8,9]. Rather than a relatively new aspect of phage study, however, the exploration of phage infection of spatially structured bacterial populations goes back to the beginning of the phage era. The first generally recognized bacteriophage study [10], that of Twort [11,12], considered in particular the phage impact on bacteria growing as colonies. Though subsequent studies of phage interaction with macroscopic bacterial colonies have been relatively rare, observation of phage-induced lysis of microscopic colonies has been a routine facet of phage biology, with a growing literature considering explicitly the dynamics of bacterial lysis within the context of phage plaque formation [9,13–17]. Given the ubiquity of biofilms within natural environments, phage interaction with spatially structured bacterial populations should be somewhat relevant to our appreciation of phage environmental microbiology in general [13,18]. Similarly, improved understanding of such interactions may possess applied significance, such as helping to inform phage choice as anti-biofilm agents [19,20] or phage modification to improve anti-biofilm properties [21,22].

Here I explore the costs to bacteria of 'group living' that can result from exposure to phages. This I term a 'spatial vulnerability' because bacteria that are physically attached together—as arrangements, microcolonies, "macrocolonies" [23], or otherwise within biofilms—display less mobility relative to each other than do equivalent bacteria found as physically isolated cells. The result, if physically associated bacteria are clonally related [1], can be a greater negative impact resulting from phage exposure than if the same bacteria instead existed as free cells. I argue that benefits associated with group living therefore are accessible to bacteria only to the extent that their vulnerability to phages nevertheless is small. Murray and Jackson [24] provide comparable though more general arguments based on particle transport theory, that is, the physics of un-self-propelled movement within fluids as applied to aquatic viruses. In general, being large is possible only if the pressure of viral predation is sufficiently low.

Mechanisms that can reduce bacterial exposure to phages include existence within environments into which phage penetration is difficult or when bacteria exist at sufficiently low densities that they are unable to support phage amplification to "inundative" densities [25]. The latter can be described as an avoidance of 'kill the winner' mechanisms [26,27] or, equivalently, bacterial existence within numerical refuges [28,29]. A general implication is that biofilms may tend to persist particularly within environments in which the densities of phages targeting those bacteria are relatively low or, alternatively, that biofilm-forming bacteria *must* possess substantial phage-resistance mechanisms [30–32] in order to maintain their populations within environments where phage predation pressures are relatively high.

The common theme is that living as physically associated and therefore spatially structured clonal groups, in and of itself, should not be expected to serve bacteria as a phage-resistance mechanism. Rather, I argue from first principles that group living can result in greater bacterial vulnerability to phages than may be experienced by bacteria that instead are physically separated from their clonal relations. With this perspective in mind, the utility of phages as an anti-bacterial as well as a specifically anti-biofilm strategy may be appreciated as an explicit ecological reversal of exactly those circumstances in which biofilms otherwise may flourish: Application of sufficient densities of phages, where phages otherwise are lacking, such that uncontrolled bacterial proliferation can be reversed.

2. Results and Discussion

In this study the primary question being asked is what might be the ecological costs to bacteria, in light of phage predation, that are associated with bacterial growth as arrangements or microcolonies. To answer this question, I generally employ ecological models, arguments, and scenarios—that is, considerations of how bacteria may interact with their environments—and this is rather than primarily enlisting evolutionary approaches or perspectives. In addition, as the study represents a relatively novel exploration the ideas presented, I limit discussion to less complex scenarios, avoiding addressing for instance consideration of stochasticity, the growth in size of bacterial arrangements, or simulations of phage-bacterial ecological dynamics.

2.1. Phage Adsorption to Free Bacteria

The interaction between phages and those bacteria that exist as individual, planktonic cells—here, collectively, "free" bacteria—is fairly straightforward. Beginning with phage attachment to a susceptible bacterium, phage-genome uptake occurs, initiating the infection proper. At some point mature virions must be released from the infected bacterium, beginning an extracellular search for new bacteria to infect [33]. This search is driven by a combination of phage diffusion, fluid flow including environmental mixing, and bacterial as well as bulk environmental movement [18]. Upon sufficient mixing, all bacteria within an environment are then equally likely to encounter a particular phage that has just been released from a specific bacterium. That is, spatial structure can be said to largely *not* exist given a combination of bacteria that are "free" and substantial environmental mixing. In this section I consider the basics of phage adsorption, focusing particularly on issues of encounter rates between phages and bacteria rather than mechanisms of phage attachment or subsequent phage initiation of infection. For visualization of the spatial scale of environments in which these interactions take place, see Abedon [34].

2.1.1. Phage Movement towards Bacterial Targets

The extracellular search at its most basic consists of a process of virion diffusion. Such diffusion, due to the comparatively small size of phages, occurs at a rate that is substantially greater than the diffusion of free-floating bacteria. As a result, phage extracellular movement towards an idealized bacterium can be described [24] (p. 104) as "simple diffusion to a single sphere". The likelihood of phage-bacterium collision, even in an environment lacking in spatial structure, thus is a function of phage diffusion much more so than the diffusion of target bacteria. Given the substantially larger size of bacteria relative to phages, the likelihood of phage-bacterium encounter is governed by bacterial target size much more so than phage diameter [35]. Free phages thus can be considered to rapidly diffuse among relatively large and stationary bacterial targets (Figure 1).

Figure 1. Illustration of phage and bacterial contributions to phage adsorption rates. Generally phages are relatively small and bacteria somewhat larger. Since diffusion rates are inversely proportional to particle size, whereas target size is proportional to particle size, the result is that phage diffusion (larger arrows pointing right) is a more important contributor to phage adsorption than is bacterial diffusion (smaller arrows point left) while bacterial target size is more important than phage target size to the likelihood of phage-bacterial encounter. An approximate doubling of total bacterial size (lower right) consequently affects target size but has little relevant impact on combined diffusion rates. Note that arrow lengths reflect an assumption that phages are one-tenth the diameter of the coccus and one-twentieth the diameter of the diplococcus.

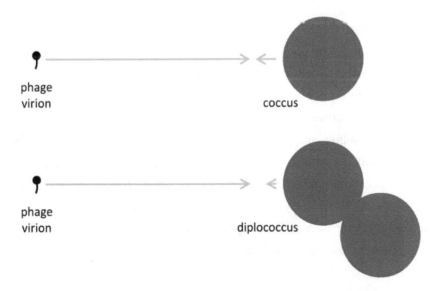

phage
virion

coccus

phage
virion

diplococcus

Rates of virus diffusion along with the size of target bacteria are not the only factors affecting rates of encounter between these entities, as so too do their environmental densities. For instance, and as considered in some detail by Murray and Jackson [24], the specific rate of virus adsorption is a function of both the size as well as number of adsorbable targets. Bacteria thus represent smaller targets relative to protozoa and as a consequence individual viruses are less likely to encounter individual bacteria in comparison to individual protozoa. Protozoa, however, tend to display lower population densities than do bacteria, resulting in lower rates of virus encounter with protozoa despite the latter's larger individual sizes. These lower rates of encounter can have the effect of keeping prey densities below 'winner' concentrations (Section 1) despite larger target sizes for individual prey organisms.

Ioannou *et al.* [36] provide both theory and experimental evidence that, in light of predation, the costs to prey of increasing in size—and therefore becoming more absolutely visible to predators—may be offset by prey decrease in abundance. In particular, larval prey that are present at lower densities will be relatively less visible due to, on average, their existing at greater distances from their three-spined stickleback predator (a type of fish), just as bacteria existing at lower densities too, on average, are present at greater distances from individual phages relative to bacteria found in higher density bacterial populations. Consistent with these considerations, in subsequent sections I suggest that bacterial existence as arrangements or as microcolonies can result in greater bacterial vulnerability to phages due to resulting increases in overall target size. I suggest in addition that such vulnerabilities may be avoided at least in part when bacterial populations are relatively rare, that is, should bacteria enter into what have been described, for free bacteria, as numerical refuges [28,29].

2.1.2. Basic Adsorption Calculations

The higher the density at which phage virions are present within environments then the more likely that a given bacterium will encounter a phage, potentially resulting in bacterial conversion from phage uninfected to phage infected. Thus,

$$N_t = N_0 e^{-kPt} \tag{1}$$

where N_t is the density of bacteria that are not phage infected at the end of some interval (t), N_0 is the density of uninfected bacteria at the beginning of that interval, P is the density of free phages (that is, phages that are both unadsorbed and no longer associated with their parental infection), and kPt represents an actual multiplicity of phage *infection* [9,19,25,37], that is, MOI_{actual} as defined by Kasman *et al.* [38]. The phage adsorption rate constant, k, is the probability that a single phage within a specified volume will encounter and then adsorb a single bacterium. This value is based in part on the rate of phage diffusion along with the size of bacterial targets (Section 2.1.1). Note in this equation that phage densities are presumed to remain constant over the course of the interval, t, a situation that may be readily approximated particularly when bacterial densities are low [39].

The more bacteria that are present within a given environment then the more likely that a specific phage will encounter some bacterium, such that,

$$P_t = P_0 e^{-kNt} \qquad\qquad (2)$$

where P_0 is the initial phage density and P_t is that density after time, t. This equation in particular describes the loss of free phages as a function of bacterial adsorption. Substantial declines in phage titers will occur due to bacterial adsorption, however, only if bacterial densities are relatively high or t is relatively large. Consequently, and as is true also with Equation (1), for this study I employ the simplifying assumption that phage densities do not vary over time. Operationally, this means that I am placing greater emphasis on consideration of bacterial vulnerability to phages than I am on the dynamics of phage generation and loss.

2.2. Phage Interaction with Bacterial Arrangements

We can increase the complexity of phage-bacterial interactions by considering bacteria that are found predominantly as arrangements rather than as otherwise "free" cells (note that generally, in using the term "arrangement", I am implying microcolony as well, that is, clonally related bacteria that by some means are found attached to one another). For example, bacteria can be arranged as doublets of cells (such as diplococci), strings of bacteria (streptococci or streptobacilli), or other, more complicated forms (staphylococci, tetrads, or sarcinae), and even, as indicated above, as microcolonies as well as biofilms. These arrangements are formed in the course of bacterial division and they differ in terms of the number of divisions that take place prior to cell separation as well as in terms of the planes of those divisions. Forming into arrangements presumably provides bacteria with selective advantages, as considered in Section 2.3.1. This is just as the specific shapes that different bacterial strains and species display, such as coccus *versus* bacillus *versus* spirillum, can be viewed as presumptively adaptive [40,41] or biofilm phenotypes can be seen as improvements in some manner upon the planktonic state [23].

Existing as arrangements, or as microcolonies, may be costly in the face of phage-mediated predation. We can consider this proposed elevation in costs as a consequence of increases in the overall target size of arrangements relative to individual bacteria, which is relevant especially in combination with increased potential for phage propagation within arrangements. At an extreme, arrangement target size could increase directly as a multiple of the number of bacteria found within an arrangement (*i.e.*, ten bacteria as a single target could be ten-times as likely to become phage adsorbed as a single bacterium). Again at an extreme, once an arrangement has become phage infected, then complete loss of all bacteria found in that arrangement could occur. In this section, I consider limitations on these extremes. I nevertheless retain the general conclusion that group living could increase bacterial vulnerability to phages.

2.2.1. Increased Target Size

The likelihood of a bacterium encountering a phage, as indicated in Equation (1), is kPt, where k is a function in part of the bacterium's target size [24,35]. If bacteria form into arrangements, then the likelihood that a specific bacterium encounters a phage may be lower due to partial shading of bacteria by other bacteria [42] or, alternatively, because of shading that results from bacterial association with surfaces. To reflect these issues, I will use the term $\ddot{k}$ to describe reductions in phage adsorption rates to bacteria that stem from shading, such that $\ddot{k} < k$. Note that the umlaut's intention is to imply a description of properties associated with bacterial arrangements, with the double dots literally suggestive of a diplococcus.

The rate of phage adsorption to a bacterial arrangement can be described as nkP, where n is the number of bacteria making up an individual arrangement. That is, the *target size* of an arrangement increases by a factor of n relative to free bacteria while at the same time decreases by a factor of $\ddot{k}/k$. The increase due to n, however, likely is greater than the decrease described by $\ddot{k}/k$, at least so long as arrangements are not sequestered within phage-excluding volumes such as (perhaps) defects in the glass walls of chemostats [29]. Larger arrangements, in other words, almost inevitably will tend to serve as larger targets for phage encounter than will either individual bacteria or smaller arrangements.

Figure 2. Shading of bacteria by bacteria. Shown is a progression starting with two "free" coccus-shaped bacteria (left) which is followed by a diplococcus displaying some degree of attachment (middle) that in turn is followed by a diplococcus displaying maximal attachment as well as minimized surface-to-volume ratio (right), *i.e.*, existing as a combined-volume sphere of $2^{1/3}$-fold increased radius over an individual cell (see calculation, below). The left-hand lack of arrangement shows no shading whereas the right-hand arrangement shows an approximation of maximal shading for a combined spherical shape. The middle arrangement displays some intermediate degree of shading and therefore some intermediate overall target size between maximal and minimal (holding cell volumes constant). Note that the volume of a sphere, V_1, is equal to $\left(\frac{4}{3}\right)\pi r_1^3$. Twice its volume ($V_2$) therefore is $\left(\frac{8}{3}\right)\pi r_1^3$, which as a sphere is equal to $\left(\frac{4}{3}\right)\pi r_2^3$. For $\left(\frac{4}{3}\right)\pi r_2^3 = \left(\frac{8}{3}\right)\pi r_1^3$, then $r_2 = 2^{1/3}r_1$. With such shading, then, diameter increases by only $2^{1/3} = 1.26$ fold.

Generally it is the diameter or 'breadth' of bacterial targets that is crucial to determining viral contact rates [24,35]. For example, the target size of paired, spherical bacteria (diplococci) will range between ~1.26 (= $2^{1/3} = n\ddot{k}$), which is the increased diameter of a two-fold larger volume, and approximately two (n) times larger than the target size of individual cocci. These values in other words range from where shading is substantial (1.26 times) to where shading instead is minimal (~2 times; for illustration, see Figure 2). Given diversity in arrangement shape it is clear that using arrangement diameter as a proxy for target size is a simplification, though one which I retain both for the sake of mathematical convenience and because assuming that targets are spherical may be the most reasonable of default assumptions. Clearly though, and as indicated in the above calculation (Figure 2), surface area (as equivalent to the "~2 times" calculation) provides a more intuitive perspective on target size and particularly so given non-spherical as well as relatively immobile targets. The larger and more important point, however, is that in terms of target size, arrangements should be inherently more vulnerable to phage encounter than individual bacteria.

2.2.2. Increased Multiplicity of Adsorption

To visualize the impact of forming into arrangements, with shading affecting phage adsorption rates, compare Equation (1) with

$$A_t = A_0 e^{-n\ddot{k}Pt} \tag{3}$$

Here A stands for arrangement and A_t is the number of arrangements that have *not* been phage adsorbed over an interval, t, given a constant phage density, P. So long as $n\ddot{k} > k$ holds, then $N_t/N_0 > A_t/A_0$. That is, fewer arrangements will remain fractionally unadsorbed (A_t/A_0) than would individual, free bacteria (N_t/N_0), holding bacterial size and adsorption susceptibility otherwise constant. Here $n\ddot{k}Pt$ is equivalent to MOA$_{actual}$ for arrangements. Note though that it is my preference to instead use the term multiplicity of *adsorption*, i.e., MOA, rather than multiplicity of *infection* because while an arrangement can be wholly adsorbed by a phage, subsequent infection of the whole arrangement is a more complicated process *versus* the infection of individual phage-adsorbed bacteria.

A complementary perspective on the above assertion—that is, that fewer arrangements will remain unadsorbed by phages relative to free bacteria, $N_t/N_0 > A_t/A_0$—is that MOA for arrangements can be up to n-fold higher than that for individual cells. A quantity that I will call MOA$_{input}$ (M) can, after Kasman *et al.* [38], be set equal to the density of phages divided by the density of phage targets. The density of arrangements (A_0), as phage targets, is expected to be n-fold lower than that of free bacteria, i.e., $A_0 = N_0/n$, assuming a constancy in both cell size and total species biomass [36,43]. Holding phage numbers constant, then M for arrangements (M_A) is expected to be n-fold higher than M for free bacteria (M_N), since $M_A = P/(N_0/n)$ whereas $M_N = P/N_0$. The fraction of targets expected to remain unabsorbed, in turn, is readily calculated as e^{-M}, which is the frequency of the zero category—bacteria (N_t/N_0) or arrangements (A_t/A_0) experiencing no phage adsorption—given a Poisson distribution of phages adsorbing to targets. The larger M then the smaller the fraction of cells or arrangements remaining unadsorbed, and therefore $N_t/N_0 > A_t/A_0$ if $M_A > M_N$. More precisely, we can consider instead MOA$_{actual}$, which are $M_A = n\ddot{k}Pt$ *versus* $M_N = kPt$. With M defined in this manner, then the fraction of phage targets expected to remain unadsorbed is equal to $e^{-n\ddot{k}Pt}$ ($= A_t/A_0$) and e^{-kPt} ($= N_t/N_0$), respectively, which are restatements of Equations (3) and (1), respectively. Note that $e^{-n\ddot{k}Pt} < e^{-kPt}$ if as expected $n\ddot{k}Pt > kPt$, implying that $A_t/A_0 < N_t/N_0$.

These considerations come with the caveat that increases in the likelihood of arrangement adsorption that occur as a function of n, that is, as n contributes to arrangement diameter and therefore to target size, may be slowed to the extent that adsorption rates to the individual cells making up an arrangement, $\ddot{k}$, also may decline as n increases. It may be harder, that is, for environmental phages on average to encounter *individual* bacteria that are found within larger arrangements (such as a large microcolony) *versus* individual bacteria that are found in smaller arrangements (such as a diplococcus). Arrangement vulnerability to adsorption, given this tendency, therefore might increase *less rapidly* as a function of the number of bacteria that they contain.

2.2.3. Phage Propagation within Arrangements

The expression $n\ddot{k}Pt$ should adequately describe the likelihood of arrangement encounter with a phage. Further, $n\ddot{k}Pt$ defines the actual multiplicity, a.k.a., multiplicity of adsorption, of a bacterial arrangement with the value $\ddot{k}$ specifying inefficiencies in this phage adsorption process in comparison to free bacteria. In addition, phage adsorption to arrangements is not identical to phage adsorption to free bacteria because only a fraction of the number of bacteria making up an arrangement become initially phage adsorbed (*i.e.*, $1/n$) rather than all of the bacteria making up a free bacterium ($1/1$). Furthermore, with bacterial arrangements an initial ("primary") phage adsorption could give rise to a variety of subsequent outcomes including infection of only the adsorbed bacterium, subsequent infection of a fraction of the bacteria found within the arrangement, or indeed subsequent infection of all of the bacteria making up an arrangement. The latter result, of course, is the more costly of outcomes to the affected bacteria, just as it is in terms of prey aggregation more generally [36]. The result is that an additional assumption must be made to argue that the vulnerability of bacterial arrangements to phages can be greater than that seen for individual bacteria. This assumption, in particular, is that the efficiency of phage propagation among bacteria found within arrangements must be greater than that which can be sustained among free bacteria.

When phage densities are higher, then the likelihood of phage adsorption of a given bacterium also should be higher, as described by Equation (1). Importantly, then, the density of phages immediately surrounding a lysing bacterium, that is, as made up predominantly of those phages released from that bacterium, will be the highest phage densities that can be readily attained within a given environment. The phages released in this burst will then diffuse outward, declining in density as they do. The result should be a higher rate of phage adsorption to any susceptible bacterium found within the immediate vicinity of a lysing bacterium, but lower rates at increasing distances (assuming that phage intrinsic adsorption ability does not substantially increase over the course of environmental diffusion).

While free bacteria can randomly find themselves in the vicinity of a lysing bacterium, bacteria that are found in arrangements can be spatially constrained to that vicinity. The result is a higher likelihood of phage infection of other bacteria found within the same phage-infected arrangement than to other environmental bacteria (Figure 3). This argument is similar to an observation made by Babic *et al.* [44] that transfer of conjugative transposons among bacteria found in arrangements (chains) too can be quite efficient and this is for similar reasons, *i.e.*, a constraining of bacterial location in arrangements to within the vicinity of agents infecting the same arrangement (p. 1): "Since many bacterial species grow naturally in chains, this intrachain transfer is likely a common mechanism for accelerating the spread of conjugative elements within microbial communities."

Figure 3. Illustration of the tendency of phages to display biases towards acquisition of locally available bacteria. Here shown to the right is phage acquisition of a bacterium (blue) that is found as part of the same arrangement as a lysing bacterium (red with dashed border). The green arrows represent outwardly diffusing phage progeny released upon bacterial lysis while the shorter, gray arrows illustrate the tendency of those phages that are released immediately adjacent to an uninfected bacterium to encounter that bacterium. Contrasting this second bacterium looming large in the vicinity of an adjacent phage burst, even at a high plankton bacterial density of 10^8 per mL, each free bacterium (left) occupies a total environmental volume of 10^4 μm^3 (1 cm = 10^4 μm, meaning that 1 mL = 1 cm^3 = 10^{12} μm^3, where 10^{12} $\mu m^3/10^8$ bacteria = 10^4 μm^3/bacterium). This density in turn implies an average distance between bacteria of about $10^{4/3}$ (*i.e.*, the cube root of 10^4 μm^3), or more than 10 μm, which one may compare with a typical bacterium diameter of about 1 μm. Thus, bacteria in arrangements can be not-unreasonably described as having local densities that should encourage phage adsorption with higher likelihood than that seen among planktonic, individual bacteria.

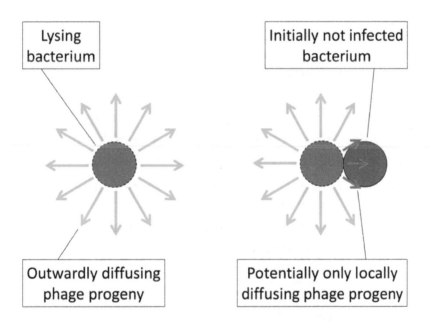

Analogously, greater bacterial densities found within bacterial arrangement can be viewed as possessing a higher "mass" relative to that associated with planktonic bacteria, where by "mass"—potentially confusingly—I am referring to immediately local *densities* of bacteria. This "mass" of bacteria within a bacterial arrangement would be more likely to exceed a "critical" level, such that phage propagation can be sustained, than may be achieved by an equivalent number of more locally dilute, free bacteria. This perspective is just as one can consider for nuclear fission, *i.e.*, radioactive decay, and associated chain reactions [45], which occurs more rapidly given higher "fuel" densities. Indeed, the idea of a "critical mass" can be directly equated with phage proliferation thresholds, that is, those bacterial densities at which rates of phage population growth are perfectly balanced by rates of phage loss [19,25]. Each is a description of target densities (atoms or bacteria) that can be sufficiently high that collision with targets (by neutrons or phages), in combination with subsequent proliferation (via fission or infection), balances any losses that can occur due either to interactions with nontargets or movement away from the focus "mass".

A target bacterium that is found in the immediate vicinity of a phage, in other words, displays a much higher local density from the perspective of that phage than may be the case for bacteria that are randomly dispersed throughout an environment. A local concentration of bacterial "mass" thus can result in a high propensity for bacterial adsorption by phages that have been generated within the same "mass". See Abedon [34], by contrast, for illustration of the relatively low propensity for phages to randomly encounter free-floating bacteria found within fluid environments; see too Hagens and Loessner [46] as well as Goodridge [47]. As default assumptions, therefore, we can view bacterial arrangements as both larger targets for phage adsorption than individual bacteria and as locally higher bacterial densities, densities that may be better able to support local phage propagation and population growth than more diffuse populations of free bacteria. Consistent with the analogy with nuclear fission, which can be controlled by the insertion of neutron-absorbing substances, we can question the efficiency of phage acquisition and then infection of seemingly adjacent bacteria, e.g., perhaps as may be separated by extracellular matrix.

2.2.4. Inefficiencies in Phage Propagation

Notwithstanding proposed tendencies for phages to more readily acquire bacteria that are within their immediate vicinity, it is as noted possible for inefficiencies to exist in the sequential phage infection of bacteria co-occupying the same arrangements. To incorporate such inefficiencies into models of arrangement vulnerability to phages, I employ the term, $\ddot{n}$. This represents the number of bacteria within an arrangement that will be lost, on average, as a consequence of phage adsorption of a single bacterium within that arrangement. This number, $\ddot{n}$, can range up to the total number of bacteria making up an arrangement (n). Another way of viewing $\ddot{n}$, however, is that $\ddot{n} > 1$ implies a phage reproductive number within a bacterial arrangement that is greater than 0 such that some degree of phage propagation within an arrangement occurs along with consequent bacterial death. If insufficient phage release from infected bacteria and/or insufficient subsequent bacterial infection occurs within an arrangement, or subsequent infections don't happen fast enough, then complete eradication of an arrangement by an infecting phage may not happen, such that $\ddot{n} < n$.

These ideas can be expressed as,

$$N_t = N_0 - \left(1 - e^{-n\ddot{k}Pt}\right)\ddot{n}A_0 = N_0 - \left(1 - e^{-n\ddot{k}Pt}\right)\ddot{n}N_0/n \tag{4}$$

where $(1 - e^{-n\ddot{k}Pt})$ is the fraction of arrangements that become individually phage adsorbed over some interval, t, and $\ddot{n}$ is the number of bacteria per arrangement that are lost to this adsorption (assuming, for simplicity, that $\ddot{n}$ is independent of the actual multiplicity of phage adsorption to a given arrangement). Were $\ddot{n} = 1$, then though arrangements are more likely to be adsorbed than free cells, nevertheless no more bacteria would be lost per arrangement adsorption. Indeed, to the extent that the initial bacterial infection is less likely, that is, given $\ddot{k} < k$ along with $\ddot{n} = 1$, then overall existence as an arrangement could result in *less* vulnerability to phages rather than more. Such a situation would occur, for example, were phage infections abortive or perhaps could result instead were infections considerably reduced in burst size or extended in latent period such that phage propagation through an arrangement were substantially impaired. Alternatively, the equality $\ddot{n} = n$ would imply complete arrangement loss following each phage adsorption of an arrangement. The parameter, $\ddot{n}$, is thus a description of arrangement vulnerability to phages post-adsorption, ranging from minimal ($\ddot{n} = 1$, or even $\ddot{n} = 0$) to maximal ($\ddot{n} = n$). A visual summary of the models represented by Equations (1) and (4) is presented in Figure 4.

Figure 4. The model. Parameters include P (density of phages in environment), k (phage adsorption constant), $\ddot{k}$ (phage adsorption constant considering reductions due to shading of bacteria by bacteria found within bacterial arrangements), n (number of bacteria found per arrangement), N (bacterial density of overall environment), L (phage latent period, which is the duration of a phage infection), and $\ddot{n}$ (number of bacteria per arrangement lost subsequent to phage infection of one cell in the arrangement). Likelihood of phage adsorption of bacterial arrangements is $n\ddot{k}$ and density of arrangements within environments is equal to $N/n = N_0/n$ (or indeed $n_0\ddot{k}$ and N_0/n_0, respectively, to reflect that n changes as a function of time in the figure). The inequality $t \geq 2L$ indicates how phage acquisition of bacteria within a bacterial arrangement, according to this model, involves at least two sequential rounds of phage infection. The absence of cells in the lower right is intentional as too is the reduction in cell number to n_t in the lower left. Both of these reductions in cell number, going from middle to bottom, indicate phage-induced bacterial lysis.

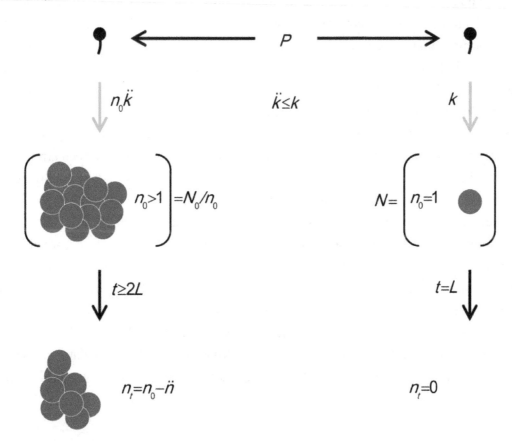

2.2.5. An Important Special Case

Equation (4) is less applicable given higher levels of phage adsorption, as can occur over longer periods. This is because, as noted, Equation (4) fails to take into account the impact, on the overall fraction of bacteria that are lost, of multiple phage adsorptions of individual arrangements. That is, for instance, we might have an expectation of greater bacterial loss with greater levels of arrangement adsorption by environmental phages under conditions where otherwise $\ddot{n} < n$, but Equation (4) does not reflect the possibility that if one adsorbing phage fails to clear an arrangement then perhaps more than one adsorbing phage will, with greater likelihood, succeed in doing so. One way to address this concern is simply by setting $\ddot{n} = n$, that is, an assumption of complete arrangement loss per primary phage adsorption.

Alternatively, one can limit one's considerations to circumstances in which phage multiplicities of adsorption to arrangements are relatively low (<<1). In such cases the equation can be simplified as

$$N_t = N_0 - \left(n\ddot{k}Pt\right)\ddot{n}A_0 = N_0 - \ddot{k}Pt\ddot{n}N_0 \tag{5}$$

which compares, for a free bacterium, with,

$$N_t = N_0 - kPtN_0 \tag{6}$$

Note that N_t (as defined by Equation (5)) is smaller than N_t (as defined by Equation (6)) when $\ddot{n} > k/\ddot{k}$. That is, when this inequality holds then bacteria found within bacterial arrangements are more vulnerable to phages than are phage-susceptible bacteria that are "free". In words: Bacteria found within arrangements are more susceptible to phage attack if losses due to existence within a phage-adsorbed arrangement, $\ddot{n}$, are greater than increases in *individual* bacterial vulnerability to "primary" adsorptions that come from *not* existing within an arrangement (that is, $k/\ddot{k}$ where $k > \ddot{k}$).

If a bacterium is half as likely to be subject to primary phage adsorption when found within an arrangement ($\ddot{k}/k = 0.5$ such that $k/\ddot{k} = 2$) but each primary adsorption results in the loss of ten bacteria ($\ddot{n} = 10$), then the resulting $10 > 2$ would imply that arrangements are more vulnerable to phages than are free bacteria. Indeed, in this example bacteria found within arrangements would be five times more vulnerable. Note further that if arrangements can reduce $\ddot{n}$ to 1, then arrangements will be expected to display lower vulnerability to phages than free bacteria so long as individual bacteria found within arrangements are less vulnerable to primary phage adsorptions than free bacteria (*i.e.*, again, such that $k > \ddot{k}$). This is an observation that could very well explain the utility of abortive infection systems to bacteria, *i.e.*, phage-resistance mechanisms in which both adsorbed bacteria and adsorbing phages die [30]. Alternatively, for $\ddot{n} = 0$, which is the case given successful bacterial display of, for example, anti-phage restriction-modification systems, then biologically the inequality, $\ddot{n} > k/\ddot{k}$, no longer holds, *i.e.*, $0 > k/\ddot{k}$. Logically, though, in this case arrangements should be no more or less vulnerable to phages than free bacteria since, in fact, neither would be vulnerable. See Table 1, under "Higher" phage densities for summary. Note, though, that calculations relevant to the "Lower" phage densities portion of the Table, *i.e.*, Equation (7), are not discussed until Sections 2.3.3 and 2.3.4.

Table 1. Summary of predictions as a function of phage densities in environments and phage potential to acquire bacteria sequentially within bacterial arrangements (recall in interpreting the table that the inequality, $N_t/N_0 > A_t/A_0$, implies greater success over time in the face of phage-mediated predation for free bacteria *versus* bacteria found within arrangements while $N_t/N_0 < A_t/A_0$ implies the opposite). Generally, $k/\ddot{k} \geq 1$. Calculations relevant to lower phage densities, as found in the bottom portion of the table, are not discussed until Section 2.3.3 and especially section 2.3.4.

Environmental Phage Density (P)	Phage Propagation Ability Through Arrangements ($\ddot{n}$)			
	Higher	**Lower**		
Higher (bacterial losses dominate dynamics)	For $\ddot{n} > k/\ddot{k}$, $\dfrac{N_t}{N_0} > \dfrac{A_t}{A_0}$ [lesser or no impediments to phage propagation within arrangements]	For $\ddot{n} < k/\ddot{k}$, $\dfrac{N_t}{N_0} < \dfrac{A_t}{A_0}$ [impediments less than absolute]	For $\ddot{n} = 1$, $\dfrac{N_t}{N_0} \leq \dfrac{A_t}{A_0}$ [e.g., abortive infections]	For $\ddot{n} = 0$, $\dfrac{N_t}{N_0} = \dfrac{A_t}{A_0}$ [e.g., phage restriction]
Lower (bacterial gains dominate dynamics)	For $\mu_A - \mu_N > P(\ddot{k}\ddot{n} - k)$, $\dfrac{N_t}{N_0} < \dfrac{A_t}{A_0}$ [which, as $P \to 0$, is more likely]	$\dfrac{N_t}{N_0} < \dfrac{A_t}{A_0}$	[assuming phage-independent advantages to arrangement formation, i.e., $\mu_A - \mu_N > 0$, and that $\mu_A - \mu_N > P(\ddot{k}\ddot{n} - k)$ holds, which is likely given both $P \to 0$ and $\ddot{n} \to 0$]	

2.3. Utility of Group Living in Light of Phages

Above I argue that the likelihood of adsorption by any given phage can be lower for individual bacteria found within arrangements ($\ddot{k}$) than for free bacteria (k) but nonetheless that arrangements can be more vulnerable to exploitation by phages. If we assume that arrangements nevertheless provide bacteria with selective benefits, then it should be possible to consider how large these benefits must be to offset costs stemming from this presumed increased susceptibility of arrangements to phage infection. Before moving on to that issue, however, I first address two underlying considerations, (1) how existence within clonal associations in fact might benefit bacteria and (2) experimental evidence that phages can propagate through bacterial arrangements and/or microcolonies.

2.3.1. Selective Benefits of Living in Arrangements

Fitness advantages accrued by bacteria from living within groups generally can result from increased short-term growth rates, greater long-term rates of reproduction, increased population resistance to extinction, or greater competitive ability in terms of more effectively sustaining population densities. The latter might be accomplished by gaining better access to nutrients.

Adherence to surfaces in conjunction with subsequent growth as a microcolony, for instance, can retain a bacterial population within the vicinity of key resources, such as flowing water [48].

Continued association of cells following division similarly may allow for more effective penetration into resource-supplying substrates, *i.e.*, burrowing, particularly as seen among filamentous microorganisms growing in spatially structured environments such as soils. Aggregated cells also will tend to have larger collective activity domains [49]. This may be beneficial to bacteria by allowing better retention of extracellular regulatory molecules, particularly such as towards quorum sensing [1,50], or for the concentrating of other beneficial extracellular factors such as exoenzymes or exotoxins. Reduced levels of sharing with unrelated organisms of any molecules generated by these extracellular factors may be possible given bacteria association within arrangements and this may be so simply given greater densities of those cells within a given microvolume.

Growing in a single location can allow for more effective interspecific interactions [49] including crossfeeding [51] as well as development of closely spatially integrated microbial consortia [52,53]. Such associations might be enhanced by cells that grow to higher densities within a specific location, including as cell arrangements or instead as cells that simply fail to disperse and thereby form into microcolonies (or, instead, simply don't disperse very far). More generally, the development of favorable physiochemical gradients might be more readily achieved when cells are living within groups consisting at least in part of related individuals, *i.e.*, as consistent with the above-noted idea of cellular aggregates possessing larger collective activity domains *versus* bacteria that are not found in association with related bacteria. The result may be a higher potential to contribute to synergistic interspecific interactions. Indeed, "Any transfer between two such organisms will be immensely inefficient unless the concentrations of cells and substrates are high… Such interactions can take place between groups of bacteria where the size of the group means an enlarged activity domain and the retention of reactants in the vicinity of the group." [49] (pp. 56 and 58).

Bacteria can display movement when found within disfavorable environments, such as can be associated with planktonic or actively dispersing cells, and then cease such movement, such as through adherence to a surface, when found within a favorable environment. This is a behavior that, minimally, is equivalent to the concept of orthokinesis, that is, where the speed of an organism is modified by external cues or, indeed, thigmokinesis, where speed is modified by degree of physical contact with something else. Included among favorable environments and external cues can be other microbial species from which a given bacterium can derive benefits. Association with other microorganisms on a surface, rather than being stable, can as a consequence of thigmokinesis instead be associated with the invasion of later-arriving organisms. The result can be an interspecific competition for space or resources that can put a premium on competitive ability, which in turn may benefit from more effective arrangement or microcolony formation [50,54,55]. Full or partial replacement of one microbial species with another on surfaces more generally is an example of ecological succession, as has been documented particularly well in terms of dental biofilms [49,56].

Cellular clones growing in a single location have the potential to interact with each other cooperatively towards mutual benefit and do so in ways that are less available to free bacteria. This can include elevation of colonies into more nutrient-rich microenvironments [48], certain degrees of cellular differentiation and/or physiological specialization, fruiting body formation, and even seemingly intentional cell cannibalism under starvation conditions [1,57]. Perhaps consistent with this idea of bacteria being able to mutually cooperate when living within single-species groups, cells growing as biofilms are known to display an increased resistance to toxic chemicals found in their environments, which clinically can include both antibiotics and disinfectants [3,4]. Increased resistance may also be seen given association other microbial species [58]. Extracellular polymer production itself is often described as a potentially cooperative activity among related, physically associated bacteria [55].

Growth as arrangements may give rise to a greater resistance to engulfment by protozoa, and biofilms otherwise may be more resistant to protozoa-mediated predation than planktonic bacteria [59]. The resulting adaptations, such as "resisting ingestion, by becoming too large or too long... making themselves inaccessible, by growing in aggregates or biofilms", [41] (p. 3), however, may conflict with avoidance of phages by these same bacteria. Biofilm formation can also serve as a means of immune system resistance during infections by bacterial pathogens and this is due at least in part to interference with the action of phagocytes, though not necessarily solely via a direct blocking of bacterial engulfment [60,61,62]. More generally, utility that comes with being larger, whether individually per cell or instead as a consequence of cell-to-cell associations, is exploited by most non-bacterial cellular organisms including animals, plants, fungi, and protists.

2.3.2. Susceptibility of Bacterial Arrangements and Microcolonies to Phage Exploitation

To what degree are bacteria living in arrangements in fact phage susceptible? This question must be addressed particularly since assertions have been made that biofilm formation serves bacteria as an inherently phage-resistant state, as I review elsewhere [9]. Costerton *et al.* [63] in particular noted that (p. 440), "the gellike state of the predominantly polysaccharide biofilm matrix limits the access of antibacterial agents, such as... bacteriophage... Therefore, biofilm bacteria are substantially protected from... bacteriophage..." Elsewhere [8,9], however, I review the substantial potential, under experimental conditions, for phages to in fact considerably impact biofilms. Phages may not be well equipped to drive biofilm bacteria completely to extinction in the course of propagating on those bacteria, and phages that specialize in targeting biofilm bacteria may not always be highly prevalent in environments. Nevertheless, there is no evidence that I know of that would appear to indicate that biofilms are *not* relatively susceptible, or in many cases even highly susceptible to exploitation by phages.

As with the phage potential to clear bacterial biofilms, evidence also exists that phages can propagate through bacterial arrangements. Barron *et al.* [64], for example, state that "the phage released from a single coccus may infect other cocci in that chain." The evidence supporting that claim is an observation by Friend and Slade [65]. They found that one-step growth curves on a group A streptococcal strain in fact were two-step in practice, implying an initial arrangement adsorption followed by post-burst adsorption within the same arrangement; see also earlier work by Kjem, 1958 and 1964, as cited by Friend and Slade, as well as work by Fischetti *et al.* [66]. In the Friend and Slade study, two-step curves were then reduced to one-step curves through the separation of streptococci into individual, that is, "free" bacteria using sonication.

The tendency for phage plaques to be clear, particularly in their centers, also can be viewed as an indication of the tendency for phages to prevent phage-sensitive bacteria from propagating in the immediate vicinity of phage bursts. Indeed, plaque formation explicitly occurs within a context of phage penetration into and subsequent clearance of the bacterial microcolonies that make up bacterial lawns [67]. Consistently, Doolittle *et al.* [68] observed phage propagation within single-species bacterial biofilms, describing the dynamics as plaque-like. Conversely, plaque cloudiness can signify limitations on the ability of phages to clear bacterial microcolonies [9,13,69]. It should *not* be controversial therefore that phages can propagate to at least some extent into bacterial arrangements or microcolonies. Indeed, in phage therapy it is often assumed that phages can be quite adept at propagating through bacterial biofilms, a process that I have described elsewhere as an 'active penetration' [19]. These claims all come with the caveat, however, that such active penetration is not necessarily going to be the case for every combination of phage, bacterial arrangement, and circumstance.

2.3.3. Phage-Mediated Costs of Existing as Arrangements

For the sake of mathematical convenience, I consider especially the phage impact on instantaneous bacterial population growth rates rather than other aspects of bacterial fitness in addressing the extent to which phages might affect bacterial arrangements. Focusing on instantaneous growth rates, in particular, greatly simplifies the mathematics while at the same time avoids consideration of the difficult issue of exactly how bacterial arrangements propagate over longer time periods. The conclusions I reach, however, should be qualitatively applicable to other situations.

These considerations of phage impact on the instantaneous population growth of bacteria existing within arrangements can be easily formulated as a differential equation in which changes in bacterial density (N) are considered in terms of rates of cell division (μ) *versus* declines due to phage adsorption,

$$dN/dt = \mu_A N - n\ddot{k}P\ddot{n}N/n = N(\mu_A - \ddot{k}P\ddot{n}) \tag{7}$$

which compares with

$$dN/dt = \mu_N N - kPN = N(\mu_N - kP) \tag{8}$$

for free bacteria. Note that μ_A and μ_N are growth rates of bacteria associated with arrangements and free bacteria, respectively.

The expression $n\ddot{k}P\ddot{n}N/n$ in Equation (7), or simply kPN in Equation (8), describes those bacteria that have been lost from the unadsorbed bacterial pool, N, as a consequence of phage adsorption either of themselves (Equation (7) and Equation (8)) or of the arrangement in which those bacteria are located (Equation (7)). As noted above (*i.e.*, $A_0 = N_0/n$ in Section 2.2.2), the expression N/n as found in Equation (7) is a description of the density of arrangements consisting of n bacteria that are found in the environment in question. Also as above, the parameter $\ddot{n}$ describes the number of bacteria that will be lost to phage infection given phage adsorption of an arrangement. Lastly, $n\ddot{k}P$ is a description of the per-arrangement rate of bacterial adsorption by phages given an environmental phage density of P.

Per bacterium, Equation (7) thus describes adsorptions that occur at a rate of $\ddot{k}P$ and each of those adsorptions results in a loss of $\ddot{n}$ bacteria. This compares with the rate of loss of individual bacteria as described by kP in Equation (8). The expression, μN with either subscript, by contrast, describes in both equations the gains in bacterial density that occur as a consequence of bacterial replication. Thus, if $\ddot{k}P\ddot{n} > \mu_A$, then the bacterial population will experience a net decline in number whereas $\ddot{k}P\ddot{n} < \mu_A$ indicates net gains and $\ddot{k}P\ddot{n} = \mu_A$ defines a steady state. For Equation (8) the equivalent expressions instead are respectively $kP > \mu_N$, $kP < \mu_N$, and $kP = \mu_N$, where P in the latter can be described as an inundation threshold or even phage minimum inhibitory concentration, that is, of bacteria [25]. In the absence of phages, the bacterial population will simply grow at rates specified by μ_A or μ_N.

In Equation (7) the dynamics of microcolony formation are not considered and nor are various other complications such as multiple adsorption by "environmental" phages of individual bacterial arrangements, where by "environmental" I am distinguishing those phages defining P (=environmental) from phages that instead are explicitly propagating through bacterial arrangements. What Equation (7) nevertheless indicates is that the phage impact on bacterial arrangements varies as a function of phage density (P) in combination with the susceptibility of individual bacteria making up an arrangement to phages ($\ddot{k}\ddot{n}$). Specifically, the larger $\ddot{k}P\ddot{n}$ then the more bacteria found within arrangements that are lost to phage infection. For example, twice as many bacteria will be lost per unit time for bacterial arrangements ($\ddot{k}P\ddot{n}$ from Equation (7)) *versus* free bacteria (kP from Equation (8)) if $\ddot{k}\ddot{n}/k = 2$.

Alternatively, a total of $\ddot{k}\ddot{n}/k$-fold more arrangement-associated bacteria will be lost *versus* free bacteria for any given phage density, P. Forming into an arrangement, and thereby incurring costs of additional vulnerability to phages, therefore should be worthwhile to bacteria *only* to the extent that μ_A, or some other measure of bacterial fitness, increases as a consequence of group living to a larger extent than bacterial fitness decreases as a result of incurring a greater spatial vulnerability to phages, *i.e.*, as described by $\ddot{k}\ddot{n}/k$.

2.3.4. Importance of Reduced Vulnerability to Phages

The extreme situations with regard to Equation (7) are as follows: (i) If no phages are present in an environment ($P = 0$) then there will be no phage-associated cost to arrangement, microcolony, or biofilm formation (in which case no advantage is required from group living to offset costs of phage adsorption) or (ii) if phages are present at effectively infinite densities ($P = \infty$), or simply sufficiently high densities, then potentially no amount of phage-independent benefit to group living could offset increases in phage-associated vulnerabilities. The latter situation is just as dire for free bacteria, however, *i.e.*, for which $\ddot{n} = 1$ but nonetheless where it is possible for $\ddot{k}P\ddot{n} \approx kP \approx \infty$ (Equation (7) *versus* Equation (8)). At a minimum, however, $\mu_A = \ddot{k}P\ddot{n}$ is necessary for bacterial fitness, as measured here in terms of increases in bacterial growth rates, to offset costs due to phage adsorption, and this compares with $\mu_N = kP$ for free bacteria; see Abedon and Thomas-Abedon [19] along with references cited, or Abedon [25], for derivation of the latter.

For circumstances in which $kP\ddot{n} > kP$, then the growth rate or other measure of the fitness of bacterial arrangements must be greater than that of free bacteria by that amount, e.g., $\mu_A - \mu_N > \ddot{k}P\ddot{n} - kP = P(\ddot{k}\ddot{n} - k)$, to offset increased phage-associated costs that are borne by bacterial arrangements. This fitness improvement, however, need not be substantial unless phage densities (P) are also substantial. Thus, as $\ddot{n}$ or $\ddot{k}$ increase so too does the potential for phages to block the evolution of bacterial arrangements, but at the same time such increases do not serve as absolute blocks on this evolution. The alternative perspective is that given sufficiently high phage densities—but not too high, as indicated in the previous paragraph—then evolution could tend to favor reductions in $\ddot{n}$ even if bacteria otherwise experience benefits from forming into arrangements, that is, reduced formation of arrangements could serve as a bacterial anti-phage strategy. In simple terms, a coccus might encounter a phage approximately half as often as a diplococcus.

A generalization on these considerations is that bacteria that have formed into arrangements will have to avoid, on a per-bacterium basis, equivalent increases in vulnerability to phages in order to partake of whatever net advantages may be associated with forming into arrangements. This reduced vulnerability, furthermore, can result either from existing in environments in which phage densities are low (where phage-independent bacterial fitness "gains" can dominate competitive dynamics) or, alternatively, from greater resistance by individual bacteria to phage attack when phage densities are high, that is, when phage-mediated losses dominate competitive dynamics [30,31]. This idea of predator-independent aspects to fitness dominating prey evolution when predator densities are low (here bacteria and phages, respectively) but predator-resistance dominating when prey densities are high is a standard conclusion from community ecology. It is also seen in phage-bacterial chemostat studies as phage-sensitive bacteria, if they possess a growth-rate advantage to phage-resistant bacteria, out-compete phage-resistant bacteria when phage densities are lower but those same phage-sensitive bacteria are at a competitive disadvantage if phages instead are more numerous [70,71]. See Table 1 for a general summary of fitness expectations for arrangements *versus* free bacteria given higher *versus* lower arrangement vulnerability to phages.

2.3.5. Reduced Bacterial Densities as Phage-Resistance Strategy

Perhaps the simplest approach to reducing arrangement vulnerability to phages would be for those bacteria to exist at lower population densities, population densities, that is, which are insufficient to support phage population growth to levels at which those phages can substantially impact arrangement fitness. In other words, bacteria that are *less* able to attain "winner" densities within a given environment—as in "kill the winner" (Section 1)—may as a consequence be better able to exploit biofilm niches. This idea is similar to the conclusions of Ioannou *et al.* [36], as described in Section 2.1.1, who suggest that one means by which prey species can offset the costs of possessing greater individual sizes is by displaying lower population densities such that, as a prey type, they effectively are a greater distance, on average, from their predator and therefore less likely to be desirable to the predator. A major difference between the scenario presented by Ioannou *et al.* and that of kill the winner, however, is seen in terms of the degree of specialization of the predator species. For phages, specialization can be extreme such that bacteria phage-susceptibility types can "hide" by failing to support phage replication to inundative densities, that is, by not being "winners". For the visual predators considered by Ioannou *et al.*, by contrast, it is not that prey avoid predation by impacting predator densities but instead that they avoid being consumed by being less visible to predators that already exist at some more or less fixed density.

Biofilms often can occupy only a small fraction of total environmental volumes, such as in aquatic environments. In such circumstances, biofilm bacteria as a consequence may be inherently unable to achieve winner-level densities across an environment. This low potential, furthermore, may be particularly the case to the extent that biofilms consist of mixed populations of bacteria rather than monocultures of specific phage susceptibility types, thereby implying even lower densities of individual bacterial phage susceptibility types than available surfaces might maximally hold. Alternatively, biofilms found within highly structured environments, such as undisturbed soils [18], may over time tend to be mostly sequestered within micro-localities away from phages to which they are susceptible. Physically associated clonal groups of bacteria, such as can make up biofilms, in other words, may not be so much inherently resistant to phage attack as either relatively unexposed to specific phages or inherently less able to support the population growth of those phages to inundative densities [9].

An additional possibility may hold: Those bacteria which as a matter of luck are resistant to all phages present within environments might be able to attain and then sustain "winner" densities, either as free bacteria or as arrangements. This latter idea is ecologically similar to the domination of animal-rich ecosystems by plants that are resistant to forage by those herbivores which happen to be present within an environment [72]. That is, bottom-up control on bacterial density (nutrient availability) rather than top-down control (predator prevalence) would be expected to operate within ecosystems where predators are lacking whereas a combination of bottom-up and top-down control may hold when differences exist among prey in terms of their resistance to predation [70]. We might therefore predict a surfeit of bacteria existing as arrangements under three distinguishable circumstances: (i) where the abundance of specific phage susceptibility types of bacteria generally is low (potentially top-down control, 'kill the winner', and high bacterial diversity); (ii) where phage-resistant organisms dominate (combination of bottom-up and top-down control, and potentially lower bacterial diversity); or (iii) where phages simply are absent (bottom-up control with the diversity of bacteria therefore determined by factors other than phage-mediated predation).

3. Experimental Section

See Results and Discussion.

4. Conclusions

Group living bestows benefits—else why live in groups?—but also engenders costs. One cost comes from an increased vulnerability to exploitation that group living creates, such as a greater potential for infection of individual bacteria by bacteriophages. Here I have provided a simple model for quantifying those costs, one that points to the idea that bacterial arrangements along with microcolonies and biofilms may persist particularly under circumstances where high rates of lytic phage infection are unlikely. This conclusion is broadly consistent with Murray and Jackson's [24] suggestion, pp. 113 and 114, that "…for any given viral concentration, a large particle is more likely to have a virus reach its surface… Any given virus, however, is far more likely to reach a small, presumably bacterial, particle than a larger one in the ocean because of the greater abundance of small particles…" That is, individual prey size and vulnerability are not the only variables controlling the susceptibility of prey *populations* to predators, with prey population density also playing a key ecological role.

This perspective contrasts with notions that group living among bacteria might *directly* serve as a means of avoiding phage-mediated predation. Observation of the existence of specific bacterial types at high environmental densities and particularly as arrangements, microcolonies, or biofilms, however, would suggest the existence of effective mechanisms by which their vulnerability to phages has been reduced, just as pathogen-resistance mechanisms have, of course, evolved in multicellular organisms such as animals [32]. One reasonable scenario explaining such a situation is that diversity may exist within bacterial communities in terms of phage susceptibility even when that diversity is not superficially obvious. In this case, densities of individual bacteria types wouldn't be as high as they would appear—perhaps as a consequence of "kill the winner" (loss of bacterial populations existing at higher environmental densities)—because densities of specific phage susceptibility types would not be appreciably high. That situation represents the default assumption that one might make with regard to large, relatively open systems, that is, stabilizing frequency-dependent selection acting on rare phage susceptibility types that results in substantial bacterial diversity particularly in terms of phage resistance [73].

Alternatively, phages to which high-density bacterial populations are sensitive simply may not have reached those bacterial populations. That situation represents the default assumption that one might make with regard to small, relatively closed systems. The utility of phage therapy as an antibacterial strategy is that in many instances the infection to be treated can be described in the latter terms: small, relatively closed systems, such as localized or even systemic bacterial infections. Efficient bacterial eradication, including of bacterial biofilms, often can be achieved, therefore, simply by "opening" these systems sufficiently to phages, and the phage therapy literature, such as in terms of successful clinical treatment of infections in humans [74,75], appears to be consistent with that scenario.

Acknowledgments

Thank you to Vince Fischetti for helpful discussion and Bob Blasdel as well as two anonymous reviewers for helpful comments.

References

1. Stoodley, P.; Sauer, K.; Davies, D.G.; Costerton, J.W. Biofilms as complex differentiated communities. *Ann. Rev. Microbiol.* **2002**, *56*, 187–209.

2. Kjelleberg, S.; Givskov, M. *The Biofilm Mode of Life: Mechanisms and Adaptations*; Horizon Biosciences: Norfolk, UK, 2007.

3. Ramage, G.; Culshaw, S.; Jones, B.; Williams, C. Are we any closer to beating the biofilm: Novel methods of biofilm control. *Curr. Opin. Infect. Dis.* **2010**, *23*, 560–566.

4. Gino, E.; Starosvetsky, J.; Kurzbaum, E.; Armon, R. Combined chemical-biological treatment for prevention/rehabilitation of clogged wells by an iron-oxidizing bacterium. *Environ. Sci. Technol.* **2010**, *44*, 3123–3129.

5. Cos, P.; Tote, K.; Horemans, T.; Maes, L. Biofilms: An extra hurdle for effective antimicrobial therapy. *Curr. Pharm. Des.* **2010**, *16*, 2279–2295.

6. Abedon, S.T. Kinetics of phage-mediated biocontrol of bacteria. *Foodborne Pathog. Dis.* **2009**, *6*, 807–815.

7. Loc-Carrillo, C.; Abedon, S.T. Pros and cons of phage therapy. *Bacteriophage* **2011**, *1*, 111–114.

8. Abedon, S.T. Bacteriophages and biofilms. In *Biofilms: Formation, Development and Properties*; Bailey, W.C., Ed.; Nova Science Publishers: Hauppauge, NY, USA, 2010; Chapter 1, pp. 1–58.

9. Abedon, S.T. *Bacteriophages and Biofilms: Ecology, Phage Therapy, Plaques*; Nova Science Publishers: Hauppauge, NY, USA, 2011.

10. Abedon, S.T.; Thomas-Abedon, C.; Thomas, A.; Mazure, H. Bacteriophage prehistory: Is or is not Hankin, 1896, a phage reference? *Bacteriophage* **2011**, *1*, 174–178.

11. Twort, F.W. An investigation on the nature of ultra-microscopic viruses. *Lancet* **1915**, *ii*, 1241–1243.

12. Twort, F.W. An investigation on the nature of ultra-microscopic viruses. *Bacteriophage* **2011**, *1*, 127–129.

13. Abedon, S.T.; Yin, J. Impact of spatial structure on phage population growth. In *Bacteriophage Ecology*; Abedon, S.T., Ed.; Cambridge University Press: Cambridge, UK, 2008; Volume 15, Chapter 4, pp. 94–113.

14. Krone, S.M.; Abedon, S.T. Modeling phage plaque growth. In *Bacteriophage Ecology*; Abedon, S.T., Ed.; Cambridge University Press: Cambridge, UK, 2008; Volume 15, Chapter 16, pp. 415–438.

15. Abedon, S.T.; Yin, J. Bacteriophage plaques: Theory and analysis. *Meth. Mol. Biol.* **2009**, *501*, 161–174.

16. Gallet, R.; Shao, Y.; Wang, I.N. High adsorption rate is detrimental to bacteriophage fitness in a biofilm-like environment. *BMC Evol. Biol.* **2009**, *9*, 241.

17. Gallet, R.; Kannoly, S.; Wang, I.N. Effects of bacteriophage traits on plaque formation. *BMC Microbiol.* **2011**, *11*, 181.

18. Abedon, S.T. Communication among phages, bacteria, and soil environments. In *Biocommunication of Soil Microorganisms*; Witzany, G., Ed.; Springer: New York, NY, USA, 2011; Volume 23, Chapter 2, pp. 37–65.

19. Abedon, S.T.; Thomas-Abedon, C. Phage therapy pharmacology. *Curr. Pharm. Biotechnol.* **2010**, *11*, 28–47.

20. Gill, J.J.; Hyman, P. Phage choice, isolation and preparation for phage therapy. *Curr. Pharm. Biotechnol.* **2010**, *11*, 2–14.

21. Lu, T.K.; Collins, J.J. Dispersing biofilms with engineered enzymatic bacteriophage. *Proc. Natl. Acad. Sci. USA* **2007**, *104*, 11197–11202.

22. Goodridge, L.D. Designing phage therapeutics. *Curr. Pharm. Biotechnol.* **2010**, *11*, 15–27.

23. Mondes, R.; O'Toole, G.A. The developmental model of microbial biofilms: Ten years of a paradigm up for review. *Trends Microbiol.* **2009**, *17*, 73–87.

24. Murray, A.G.; Jackson, G.A. Viral dynamics: A model of the effects of size, shape, motion, and abundance of single-celled planktonic organisms and other particles. *Mar. Ecol. Prog. Ser.* **1992**, *89*, 103–116.

25. Abedon, S. Phage therapy pharmacology: Calculating phage dosing. *Adv. Appl. Microbiol.* **2011**, *77*, 1–40.

26. Thingstad, T.F. Elements of a theory for the mechanisms controlling abundance, diversity, and biogeochemical role of lytic bacterial viruses in aquatic systems. *Limnol. Oceanogr.* **2000**, *45*, 1320–1328.

27. Thingstad, T.F.; Bratbak, G.; Heldal, M. Aquatic Phage Ecology. In *Bacteriophage Ecology*; Abedon, S.T., Ed.; Cambridge University Press: Cambridge, UK, 2008; Chapter 10, pp. 251–280.

28. Chao, L.; Levin, B.R.; Stewart, F.M. A complex community in a simple habitat: An experimental study with bacteria and phage. *Ecology* **1977**, *58*, 369–378.

29. Schrag, S.; Mittler, J.E. Host-parasite persistence: The role of spatial refuges in stabilizing bacteria-phage interactions. *Am. Nat.* **1996**, *148*, 348–377.

30. Hyman, P.; Abedon, S.T. Bacteriophage host range and bacterial resistance. *Adv. Appl. Microbiol.* **2010**, *70*, 217–248.

31. Labrie, S.J.; Samson, J.E.; Moineau, S. Bacteriophage resistance mechanisms. *Nat. Rev. Microbiol.* **2010**, *8*, 317–327.

32. Abedon, S.T. Bacterial 'immunity' against bacteriophages. *Bacteriophage* **2012**, *2*, 50–54.

33. Abedon, S.T. Lysis of lysis inhibited bacteriophage T4-infected cells. *J. Bacteriol.* **1992**, *174*, 8073–8080.

34. Abedon, S.T. Envisaging bacteria as phage targets. *Bacteriophage* **2011**, *1*, 228–230.

35. Stent, G.S. *Molecular Biology of Bacterial Viruses*; WH Freeman and Co.: San Francisco, CA, USA, 1963.

36. Ioannou, C.C.; Bartumeus, F.; Krause, J.; Ruxton, G.D. Unified effects of aggregation reveal larger prey groups take longer to find. *Proc. Biol. Sci* **2011**, *278*, 2985–2990.

37. Abedon, S.T. Phage Population Growth: Constraints, Games, Adaptation. In *Bacteriophage Ecology*; Abedon, S.T., Ed.; Cambridge University Press: Cambridge, UK, 2008; Volume 15, Chapter 3, pp. 64–93.

38. Kasman, L.M.; Kasman, A.; Westwater, C.; Dolan, J.; Schmidt, M.G.; Norris, J.S. Overcoming the phage replication threshold: A mathematical model with implications for phage therapy. *J. Virol.* **2002**, *76*, 5557–5564.

39. Abedon, S.T. Bacteriophage T4 resistance to lysis-inhibition collapse. *Genet. Res.* **1999**, *74*, 1–11.

40. Young, K.D. The selective value of bacterial shape. *Microbiol. Mol. Biol. Rev.* **2006**, *70*, 660–703.

41. Young, K.D. Bacterial morphology: Why have different shapes? *Curr. Opin. Microbiol.* **2007**, *10*, 596–600.

42. Azeredo, J.; Sutherland, I.W. The use of phages for the removal of infectious biofilms. *Curr. Pharm. Biotechnol.* **2008**, *9*, 261–266.

43. Turner, G.F.; Pitcher, T.J. Attack abatement: A model for group protection by combined avoidance and dilution. *Am. Nat.* **1986**, *128*, 228–240.

44. Babic, A.; Berkmen, M.B.; Lee, C.A.; Grossman, A.D. Efficient gene transfer in bacterial cell chains. *MBio* **2011**, *2*, e00027-11.

45. Abedon, S.T. Phages, ecology, evolution. In *Bacteriophage Ecology*; Abedon, S.T., Ed.; Cambridge University Press: Cambridge, UK, 2008; Volume 15, Chapter 1, pp. 1–28.

46. Hagens, S.; Loessner, M.J. Bacteriophage for biocontrol of foodborne pathogens: Calculations and considerations. *Curr. Pharm. Biotechnol.* **2010**, *11*, 58–68.

47. Goodridge, L.D. Phages, bacteria, and food. In *Bacteriophage Ecology*; Abedon, S.T., Ed.; Cambridge University Press: Cambridge, UK, 2008; Chapter 12, pp. 302–331.

48. Xavier, J.B.; Foster, K.R. Cooperation and conflict in microbial biofilms. *Proc. Natl. Acad. Sci. USA* **2007**, *104*, 876–881.

49. Wimpenny, J. Ecological determinants of biofilm formation. *Biofouling* **1996**, *10*, 43–63.

50. Nadell, C.D.; Xavier, J.B.; Levin, S.A.; Foster, K.R. The evolution of quorum sensing in bacterial biofilms. *PLoS Biol.* **2008**, *6*, e14.

51. Crombach, A.; Hogeweg, P. Evolution of resource cycling in ecosystems and individuals. *BMC Evol. Biol.* **2009**, *9*, 122.

52. Hoffmeister, M.; Martin, W. Interspecific evolution: Microbial symbiosis, endosymbiosis and gene transfer. *Environ. Microbiol.* **2003**, *5*, 641–649.

53. Searcy, D.G. Metabolic integration during the evolutionary origin of mitochondria. *Cell Res.* **2003**, *13*, 229–238.

54. Kreft, J.U. Biofilms promote altruism. *Microbiology* **2004**, *150*, 2751–2760.

55. Nadell, C.D.; Bassler, B.L. A fitness trade-off between local competition and dispersal in *Vibrio cholerae* biofilms. *Proc. Natl. Acad. Sci USA* **2011**, *108*, 14181–14185.

56. Takahashi, N.; Nyvad, B. The role of bacteria in the caries process: Ecological perspectives. *J. Dent. Res.* **2011**, *90*, 294–303.

57. Kolter, R. Biofilms in lab and nature: A molecular geneticist's voyage to microbial ecology. *Int. Microbiol.* **2010**, *13*, 1–7.

58. Whiteley, M.; Ott, J.R.; Weaver, E.A.; McLean, R.J.C. Effects of community composition and growth rate on aquifer biofilm bacteria and their susceptibility to betadine disinfection. *Environ. Microbiol.* **2001**, *3*, 43–52.

59. Matz, C. Biofilms as refuge against predation. In *The Biofilm Mode of Life: Mechanisms and Adaptations*; Kjelleberg, S., Givskov, M., Eds.; Horizon Bioscience: Norfolk, UK, 2007; Chapter 11, pp. 195–213.

60. Thurlow, L.R.; Hanke, M.L.; Fritz, T.; Angle, A.; Aldrich, A.; Williams, S.H.; Engebretsen, I.L.; Bayles, K.W.; Horswill, A.R.; Kielian, T. *Staphylococcus aureus* biofilms prevent macrophage phagocytosis and attenuate inflammation *in vivo*. *J. Immunol.* **2011**, *186*, 6585–6596.

61. Arciola, C.R. Host defense against implant infection: The ambivalent role of phagocytosis. *Int. J. Artif. Organs* **2010**, *33*, 565–567.

62. Jensen, P.O.; Givskov, M.; Bjarnsholt, T.; Moser, C. The immune system *vs. Pseudomonas aeruginosa* biofilms. *FEMS Immunol. Med. Microbiol.* **2010**, *59*, 292–305.

63. Costerton, J.W.; Cheng, J.-J.; Geesey, G.G.; Ladd, T.I.; Nickel, J.C.; Dasgupta, M.; Marrie, T.J. Bacterial biofilms in nature and disease. *Ann. Rev. Microbiol.* **1987**, *41*, 435–464.

64. Barron, B.A.; Fischetti, V.A.; Zabriskie, J.B. Studies of the bacteriophage kinetics of multicellular systems: A statistical model for the estimation of burst size per cell in streptococci. *J. Appl. Bacteriol.* **1970**, *33*, 436–442.

65. Friend, P.L.; Slade, A.D. Characteristics of group A streptococcal bacteriophages. *J. Bacteriol.* **1966**, *92*, 148–154.

66. Fischetti, V.A.; Barron, B.; Zabriskie, J.B. Studies on streptococcal bacteriophages. I. burst size and intracellular growth of group A and group C streptococcal bacteriophages. *J. Exp. Med.* **1968**, *127*, 475–488.

67. Kaplan, D.A.; Naumovski, L.; Rothschild, B.; Collier, R.J. Appendix: A model of plaque formation. *Gene* **1981**, *13*, 221–225.

68. Doolittle, M.M.; Cooney, J.J.; Caldwell, D.E. Tracing the interaction of bacteriophage with bacterial biofilms using fluorescent and chromogenic probes. *J. Indust. Microbiol.* **1996**, *16*, 331–341.

69. Abedon, S.T. Bacteriophage Intraspecific Cooperation and Defection. In *Contemporary Trends in Bacteriophage Research*; Adams, H.T., Ed.; Nova Science Publishers: Hauppauge, NY, USA, 2009; Chapter 7, pp. 191–215.

70. Bohannan, B.J.M.; Lenski, R.E. Effect of prey heterogeneity on the response of a food chain to resource enrichment. *Am. Nat.* **1999**, *153*, 73–82.

71. Bohannan, B.J.M.; Lenski, R.E. Linking genetic change to community evolution: Insights from studies of bacteria and bacteriophage. *Ecol. Lett.* **2000**, *3*, 362–377.

72. Terborgh, J.; Lopez, L.; Nunez, V.; Rao, M.; Shahabuddin, G.; Orihuela, G.; Riveros, M.; Ascanio, R.; Adler, G.H.; Lambert, T.D.; *et al.* Ecological meltdown in predator-free forest fragments. *Science* **2001**, *294*, 1923–1926.

73. Abedon, S.T. Phage evolution and ecology. *Adv. Appl. Microbiol.* **2009**, *67*, 1–45.

74. Kutter, E.; De Vos, D.; Gvasalia, G.; Alavidze, Z.; Gogokhia, L.; Kuhl, S.; Abedon, S.T. Phage therapy in clinical practice: Treatment of human infections. *Curr. Pharm. Biotechnol.* **2010**, *11*, 69–86.

75. Abedon, S.T.; Kuhl, S.J.; Blasdel, B.G.; Kutter, E.M. Phage treatment of human infections. *Bacteriophage* **2011**, *1*, 66–85.

Utility of the Bacteriophage RB69 Polymerase gp43 as a Surrogate Enzyme for Herpesvirus Orthologs

Nicholas Bennett [1] and Matthias Götte [1,2,3,*]

[1] Department of Microbiology and Immunology, McGill University, 3775 University Street, Montreal, Quebec H3A 2B4, Canada

[2] Department of Biochemistry, McGill University, 3655 Sir William Osler Promenade, Montreal, Quebec H3G1Y6, Canada

[3] Department of Medicine, Division of Experimental Medicine, McGill University, 1110 Pine Avenue West, Montreal, Quebec H3A 1A3, Canada

* Author to whom correspondence should be addressed; E-Mail: matthias.gotte@mcgill.ca

Abstract: Viral polymerases are important targets in drug discovery and development efforts. Most antiviral compounds that are currently approved for treatment of infection with members of the *herpesviridae* family were shown to inhibit the viral DNA polymerase. However, biochemical studies that shed light on mechanisms of drug action and resistance are hampered primarily due to technical problems associated with enzyme expression and purification. In contrast, the orthologous bacteriophage RB69 polymerase gp43 has been crystallized in various forms and therefore serves as a model system that provides a better understanding of structure–function relationships of polymerases that belong the type B family. This review aims to discuss strengths, limitations, and opportunities of the phage surrogate with emphasis placed on its utility in the discovery and development of anti-herpetic drugs.

Keywords: DNA polymerase; T4 DNA polymerase; gp43; *herpesviridae*; UL30; UL54; HSV1; HCMV; RB69 DNA polymerase

Nomenclature

HSV1	Herpes Simplex Virus 1
HSV2	Herpes Simplex Virus 2
VZV	Varicella Zoster Virus
EBV	Epstein–Barr Virus
HCMV	Human cytomegalovirus
HHV6	Human Herpesvirus 6
HHV7	Human Herpesvirus 7
KSHV	Kaposi's sarcoma-associated herpesvirus
PFA	Phosphonoformic acid
PAA	Phosphonoacetic acid
ACV	Acyclovir
GCV	Ganciclovir
CDV	Cidofovir

1. Introduction

The eukaryotic viruses of the *herpesviridae* family are important human pathogens. In all, there are eight different human herpesviruses; Herpes simplex virus 1 and 2 (HSV1: HHV1 and HSV2: HHV2), varicella zoster virus (VZV: HHV3), Epstein-Barr virus (EBV: HHV4), Human cytomegalovirus (HCMV: HHV5), Human Herpes virus 6 and 7 (HHV6 and HHV7) and Kaposi's sarcoma-associated herpesvirus (KSHV, HHV8). Human herpesviruses cause a spectrum of diseases ranging from relatively benign cutaneous lesions to serious conditions like encephalitis and cancer. Viruses that belong to the *herpesviridae* family are characterized by their ability to establish lifelong, latent infections. Thus, a *substantial proportion of the global population is seropositive for one or more herpesviridae viruses. Although individuals with a functioning immune system can generally keep the virus suppressed, the ability to form latent infections, and the fact that the virus is widespread in the human population means that herpesvirus reactivation is a major source of disease and morbidity in immunocompromised individuals.*

The majority of approved antiviral drugs have been shown to inhibit the herpesvirus-specific DNA polymerase, reducing viral DNA replication, and, in turn, viral load [1]. However, although herpesvirus polymerases are all structurally related, they are not highly homologues. As a consequence, most drugs do not show broad antiviral activities against the various members of the *herpesviridae*. The nucleoside analog acyclovir (ACV) and its pro-drug valacyclovir are utilized to treat infection with HSV1, HSV2 or VZV, while the nucleotide analog ganciclovir (GCV) (or valganciclovir) and cidofovir (CDV) are approved to manage HCMV infection. The pyrophosphate analog phosphonoformic acid (PFA, foscarnet) provides an option to treat HSV1, HSV2, VZV and HCMV, if first-line drugs have failed to lower the viral burden. Like all current antiviral treatments, long-term treatment can lead to the development of drug resistance. Severe side effects and complicated treatment schedules represent other problems in the management of herpesvirus infection.

Unfortunately, the development of assays to screen for novel anti-herpetic DNA polymerase inhibitors has been limited by technical problems. For the purpose of biochemical screens, *herpesviridae* DNA polymerases are difficult to overexpress in heterologous expression systems and have limited solubility. Hence, it has been difficult to characterize structural and functional details of these polymerases [2–8]. Of the eight human herpesvirus DNA polymerases, the best-studied is perhaps UL30 from HSV1. This enzyme has been characterized extensively biochemically and has been successfully crystallized [9]. Progress has also been made in characterizing HCMV UL54 [10,11]

In contrast to *herpesviridae* DNA polymerases, the orthologues enzymes of bacteriophage T4 (T4gp43) and "T4 like" bacteriophage RB69 (RB69gp43) are well studied. T4gp43 has been studied extensively using genetic, molecular biology, and biochemistry. Research into T4gp43 has been key to our current understanding of the dynamics of DNA replication [12]. RB69gp43 has been crystallized in various forms and therefore provides an important structural model for polymerases that belong to the same family [13–16]. It is here attempted to discuss the general aspects of structure and function of these related enzymes and the utility of RB69gp43 as a surrogate system for *herpesviridae* DNA polymerases in efforts to provide a better understanding of mechanisms of drug action and resistance.

2. Structure and Function of B Family Polymerases

DNA dependant DNA polymerases can be subdivided into five different families based on sequence and structural homology [17]. The DNA polymerases of bacteriophage RB69 and the *herpesviridae* are classified as B family polymerases (Figure 1a,b) [18]. B family polymerases have been identified in all domains of life and are primary involved in genome replication [19]. Unlike other polymerase families, the B family polymerases form part of a multi-subunit complexes, sometime referred to as the DNA replisome, which can co-ordinate both leading and lagging strand replication [17]. However, the polymerase catalytic activity of B family DNA polymerases is encoded by a single gene, which is sometimes referred to as the DNA polymerase catalytic subunit [20]. The catalytic subunit also often encodes an intrinsic 3'–5' exonuclease activity which provides proofreading. This substantially increases the accuracy of DNA synthesis [21,22]. The B family catalytic subunit, in the presence of the polymerase accessory proteins, is both high faithful in replicating DNA and are highly processive [12].

RB69 and each of the members of the *herpesviridae* family encode a B family polymerase (Figures 1 and 2). The virally encoded polymerase serves to replicate the viral genome. Both RB69gp43 [13] and HSV1 UL30 [9] have been studied using X-ray crystallography. Both polymerases are composed of five conserved structural domains, referred to as N-terminal, 3'–5' exonuclease, palm, fingers and thumb subdomains. In addition to these five conserved domains, the x-ray crystal structure of HSV1 UL30 showed an extra domain at the N-terminal end of the protein, which is called the pre N-terminal domain (Figure 1b).

Figure 1. (a) Domain structure of HSV1 UL30 (pdb 2GV9) [9]. The pre N-terminal domain is shown in white, the N-terminal domain is yellow, the exonuclease domain is red, the palm domain is magenta, the fingers domain is blue and the thumb domain is green. (b) The structure of the RB69gp43 apo form (pdb file 1IH7) [15]. Both structures show the fingers subdomain the open conformation. Images were generated using Pymol [23].

(a)

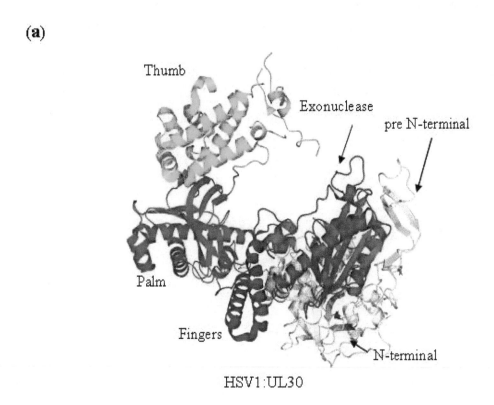

HSV1:UL30

(b)

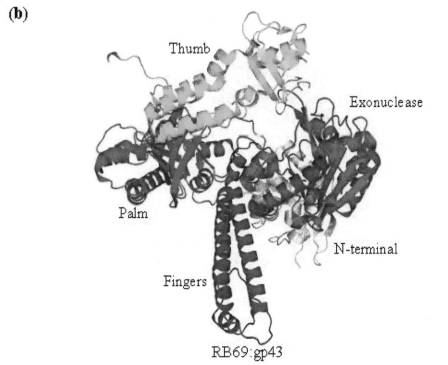

RB69:gp43

Figure 2. Protein sequence alignment of herpesvirus and bacteriophage polymerase finger domains. Herpesvirus and bacteriophage polymerase were aligned individually using the Muscle algorithm within Geneious [24,25]. The bacteriophage and herpesvirus sequences were then structurally aligned by RAPIDO [26]. Blocks above sequence highlight structural domains of polymerase. The palm domain is in pink. The fingers domain is in blue. The known conserved regions are shown in magenta blocks above sequence [27–31]. Secondary structural elements of HSV1 UL30 are indicated and are number according to Liu *et al.* (2006). Secondary structural elements of RB69 gp43 are indicated and are numbered according to Wang *et al.* (1997). Structural motifs are highlighted in sequence. N helix residues are in yellow. Motif B is in green. Mutations that have been associated with resistance to current anti-herpetic drugs are shown below the corresponding residue [32,33]. Resistance mutations are colored using the following scheme: Red: Pyrophosphate^R (Resistant), Blue: Nucleotide^R, Green: Pyrophosphate^R and Nucleotide^R. Purple: Pyrophosphate^{HS} (Hypersensitive), Brown: Nucleotide^R but Pyrophosphate^{HS}.

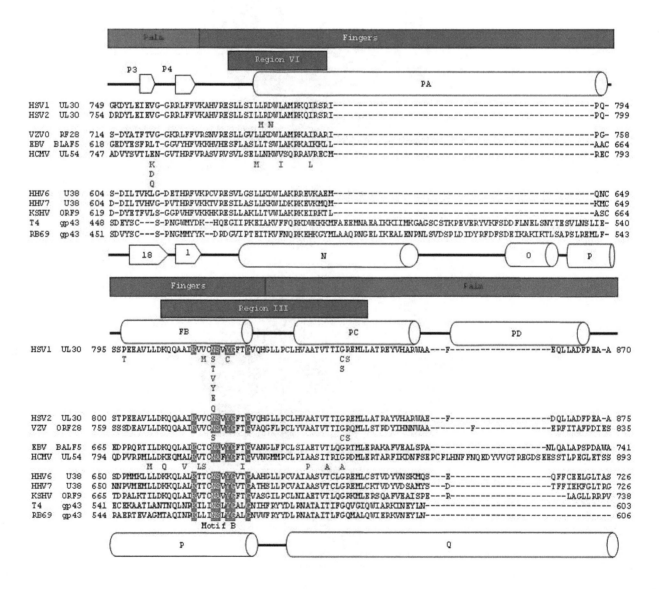

Structures of RB69gp43 in various forms in the absence and presence of substrates provided a detailed insight into distinct events involved in nucleotide incorporation, as well as the dynamics of exonuclease function. The structure of the HSV1 UL30 apo enzyme shows a similar domain structure as seen with RB69gp43; however, the exact structural requirements for DNA and nucleotide binding have yet to be established in this case. HSV1 UL30 contains the particular conserved motifs shared by all B family polymerases (Appendix Figure A1).

3. Subdomains of B Family Polymerase

Protein sequence alignments of HSV1 UL30 to other *herpesviridae* DNA polymerase show that it is likely that all Human *herpesviridae* DNA polymerases contain a pre-N terminal domain [9]. The exact function of the pre-N terminal this domain remains elusive. The pre-N terminal domain also contains the FYNPYL motif specific to *herpesviridae* family polymerases (Appendix Figure A1) [9]. It has recently been shown that this motif is required for efficient replication of viral DNA synthesis *in vivo*; a mutant polymerase lacking the FYNPYL motif showed a substantial reduction in viral DNA synthesis [34]. However, the purified FYNPYL deletion mutant showed no reduction in polymerase activity, suggesting that this motif may have a function in the formation of the viral DNA replisome.

The N-terminal domain shows a βαββαβ fold, which has been found in some RNA binding proteins [9,35]. In addition, the crystal structure of the RB69 gp43 contains a rGMP bound to the N-terminal domain [15], and some mutations in the N-terminal domain of T4gp43 decrease the expression of the polymerase leading to the suggestion that the N-terminal domain may be involved in expression regulation [36]. However, in spite of these observations, the functional role of this domain remains to be defined.

The sequence of the 3'–5' exonuclease domain of B family polymerases is not highly conserved. However, all 3'–5' exonuclease domains currently characterized adopt a ribonuclease H-like (RNase H-like) fold. The RNase H fold brings four highly conserved negatively charged residues together to form the active site. In both RB69gp43 and HSV1 UL30 these residues have been identified as three aspartic acids and a glutamic acid. These residues are essential for the binding of two divalent, catalytic metal ions. Structural elements, that harbor active site residues in RB69gp43, are referred to as exoI (D114, E116), exoII (D222), and exoIII (D327) (Table 1, Appendix Figure A1). In *herpesviridae* polymerases the equivalent regions are referred to as ExoI, region IV and Delta (δ) C, respectively. In HSV1 UL30 these residues are; ExoI, D368 and E370, region IV or ExoII, D471 and delta C or ExoIII, D581 (Table 1, Appendix Figure A1).

Figure 3. (**a**) Superpositioning of open and closed structures of RB69 gp43 showing finger domain movement. This diagram is composed of RB69 gp43 in the fingers closed position (pdb 3LDS) [37] and in the fingers opened position (pdb 1IH7) [15]. (**b**) Polymerase active site of RB69 gp43 showing interactions between conserved residues of motif A and C, metal ions A and B and dNTP and interactions between K560 and dNTP. This image is an aligned composite image of pdb 3LDS [37] and 3SCX [38]. (**c**) Structural alignment of the polymerase active site of RB69gp43 (pdb 3LDS) and HSV1 UL30 (pdb 2GV9). RB69 gp43 backbone is in light blue while the HSV1 UL30 backbone is in light orange. Active site residues of RB69 gp43 are indicated. RB69 gp43 motif A is in magenta, motif C is in orange and KKRY is in purple. HSV1 UL30 motif A is in pink, motif C is in light orange and KKRY is in light purple. Images were generated using Pymol [23]. (**d**) Generalized diagram of the polymerase catalytic cycle showing steps at which inhibitors can act. Nucleotide inhibitors, once incorporated, prevent further extension of the DNA primer, by inhibiting nucleotidyl transfer. Whereas, pyrophosphate inhibitors mimicking the pyrophosphate leaving group, stabilizing the pre-translocation complex and prevent translocation.

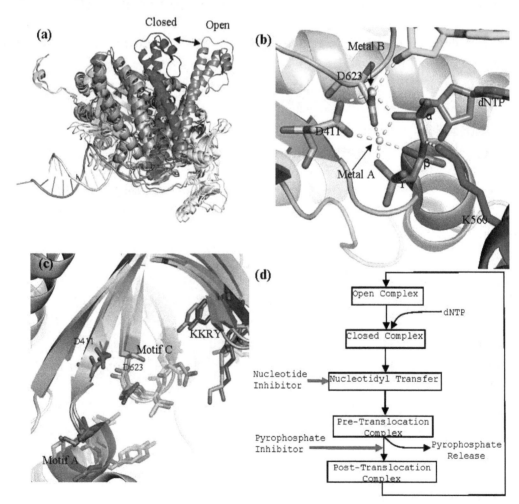

The polymerase active site of B family polymerases is made up of the three domains; the palm, fingers and thumb domains. Together they adopt the classic right hand conformation seen in all available structures of viral polymerases (Figure 1). Two highly conserved motifs in the palm domain, one in the fragment of the palm domain prior to the fingers domain called motif A and one in the fragment post fingers domain called motif C (Table 1, Appendix Figure A1) are likewise seen as signature motifs. In *herpesviridae* polymerases these domains are referred to as region II and region I, respectively. These motifs are; motif A; DXXLYPS and motif C; DTDS (Table 1, Appendix Figure A1). Structurally, these motifs fold together to form a three-strand anti-parallel β sheet (Figure 3c) [13]. Two conserved aspartic acid residues, D411 from motif A and the D625 from motif C, are required to form critical interactions which help co-ordinate the two divalent metal cation that are critical for DNA polymerization (Figure 3b) [15]. In addition, to motif A and C there are two other conserved motifs which also form parts the active site, motif B (KXXXNSXYG), which is known as Region III in *herpesviridae*, which is located on the helix P of the fingers domain, and the KKRY motif which is in the palm domain sequentially after motif C (Appendix Figure A1).

Table 1. Comparison of size, weight and position of conserved motifs or active site residues of RB69 and *herpesviridae* DNA polymerases. The position conserved motifs where assigned based on alignments generated using geneious [24].

Virus (Gene)	Amino Acids (aa)	Weight (kDa)	FYNPYL motif	Exo 1	Exo2/ region IV	Exo3/ delta (ä) C	Motif A/ region II	Motif B/ region III	Motif C/ region I	KKRY motif
RB69 (gp43)	903	104.47	N/A	113–117	222	327	411–420	560–571	621–624	804–807
HSV1 (UL30)	1236	136.42	167–173	367–371	471	581	717–726	811–822	886–889	938–941
HSV2 (UL30)	1241	137.32	166–172	368–372	472	582	722–731	816–827	891–894	943–946
VZV (Orf28)	1195	134.05	7–13	348–352	452	562	682–691	775–786	851–854	903–906
EBV (BALF5)	1016	113.43	6–12	295–299	384	497	584–593	681–692	755–758	807–810
HCMV (UL54)	1242	137.21	2–8	300–304	413	542	717–726	811–822	910–913	962–965
HHV6 (UL38)	1013	115.67	6–12	281–285	369	482	572–581	666–677	740–743	792–795
HHV7 (UL38)	1014	115.91	6–12	280–284	368	480	572–581	666–677	740–743	792–795
KSHV (Orf9)	1013	113.33	3–9	295–299	383	498	585–594	681–692	752–755	804–807

Residues K560 and N564 of motif B are important in coordinating the tri-phosphate tail of the incoming dNTP during catalysis [15]. Lysine 560 also serves as a proton donor during catalysis (Figure 3b) [39]. Residue Y567 of motif B has been shown to be involved in forming an important interaction with the minor groove of the DNA [40]. This interaction has been shown to be highly important in maintaining polymerase fidelity [40]. In addition to motif B of helix P, there are also conserved positively charged residues within helix N (R482 and K486) that form important interactions with the tri-phosphate tail of the incoming dNTP during catalysis.

The KKRY motif is primarily involved in stabilizing the B form of DNA. Residue Y708 forms a hydrogen bond with the 3' terminus of the primer while the K705 and R707 form interactions with the phosphate backbone. Together, these interactions help stabilize the interactions between base pairs of the primer and template strands [15].

4. Catalytic Cycle of B Type Polymerase RB69gp43

Both RB69gp43 and HSV1 UL30 have been crystallized in the so-called open conformation [9,13], but only RB69 has been crystallized in the closed conformation [15]. The closed conformation of a ternary complex contains a DNA primer-template pair and a trapped nucleotide. The most striking difference between the open and closed formation of RB69gp43 is the movement of the fingers domain. Structural alignment of the open and close conformation of RB69gp43 shows that, upon dNTP binding, the fingers domain rotates 60° inwards relative to the palm domain, with the tip of the fingers moving approximately 30 Å (Figure 3a) [15]. Closure of the fingers domain moves the residues on helix N and P that are involved in binding the tri-phosphate tail of the incoming dNTP, 4–8 Å closer to the polymerase active site. This action traps the dNTP in the active site and allows the nucleotidyl transfer to take place. By contrast, the overall structure of the thumb and palm domain between the open and closed conformation remains relatively unchanged. The thumb domain moves approximately 8° toward the palm domain. This action wraps the minor groove of primer-template duplex [15].

B family polymerases employ a two divalent metal ion mechanism for the nucleotidyl transfer [41], in conjunction with a concurrent two proton transfer reaction [39,42]. The catalytic ions are bound by D411 of motif A and D623 of motif C, and form an extensive network of interactions with the dNTP aligning it in the correct orientation for polymerization (Figure 3b). Metal ion A is required to activate the 3'-OH of the primer terminus. Interaction between metal ion A and the primer terminus attract the primer terminus closer to the α-phosphorus atom of the incoming dNTP. This lowers the pKa of the 3'-OH group allowing it to be deprotonated, which facilitates the nucleophilic attack on the α-phosphorous atom of the nucleotide substrate [43]. Metal ion B orientates the dNTP triphosphate tail and helps stabilizes the transition states; it has also been suggested that it assists in pyrophosphate release [38]. In RB69gp43, lysine 560 acts as a proton donor [42] and is required to protonates the pyrophosphate leaving group, which may facilitate its release from the complex. This step formally ends the catalysis, leaving the complex now in the pre-translocational state. The fingers of the polymerase rotate away from the active site allowing the release of the pyrophosphate, which allows the DNA substrate to translocate relative to the enzyme. This movement shifts the new 3'-OH terminus into the 1+ position forming a post-tranlocated complex with the polymerase reset for a new catalytic cycle. Polymerases are in general able to discriminate between correct and incorrect nucleotides [44]. Incorrect binding of a nucleotide destabilizes the closed ternary complex, which enables the fingers domain to return to the open form releasing the incorrect nucleotide. Effective discrimination against the incorrect nucleotide at the level of substrate binding raises the fidelity of DNA synthesis significantly[45]. However, in B family polymerases, if an incorrect nucleotide is indeed incorporated, these enzymes can switch into the 3'–5' exonuclease mode and remove the misincorporated base. The

3'–5' exonuclease activity has been shown to increases the fidelity of the RB69 polymerase from the Exo⁻ rate of 2.8 errors per genome (μ_g) to the Exo⁺ error rate of 4×10^{-3} μ_g [40,46]. The excision of an incorrectly incorporated terminal nucleotide is also dependant on two divalent metal ions. Structures of RB69gp43 have shown that the exonuclease active site is located approximately 40 Å from the polymerase active site [47]. Thus, for 3'–5' exonuclease activity to occur the DNA primer-template terminus must be translocated from the polymerase active site to the exonuclease site. The details of this process remain to be defined. It has been suggested that DNA replication accessory proteins, particularly the sliding clamp gp45, may be involved in the process of translocating DNA from polymerase to exonuclease active site [15]. It has also been shown that a β hairpin loop between residues 251–262 is important for exonuclease function [48,49]. This loop is involved in stabilizing the frayed base pair at the exonuclease active site allowing the removal of the incorrect nucleotide. During translocation from the polymerisation active site to the exonuclease active site the primer-template pair is partially melted, producing three unpaired bases. The three unpaired bases of the primer strand are then sequestered into the exonuclease active site, which facilitates the excision. Structures of RB69gp43 poised with its primer-template in the exonuclease mode are available [47,48].

5. Base Selectivity in RB69 DNA Polymerase

A large number of mutant RB69 and T4 DNA polymerases that affect both efficiency and fidelity of DNA replication have been isolated and characterized. In the case of RB69, X-ray crystallography has firmly established the nucleotide binding site at the atomic level. Crucial residues include K560 (Motif B), L415 and Y416 (Motif A) and L561 (Not conserved), Y567 and G568 (Motif B) [50].

Conserved residue Motif A has been shown to be important for sugar selectivity. Y416 forms a stacking interaction with deoxyribosyl moiety of the incoming dNTP [15]. It has been proposed that the 2'-OH group of a mismatched rNTP would cause a steric conflict with Y416 preventing formation of a stable complex [51]. Biochemical experiments with the Y416A mutant enzyme corroborated this notion. Unlike the wild type, the mutant was able to incorporate rCTP, ddCTP, and dCTP at similar rates [51].

Residue Y567 of motif B has been shown to be important in maintaining the fidelity of base selection [40,52]. This residue forms an interaction—via a water molecule—with the minor groove of the terminal primer-template pair. This interaction is important for sensing the geometry of the newly form base pair and thus detecting distortions caused by incorrect base pairing [15]. Mutations to residue 567 increase the size of the nascent base-pair-binding pocket allowing the misincorporation of nucleotides.

Residue L561 protrudes into the major groove of the templating base. It has been proposed that this residue is involved in detecting mismatches that lead to distortion of the major groove. L561A mutant confers a mutator phenotype [53]. Interestingly, the equivalent residue in herpesvirus-associated polymerases are not conserved (Figure 4b).

Figure 4. (**a**) Diagram of RB69 Fingers domain showing location of ABC block mutation relative to dNTP binding residues. Motif B residues are in orange, Tri-phosphate interacting residues on Helix N are in yellow, Block A is in red, Block B is in dark blue and Block C is in yellow. Aligned image of pdb 3LDS (dNTP) and pdb 3KD5 (N and P helix). (**b**) Sequence alignment of RB69 and *herpesviridae* sequences showing location of block mutations. (**c**) Diagram showing clash between W478 of Block A and W365. RB69 ABC5 is in white (3KD5), RB69 WT is in blue (pdb 1IH7) and HSV1 UL30 is in orange (pdb 2GV9). (**d**) Diagram of RB69 ABC5 block mutations active site showing phosphonoformic acid binding in β and γ phosphate position and with acyclovir in the pre-translocation position (pdb 3KD5). Images were generated using Pymol [23]. Alignment was generated using Geneious [24].

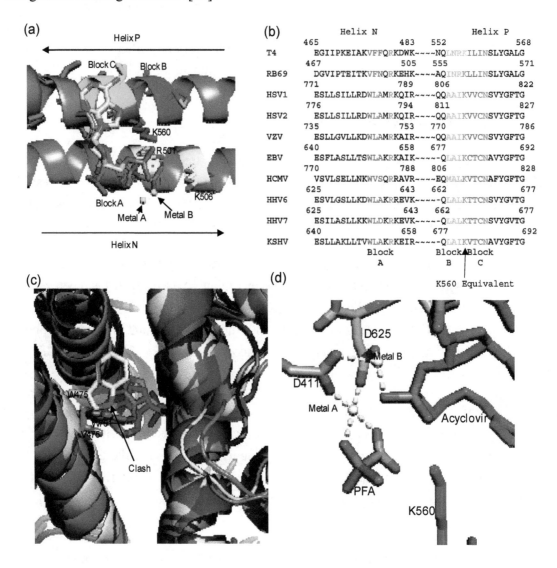

6. The Active Site of HSV1 UL30 and RB69gp43

At both the nucleotide and amino acid level, *herpesviridae polymerases* are not highly conserved (Appendix Figure A1). They range in length from 1013–1236 aa, and range in molecular weight from 113 to 137 kDa (Table 1). RB69gp43 is relatively small with 903 aa in length and a molecular weight of 104 kDa (Table 1). In terms of homology, there is very low homology between RB69gp43 and *herpesviridae* DNA polymerases sequences. However, both RB69gp43 and *herpesviridae* contain all the conserved motifs associated with B family polymerases (Appendix Figure A1).

Unlike herpesvirus DNA polymerases, there is a wealth of information on the structure and function of RB69gp43. Thus, this enzyme is often used as a model for herpesviridae DNA polymerase. Because of the shared conserved motifs between B family polymerase it is reasonable to assume that both herpesviridae and RB69gp43 have a similar, if not an identical catalytic mechanisms. The residues of motif A and C in the HSV1 UL30 and RB69gp43 structure are superimposable, supporting the notion that the function of these residues is similar (Figure 3c). In contrast to the metal cation-binding portion of the active site, the residues of helix N and P of the fingers domain vary greatly between HSV1 UL30 and RB69 gp43. On helix N, the only conserved residues are two basic amino acids R482 and K486 in RB69 gp43, and R785 and R789 in HSV1 UL30. Likewise, on the helix P, the only conserved residues are those of motif B: K560 to N565 in RB69gp43 and K811 to N815 in HSV1 UL30. Thus, the non-conserved residues of helix N and P account for the major differences in the nucleotide binding site, which could in turn account for the difference in sensitivity to antiviral drugs when herpesviridae polymerases are compared with RB69gp43. This notion is supported by biochemical studies with mutant enzymes derived from HCMV UL54 (see below) [11].

7. Mutations of Bacteriophage T4 that Induce PAA Sensitivity

There are several known natural mutations of bacteriophage T4 DNA polymerase (T4gp43) that affect sensitivity to the pyrophosphate analog phosophonoacetic acid (PAA). T4gp43 is highly homologues to RB69gp43, and, like RB69gp43, is naturally resistant to PAA. However, mutants of bacteriophage T4 with reduced plaque formation in the presence of PAA have been identified [54,55]. One of these mutations, L412M, is located in the conserved motif A. L412M confers increased sensitivity to PAA and interestingly also confers mutator properties to the polymerase [54]. Another interesting feature of the L412M mutant phage is that it can replicate in *E. coli* strains with restricted dGTP pools (*optA1*). The ability to grow under these conditions suggests that the mutant polymerase can make more efficient use of the nucleotide substrates. Because fidelity of DNA synthesis is in part controlled by the rate with which a frayed primer switches from polymerase to exonuclease activity, a bias toward stabilizing the frayed primer in the polymerase active site would cause an overall reduction in enzyme fidelity. When this hypothesis was tested with purified T4gp43 L412M enzyme, it was found that T4gp43 L412M exhibited less exonulease activity in relation to the wild type supporting the notion that the L412M mutation was in fact changing the partitioning between exonuclease and polymerase activity away from exonuclease activity. The equivalent residue to L412 in RB69gp43 is L415. Mutations at this residue show increased rates of mis-incorporation [56]. Interestingly, several suppressor mutants of L412M were also isolated. When subjected to a similar analysis, they were found to be antimutator mutations and unable to grow on *E. coli optA1* [55]. Mutations R335C and S345F are both located within the delta C region of the exonuclease domain (Appendix Figure A1). It has been proposed, that these mutation may affect pyrophosphate sensitivity by increasing the opportunity for exonuclease activity [54]. Since the primer terminus needs to be physically transferred to the exonuclease site, any increase in stability of the primer-terminus in the exonuclease active site or decrease of stability of the primer-terminus in the polymerase active site would increase the disruption of the pyrophosphate analog inhibited complex and henceforth increase the polymerase resistance to inhibition by pyrophosphate analogs. Interestingly in the RB69gp43 crystal structure the equivalent residue to R335 is R338, which is positioned at the very C terminus of the exonuclease domain and points into the cleft in which the finger domain rotates into during catalysis. Thus, the R335 mutation may not directly affect exonuclease function, but may instead inhibit finger domain movement [57].

8. Chimeric Enzymes

Tchesnokov *et al.* (2009) engineered an RB69gp43-UL54 chimeric enzyme, by mutating the active site of RB69gp43 to include the non-conserved elements from helix N and P of the HCMV enzyme. Swapping the polymerase active sites produced an enzyme that can be expressed in *E. coli*, and is soluble and easily purified. Biochemical assays have shown that the chimeric enzyme is sensitive to the nucleotide inhibitor acyclovir and the pyrophosphate analog PFA [58].

Three blocks of non-conserved amino acid residues were considered to engineer the chimera. Block A is located on helix N, and consists of residues 478–480 from RB69gp43 (VFN), these residue were replaced with equivalent residues of HCMV UL54: residues 779 to 781 (WVS). Block B and Block C are both located on Helix P. Block B consist of residues 557–559 of RB69gp43 (INR) and Block C 561–563 (LLI) of RB69gp43 where replaced with residues 808–810 (MAL) and 812–814 (VTC) from HCMV UL54, respectively. Block B and C flank residue K560, which is the conserved basic amino acid that likely donates a proton to the pyrophosphate leaving group [39,42]. Previous studies had already shown that several amino acids within this region can affect sensitivity to PFA (Figure 2) [11]. The chimeric polymerase reproduces drug sensitive and drug resistant phenotypes in cell-free biochemical assays, which validates this enzyme as a model system for polymerase active site inhibitors.

The structure of the chimera provides a detailed understanding of the mechanism of action of PFA [59]. The enzyme was co-crystallized in complex with a primer-template terminated with acyclovir in the presence and absence of PFA. PFA is bound at the polymerase active site and traps the enzyme in the pre-translocational state. The compound interacts with metal ion B and residue R482 of helix N, similar to the interactions formed by the β- and γ-phosphate of a bound dNTP in the post-translocational conformation. It appears that W478 of block A is critical in mediating sensitivity to PFA. Although no additional contacts are formed, this residue likely reduces the population of complexes that exist in the open conformation due to steric interference with W365 on helix J. Generally binary complex structures of B family polymerase are found to be in the open conformation; however, in the case of the chimeric enzyme the fingers are in the closed conformation even in the absence of PFA [59]. The predicted steric clash was confirmed with enzymes containing amino acid substitutions at residues 478 and 365 [59].

9. Resistance to Antiviral Drugs

Resistance-conferring mutations in the polymerase of HSV1, HSV2, VZV and HCMV UL54 have been identified *in vivo* and *in vitro* [10,32,33,60]. Drug resistance is measured as an increase in the inhibitory concentrations of a given drug required to block 50% viral replication (IC_{50}). In HSV1, HSV2 and VZV an $IC_{50} \geq 4.4$ µM confers significant levels of resistance to acyclovir [32], and in HCMV an $IC_{50} \geq 12$µM confers significant levels of resistance to ganciclovir, and $IC_{50} \geq 400$ µM to PFA [60]. Because of the difficulties of purifying Herpesvirus DNA polymerase the many resistance mutation have only been assessed in cell-based phenotypic assays [60].

All known resistance mutations can be roughly divided into two groups: 1. mutations within the exonuclease domain, which may affect 3'–5' exonuclease function, and 2. mutations within the domains that make up the polymerase active site, which may therefore affect polymerase function directly (Appendix Figure A1). Within the 3'–5' exonuclease domain there are several resistance mutations around the active site residues exo 1, exo2 and exo 3 respectively (Appendix Figure A1). Many of these mutations have been characterized to impair 3'–5' exonuclease activity [61,62]. Interestingly some of these resistance mutations have been characterized to confer resistance to pyrophosphate inhibitors but hypersensitivity to nucleotide inhibitors. An example of this phenotype is the HSV1 mutations Y577H and D581A [62] (Appendix Figure A1). Both mutations are located within the delta C region close to the exo 3 residue. Both mutations have been shown to impair exonuclease activity [62]. Logically a polymerase impaired in 3'–5' exonuclease activity would be unable to remove an incorporated nucleotide inhibitor from viral DNA, which would increase the viruses sensitivity to nucleotide inhibitor, thus explaining the hypersensitivity phenotype. However, the mutations also affect pyrophosphate inhibitors potency and this effect is harder to reconcile. Since pyrophosphate inhibitors mimic the pyrophosphate leaving group, and thus are not incorporated into DNA, the presence or absence of exonuclease activity should not directly affect pyrophosphate analog inhibitors susceptibility. The analysis of T4gp43 PAA sensitive mutants provides some possible insight into a mechanism [54,55]. While working with the T4gp43 L412M mutant they identified several suppressors of PAA sensitivity [54]. These suppressor mutants were shown to be antimutator polymerases, implying that L412M suppressor mutant polymerase were likely to have an altered rate of exonuclease activity compared to the L412M mutants. The authors suggested that because pyrophosphate analogs competitively inhibit polymerase activity by mimicking the pyrophosphate leaving group, that the transition to exonuclease activity could potentially bypass inhibition. Thus, altering the rate of exonuclease activity could potentially affect pyrophosphate inhibitor potency. Unfortunately this hypothesis has not fully been tested in either T4gp43 or herpesviridae DNA polymerases.

Resistance mutations within the polymerase active site can be arbitrarily split into two groups: 1. Fingers domain mutations and palm domain mutations. In HSV1, HSV2, VZV and HCMV several resistance mutations within the finger domain associated with region VI within helix PA and region III within helix PB. Helix PA and PB are equivalent to helix N and P of RB69gp43 (Figure 2) [32,33]. These two regions make up the finger domain contribution to the polymerase active site. Helix N contains several conserved residues important for nucleotide binding, while Helix P contains conserved residue K560 that is required for proton transfer during catalysis and N564, which is required for nucleotide binding. Crystal structures of RB69gp43 have shown that residue N564 interacts with the β phosphate of the nucleotide via a water molecule [52,63]. There are also several resistance mutations associated with the motif A, C and KKRY within the palm domain (Appendix Figure A1). In Motif A or region II there are several mutations prior to Motif A, which cause a pyrophosphate inhibitor hypersensitive, nucleotide inhibitor resistance phenotype. Mutations in motif C or region I and the KKRY motif or region VII (Appendix Figure A1) have all been characterized as inducing both pyrophosphate and nucleotide inhibitor resistance. Being that motif C and the KKRY motif is involved in aligning the 3'-OH nucleophile during catalysis any subtle change to the positioning of these motifs could change the binding and catalytic constants of the polymerase.

Several mutations remain to be confirmed as resistance-conferring amino acid substitutions [33]. Because of the difficulties in working with herpesviridae these mutations have not been tested in a defined genetic background to determine the phenotypic effect on resistance or susceptibility.

10. Conclusions

RB69gp43 provides an excellent model for the study of structure–function relationships of B family polymerases. However, there are limitations for the study of orthologous *herpesviridae* polymerases. Most importantly, the phage enzyme is not inhibited by approved drugs that bind to the polymerase active site in either post- or pre-translocational states. Chimeric enzymes composed of a RB69gp43 backbone and important elements of the active site of *herpesviridae* DNA polymerases can potentially address this problem. These findings warrant further investigation in such enzymes as novel tools in future drug discovery and development efforts.

Acknowledgments

M.G. is the recipient of a career award from the Fonds de la recherche en santé du Québec (FRSQ). This work was supported by funds from the Québec Consortium for Drug Discovery (CQDM).

References

1. Elion, G.B.; Furman, P.A.; Fyfe, J.A.; de Miranda, P.; Beauchamp, L.; Schaeffer, H.J. Selectivity of action of an antiherpetic agent, 9-(2-hydroxyethoxymethyl) guanine. *Proc. Natl. Acad. Sci. USA* **1977**, *74*, 5716–5720.

2. Ertl, P.F.; Thomas, M.S.; Powell, K.L. High level expression of DNA polymerases from herpesviruses. *J. Gen. Virol.* **1991**, *72*, 1729–1734.

3. Tsurumi, T.; Kobayashi, A.; Tamai, K.; Daikoku, T.; Kurachi, R.; Nishiyama, Y. Functional expression and characterization of the epstein-barr virus DNA polymerase catalytic subunit. *J. Virol.* **1993**, *67*, 4651–4658.

4. Lin, J.C.; De, B.K.; Mar, E.C. Functional characterization of partially purified epstein-barr virus DNA polymerase expressed in the baculovirus system. *Virus Genes* **1994**, *8*, 231–241.

5. Tsurumi, T.; Daikoku, T.; Nishiyama, Y. Further characterization of the interaction between the epstein-barr virus DNA polymerase catalytic subunit and its accessory subunit with regard to the 3'-to-5' exonucleolytic activity and stability of initiation complex at primer terminus. *J. Virol.* **1994**, *68*, 3354–3363.

6. De Bolle, L.; Manichanh, C.; Agut, H.; De Clercq, E.; Naesens, L. Human herpesvirus 6 DNA polymerase: Enzymatic parameters, sensitivity to ganciclovir and determination of the role of the a961v mutation in hhv-6 ganciclovir resistance. *Antivir. Res.* **2004**, *64*, 17–25.

7. Picard-Jean, F.; Bougie, I.; Bisaillon, M. Characterization of the DNA- and dntp-binding activities of the human cytomegalovirus DNA polymerase catalytic subunit ul54. *Biochem. J.* **2007**, *407*, 331–341.

8. Dorjsuren, D.; Badralmaa, Y.; Mikovits, J.; Li, A.Q.; Fisher, R.; Ricciardi, R.; Shoemaker, R.; Sei, S. Expression and purification of recombinant kaposi's sarcoma-associated herpesvirus DNA polymerase using a baculovirus vector system. *Protein Expr. Purif.* **2003**, *29*, 42–50.

9. Liu, S.; Knafels, J.D.; Chang, J.S.; Waszak, G.A.; Baldwin, E.T.; Deibel, M.R., Jr.; Thomsen, D.R.; Homa, F.L.; Wells, P.A.; Tory, M.C.; *et al.* Crystal structure of the herpes simplex virus 1 DNA polymerase. *J. Biol. Chem.* **2006**, *281*, 18193–18200.

10. Ducancelle, A.; Gravisse, J.; Alain, S.; Fillet, A.M.; Petit, F.; Pors, M.J.; Mazeron, M.C. Phenotypic characterisation of cytomegalovirus DNA polymerase: A method to study cytomegalovirus isolates resistant to foscarnet. *J. Virol. Meth.* **2005**, *125*, 145–151.

11. Tchesnokov, E.P.; Gilbert, C.; Boivin, G.; Gotte, M. Role of helix p of the human cytomegalovirus DNA polymerase in resistance and hypersusceptibility to the antiviral drug foscarnet. *J. Virol.* **2006**, *80*, 1440–1450.

12. Mueser, T.C.; Hinerman, J.M.; Devos, J.M.; Boyer, R.A.; Williams, K.J. Structural analysis of bacteriophage t4 DNA replication: A review in the virology journal series on bacteriophage t4 and its relatives. *Virol. J.* **2010**, *7*, 359.

13. Wang, J.; Sattar, A.K.; Wang, C.C.; Karam, J.D.; Konigsberg, W.H.; Steitz, T.A. Crystal structure of a pol alpha family replication DNA polymerase from bacteriophage rb69. *Cell* **1997**, *89*, 1087–1099.

14. Yang, G.; Lin, T.; Karam, J.; Konigsberg, W.H. Steady-state kinetic characterization of rb69 DNA polymerase mutants that affect dntp incorporation. *Biochemistry* **1999**, *38*, 8094–8101.

15. Franklin, M.C.; Wang, J.; Steitz, T.A. Structure of the replicating complex of a pol alpha family DNA polymerase. *Cell* **2001**, *105*, 657–667.

16. Hogg, M.; Wallace, S.S.; Doublie, S. Crystallographic snapshots of a replicative DNA polymerase encountering an abasic site. *EMBO J.* **2004**, *23*, 1483–1493.

17. Garcia-Diaz, M.; Bebenek, K. Multiple functions of DNA polymerases. *Crit. Rev. Plant Sci.* **2007**, *26*, 105–122.

18. Braithwaite, D.K.; Ito, J. Compilation, alignment, and phylogenetic relationships of DNA polymerases. *Nucleic Acids Res.* **1993**, *21*, 787–802.

19. Garg, P.; Burgers, P.M. DNA polymerases that propagate the eukaryotic DNA replication fork. *Crit. Rev. Biochem. Mol. Biol.* **2005**, *40*, 115–128.

20. De Waard, A.; Paul, A.V.; Lehman, I.R. The structural gene for deoxyribonucleic acid polymerase in bacteriophages t4 and t5. *Proc. Natl. Acad. Sci. USA* **1965**, *54*, 1241–1248.

21. Goulian, M.; Lucas, Z.J.; Kornberg, A. Enzymatic synthesis of deoxyribonucleic acid. Xxv. Purification and properties of deoxyribonucleic acid polymerase induced by infection with phage t4. *J. Biol. Chem.* **1968**, *243*, 627–638.

22. Kornberg, A. Active center of DNA polymerase. *Science* **1969**, *163*, 1410–1418.

23. Schrodinger, L.L.C. *The Pymol Molecular Graphics System*, version 1.3r1; Schrödinger K.K.; http://www.pymol.org/, 2010.

24. Drummond, A.; Ashton, B.; Buxton, S.; Cheung, M.; Cooper, A.; Duran, C.; Field, M.; Heled, J.; Kearse, M.; Markowitz, S.; *et al. Geneious v5.4*; Biomatters Ltd; http://www.geneious.com/, 2010

25. Edgar, R.C. Muscle: Multiple sequence alignment with high accuracy and high throughput./Zh *Nucleic Acids Res.* **2004**, *32*, 1792–1797.

26. Mosca, R.; Brannetti, B.; Schneider, T.R. Alignment of protein structures in the presence of domain motions. *BMC Bioinformatics* **2008**, *9*, 352.

27. Hwang, C.B.; Ruffner, K.L.; Coen, D.M. A point mutation within a distinct conserved region of the herpes simplex virus DNA polymerase gene confers drug resistance. *J. Virol.* **1992**, *66*, 1774–1776.

28. Wong, S.W.; Wahl, A.F.; Yuan, P.M.; Arai, N.; Pearson, B.E.; Arai, K.; Korn, D.; Hunkapiller, M.W.; Wang, T.S. Human DNA polymerase alpha gene expression is cell proliferation dependent and its primary structure is similar to both prokaryotic and eukaryotic replicative DNA polymerases. *EMBO J.* **1988**, *7*, 37–47.

29. Zhang, J.; Chung, D.W.; Tan, C.K.; Downey, K.M.; Davie, E.W.; So, A.G. Primary structure of the catalytic subunit of calf thymus DNA polymerase delta: Sequence similarities with other DNA polymerases. *Biochemistry* **1991**, *30*, 11742–11750.

30. Simon, M.; Giot, L.; Faye, G. The 3' to 5' exonuclease activity located in the DNA polymerase delta subunit of saccharomyces cerevisiae is required for accurate replication. *EMBO J.* **1991**, *10*, 2165–2170.

31. Blanco, L.; Bernad, A.; Blasco, M.A.; Salas, M. A general structure for DNA-dependent DNA polymerases. *Gene* **1991**, *100*, 27–38.

32. Gilbert, C.; Bestman-Smith, J.; Boivin, G. Resistance of herpesviruses to antiviral drugs: Clinical impacts and molecular mechanisms. *Drug Resist. Updates* **2002**, *5*, 88–114.

33. Lurain, N.S.; Chou, S. Antiviral drug resistance of human cytomegalovirus. *Clin. Microbiol. Rev.* **2010**, *23*, 689–712.

34. Terrell, S.L.; Coen, D.M. The pre-nh2-terminal domain of the herpes simplex virus 1 DNA polymerase catalytic subunit is required for efficient viral replication. *J. Virol.* **2012**, *86*, 11057–11065.

35. Burd, C.G.; Dreyfuss, G. Conserved structures and diversity of functions of rna-binding proteins. *Science* **1994**, *265*, 615–621.

36. Hughes, M.B.; Yee, A.M.; Dawson, M.; Karam, J. Genetic mapping of the amino-terminal domain of bacteriophage t4 DNA polymerase. *Genetics* **1987**, *115*, 393–403.

37. Hogg, M.; Rudnicki, J.; Midkiff, J.; Reha-Krantz, L.; Doublie, S.; Wallace, S.S. Kinetics of mismatch formation opposite lesions by the replicative DNA polymerase from bacteriophage rb69. *Biochemistry* **2010**, *49*, 2317–2325.

38. Xia, S.; Wang, M.; Blaha, G.; Konigsberg, W.H.; Wang, J. Structural insights into complete metal ion coordination from ternary complexes of b family rb69 DNA polymerase. *Biochemistry* **2011**, *50*, 9114–9124.

39. Castro, C.; Smidansky, E.D.; Arnold, J.J.; Maksimchuk, K.R.; Moustafa, I.; Uchida, A.; Gotte, M.; Konigsberg, W.; Cameron, C.E. Nucleic acid polymerases use a general acid for nucleotidyl transfer. *Nat. Struct. Mol. Biol.* **2009**, *16*, 212–218.

40. Bebenek, A.; Dressman, H.; Carver, G.; Ng, S.; Petrov, V.; Yang, G.; Konigsberg, W.; Karam, J.; Drake, J. Interacting fidelity defects in the replicative DNA polymerase of bacteriophage rb69. *J. Biol. Chem.* **2001**, *276*, 10387–10397.

41. Steitz, T.A. A mechanism for all polymerases. *Nature* **1998**, *391*, 231–232.

42. Castro, C.; Smidansky, E.; Maksimchuk, K.R.; Arnold, J.J.; Korneeva, V.S.; Gotte, M.; Konigsberg, W.; Cameron, C.E. Two proton transfers in the transition state for nucleotidyl transfer catalyzed by rna- and DNA-dependent rna and DNA polymerases. *Proc. Natl. Acad. Sci. USA* **2007**, *104*, 4267–4272.

43. Fothergill, M.; Goodman, M.F.; Petruska, J.; Warshel, A. Structure-energy analysis of the role of metal-ions in phosphodiester bond hydrolysis by DNA-polymerase-i. *J. Am. Chem. Soc.* **1995**, *117*, 11619–11627.

44. Patel, S.S.; Wong, I.; Johnson, K.A. Pre-steady-state kinetic analysis of processive DNA replication including complete characterization of an exonuclease-deficient mutant. *Biochemistry* **1991**, *30*, 511–525.

45. Wong, I.; Patel, S.S.; Johnson, K.A. An induced-fit kinetic mechanism for DNA replication fidelity: Direct measurement by single-turnover kinetics. *Biochemistry* **1991**, *30*, 526–537.

46. Drake, J.W. A constant rate of spontaneous mutation in DNA-based microbes. *Proc. Natl. Acad. Sci. USA* **1991**, *88*, 7160–7164.

47. Shamoo, Y.; Steitz, T.A. Building a replisome from interacting pieces: Sliding clamp complexed to a peptide from DNA polymerase and a polymerase editing complex. *Cell* **1999**, *99*, 155–166.

48. Aller, P.; Duclos, S.; Wallace, S.S.; Doublie, S. A crystallographic study of the role of sequence context in thymine glycol bypass by a replicative DNA polymerase serendipitously sheds light on the exonuclease complex. *J. Mol. Biol.* **2011**, *412*, 22–34.

49. Hogg, M.; Aller, P.; Konigsberg, W.; Wallace, S.S.; Doublie, S. Structural and biochemical investigation of the role in proofreading of a beta hairpin loop found in the exonuclease domain of a replicative DNA polymerase of the b family. *J. Biol. Chem.* **2007**, *282*, 1432–1444.

50. Zhang, H.; Beckman, J.; Wang, J.; Konigsberg, W. Rb69 DNA polymerase mutants with expanded nascent base-pair-binding pockets are highly efficient but have reduced base selectivity. *Biochemistry* **2009**, *48*, 6940–6950.

51. Yang, G.; Franklin, M.; Li, J.; Lin, T.C.; Konigsberg, W. A conserved tyr residue is required for sugar selectivity in a pol alpha DNA polymerase. *Biochemistry* **2002**, *41*, 10256–10261.

52. Yang, G.; Franklin, M.; Li, J.; Lin, T.C.; Konigsberg, W. Correlation of the kinetics of finger domain mutants in rb69 DNA polymerase with its structure. *Biochemistry* **2002**, *41*, 2526–2534.

53. Zhang, H.; Rhee, C.; Bebenek, A.; Drake, J.W.; Wang, J.; Konigsberg, W. The l561a substitution in the nascent base-pair binding pocket of rb69 DNA polymerase reduces base discrimination. *Biochemistry* **2006**, *45*, 2211–2220.

54. Reha-Krantz, L.J.; Wong, C. Selection of bacteriophage t4 antimutator DNA polymerases: A link between proofreading and sensitivity to phosphonoacetic acid. *Mutat. Res.* **1996**, *350*, 9–16.

55. Reha-Krantz, L.J.; Nonay, R.L.; Stocki, S. Bacteriophage t4 DNA polymerase mutations that confer sensitivity to the ppi analog phosphonoacetic acid. *J. Virol.* **1993**, *67*, 60–66.

56. Zhong, X.; Pedersen, L.C.; Kunkel, T.A. Characterization of a replicative DNA polymerase mutant with reduced fidelity and increased translesion synthesis capacity. *Nucleic Acids Res.* **2008**, *36*, 3892–3904.

57. Li, V.; Hogg, M.; Reha-Krantz, L.J. Identification of a new motif in family b DNA polymerases by mutational analyses of the bacteriophage t4 DNA polymerase. *J. Mol. Biol.* **2010**, *400*, 295–308.

58. Tchesnokov, E.P.; Obikhod, A.; Schinazi, R.F.; Gotte, M. Engineering of a chimeric rb69 DNA polymerase sensitive to drugs targeting the cytomegalovirus enzyme. *J. Biol. Chem.* **2009**, *284*, 26439–26446.

59. Zahn, K.E.; Tchesnokov, E.P.; Götte, M.; Doublié, S. Phosphonoformic acid inhibits viral replication by trapping the closed form of the DNA polymerase. *J. Biol. Chem.* **2011**, *286*, 25246–25255.

60. Landry, M.L.; Stanat, S.; Biron, K.; Brambilla, D.; Britt, W.; Jokela, J.; Chou, S.; Drew, W.L.; Erice, A.; Gilliam, B.; *et al.* A standardized plaque reduction assay for determination of drug susceptibilities of cytomegalovirus clinical isolates. *Antimicrob. Agents Chemother.* **2000**, *44*, 688–692.

61. Kuhn, F.J.; Knopf, C.W. Herpes simplex virus type 1 DNA polymerase. Mutational analysis of the 3'-5'-exonuclease domain. *J. Biol. Chem.* **1996**, *271*, 29245–29254.

62. Hwang, Y.T.; Smith, J.F.; Gao, L.; Hwang, C.B. Mutations in the exo iii motif of the herpes simplex virus DNA polymerase gene can confer altered drug sensitivities. *Virology* **1998**, *246*, 298–305.

63. Wang, M.; Xia, S.; Blaha, G.; Steitz, T.A.; Konigsberg, W.H.; Wang, J. Insights into base selectivity from the 1.8 a resolution structure of an rb69 DNA polymerase ternary complex. *Biochemistry* **2011**, *50*, 581–590.

Appendix Figure A1. Protein sequence alignment of herpesvirus and bacteriophage polymerases based on structural data. Herpesvirus and bacteriophage polymerase were aligned individually using the muscle algorithm within Geneious [24,25]. Then the bacteriophage and herpesvirus sequences were aligned using an alignment based on a structural alignment of HSV1 UL30 (2GV9) and RB69 gp43 (1IH7) produced by RAPIDO [26]. Blocks above sequence highlight structural domains of polymerase. Regions unresolved in the HSV1 structural model are shown in green. The pre N terminal domain is in white. The N terminal domain is in yellow. The 3'–5' exonuclease domain is in red. The palm domain is in pink. The fingers domain is in blue and thumb domain is in green. The known conserved regions are shown in magenta blocks above sequence [27–31]. Secondary structural elements of HSV1 UL30 are indicated. Elements are number according scheme provided in Liu *et al.* (2006). Secondary Structural elements of RB69 gp43 are indicated. Elements are numbered according scheme provided in Wang *et al.* (1997). Structural motifs involved in polymerase and exonuclease activity are highlighted in sequence. Herpes virus specific motif is in blue. Exonuclease conserved residues are in red. Motif A is in magenta. N helix residues are in yellow. Motif B is in green. Motif C is in brown. KKRY motif is in purple. Mutations that have been associated with resistance to anti-herpetic drugs are shown below the corresponding residue [32,33]. Resistance conferring mutations are colored using the following scheme: Red: Pyrophosphate[R] (Resistant), Blue: Nucleotide[R], Green: Pyrophosphate[R] and Nucleotide[R]. Purple: Pyrophosphate[HS] (Hypersensitive). Pink: Pyrophosphate[R] but Nucleotide[HS], Brown: Nucleotide[R] but Pyrophosphate[HS].

Appendix Figure A1. *Cont.*

```
                    Not Visible                                              Pre N Terminal
                                                                                  ▭ pN1
HSV1  UL30  1 MFSGGGGPLSPGGKSAARAASGFFAPAGPRGAS-RGPPPCLRQNFYNPYLAPVGTQQKPTGPTQR------HTYYSECDE-FRFIAPRVLDEDAPPEKRA  92
HSV2  UL30  1 MFCAAGGPASPGGKPAARAASGFFAPHNPRGATQTAPPCRRQNFYNPHLAQTGTQPKALGPAQR------HTYYSECDE-FRFIAPRSLDEDAPAEQRT  93
VZV   ORF28 1 ------------------------MALRTGFCNPFLTQASGIKYNPRTGRGSNREFLHSYKTTMSS-FQFLAPKCLDEDVPMEERK                61
EBV   BALF5 1 ------------------------MSGGLFYNPFLR-----------------------------PNKGLLKKPDKEYLRLLIPKC                32
HCMV  UL54  1 ------------------------MFFNFYLSGGVTGGAVAGG---RRQRSQPGSAQGSGKRPFQKQFLQIVPRG                            48
HHV6  U38   1 ------------------------MDSVSFFNPYLEANR-LKK------KSRSSYIRILPRG                                         31
HHV7  U38   1 ------------------------MDLVSFFNPYLENVRT-KK------KTKSTFLRIFPRG                                         31
KSHV  ORF9  1 ------------------------MDFFNPFIDPTRGGPR------NTVRQPTPSQSPT                                           29
T4    gp43    ----
RB69  gp43    ----
                                    Herpes Specific Motif
```

```
                 Pre N Terminal                                    N Terminal
                      ▭ pN2                              ▭ N1  ▭ N2  ▭ NA
HSV1  UL30  93 GVHDGHLKR---A-----P---KVYCGGDERDVLRVGS----GGFWPRRSRLWGGVDH-APAGENPTVTVFHVYDILENVEHAYGMRAAQFHARFMDAITPT  178
HSV2  UL30  94 GVHDGRLRR---A-----P---KVYCGGDERDVLRVGP----EGFWPRRLRLWGGADH-APEGFDPTVTVFHVYDILEHVEHAYSMRAAQLHEREMDAITPA  179
VZV   ORF28 62 GVHVGTLSR---------P---PKVYCNGKEVPILDFRC---SSPWPRRVNIWGEIDF-RGDKFDPRFNTFHVYDIVETTEAASNGDVS----RFATATRPL  143
EBV   BALF5 33 FQTPGAAGV--VDVRGPQ--PPLCFYQDSLTVVGGDEDGKGMWWRQRAQEGTAR--PEADTHGSPLDFHVDILETVYTHEKCAVIPS--DKQGYVVPC     123
HCMV  UL54  48 VMFDGQTGL--------I---KHKTGRLPLMFYRE--IKHLLSHDMVWPCFWRETLVGRVVGPIREFHTYDQTDAVLFFDSPENVSP--RYRQHLVPS     130
HHV6  U38   31 IMHDGAAGL-----I------KDVCDSEPR--MFYRD---RQYLLSKEMTWPSLDRVRSKDYDHTRMKFHIYDAVETLMFTDSIENLPF--QYRHFVIPS   113
HHV7  U38   31 IMHDGAPGLM-----K----TLCDSEPR--MFYQD--KQYILKNDMTWPSLSQVAEKELR-APLKFHIYDASESLLFTDSIENIPF--QYRHFVIPS     112
KSHV  ORF9  29 VPSETRVCRLIPACFQTPGRPGVVAVDTTFPTYFQGPKRGEVFAGETGSIWKTRRGQARNAPMSHLIFHVYDIVETTYTADRCEDVPFS--FQTDIIPS   127
T4    gp43  1  ------------------------MKEFYISIETVG------                                                            12
RB69  gp43  1  ------------------------MQEFYLTVEQIG------                                                            12
                                                                                            ▭ 1
```

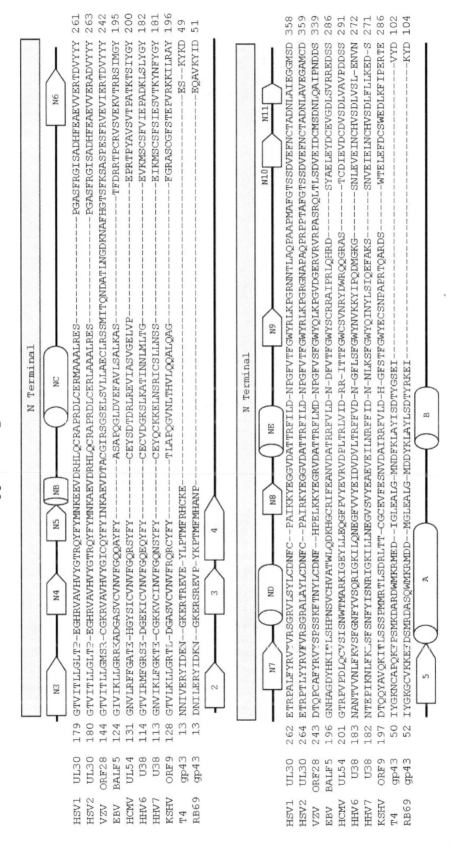

Appendix Figure A1. *Cont.*

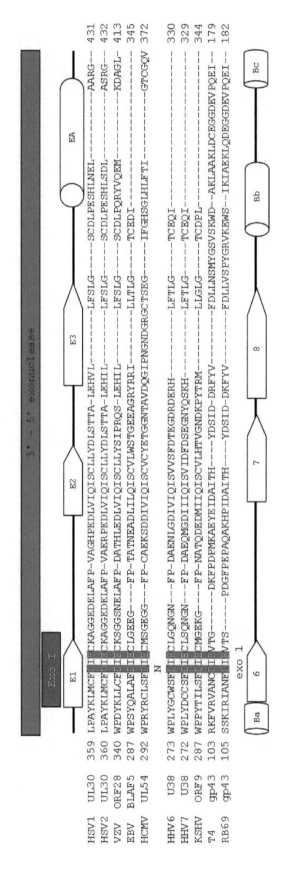

Appendix Figure A1. *Cont.*

Appendix Figure A1. *Cont.*

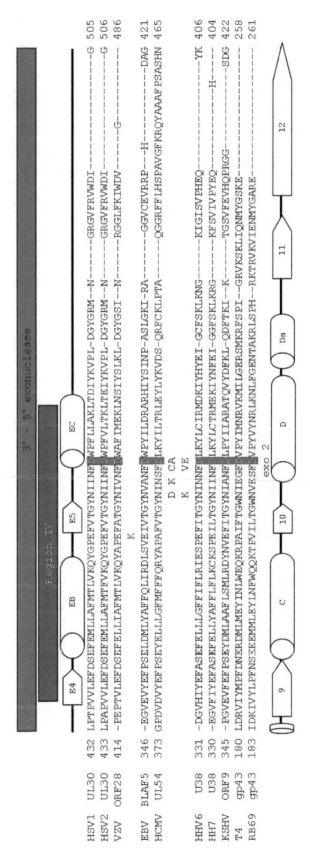

Appendix Figure A1. *Cont.*

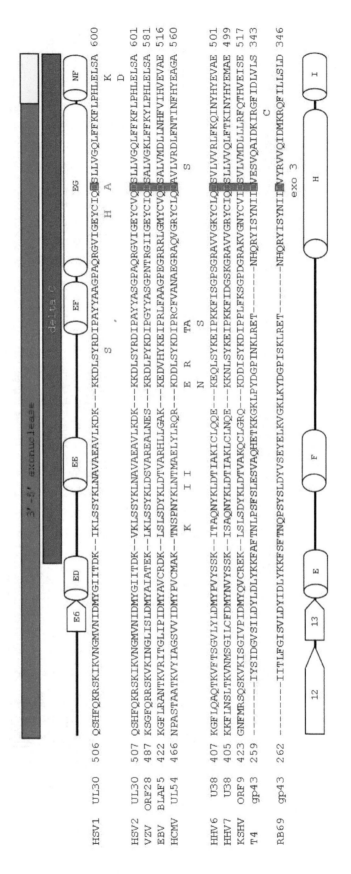

Appendix Figure A1. *Cont.*

N Terminal — delta C — Non visible

NG   NH   I   J   K

```
HSV1  UL30  601  VARLAGINITRTIYDGQQIRVFTCLLRLADQKGFILPDTQ--GRFRGAGGEAPKRPAA----------A-----------REDEERPEEGE  669
HSV2  UL30  602  VARLAGINITRTIYDGQQIRVFTCLLRLAGQKGFILPDTQ--GRFRGLDKEAPKRPAV----------P-----------RGEGERPGDGNG 670
VZV   ORF28 582  VARLARITLTKAIYDGQQVRIYTCLLGLASSRGFILPDGGYPATFEYKDVIPDVGDVE-------------EEMDEDESVSPTGTSS        655
EBV   BALF5 515  IAKIAHIPCRRVLDDGQQIRVFSCLLAAAQKENFILP-M---P-------SASDRD--                                    561
HCMV  UL54  561  IARIAKIPLRRVIFDGQQIRIYTSLLDECACRDFILPNHYSKGTTVPETNSVAVSPNAAIISTAAVPGDAGSVAAMFQMSPPLQSAPSSQDGVLPGSGSN 660
                                                   N
                                                   E
HHV6  U38   502  VARLAHVTARCVVFEGQQKKIFPCILTEAKRRNMILPSM----------V-----------SSHNRQG--                        548
HHV7  U38   498  VASLAYITRCFVFEGQQKKIFPCILHEAKNLNMILPSM----------N-----------TNFNKGKEN-                        548
KSHV  ORF9  518  IAKLAKIPTRRVLTDGQQIRVFSCLLEAAATEGYILPVP----------K-----------GDAVS----                        562
T4    gp43  344  MSYYAKMPFS-GVMS-PIKTWDAIIFNSLKGEHKVIPQ------------------------                                 379
                  F
RB69  gp43  347  MGYYAKMQIQ-SVFS-PIKTWDAIIFNSLKEQGRVIPQ------------------------                                 382
```

Appendix Figure A1. *Cont.*

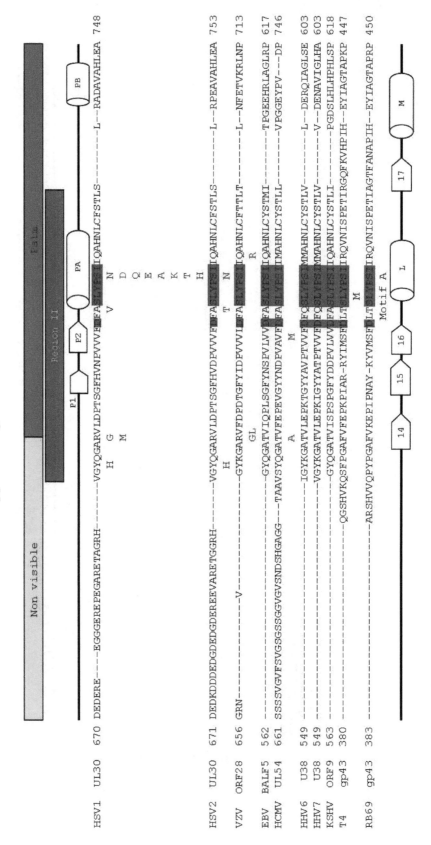

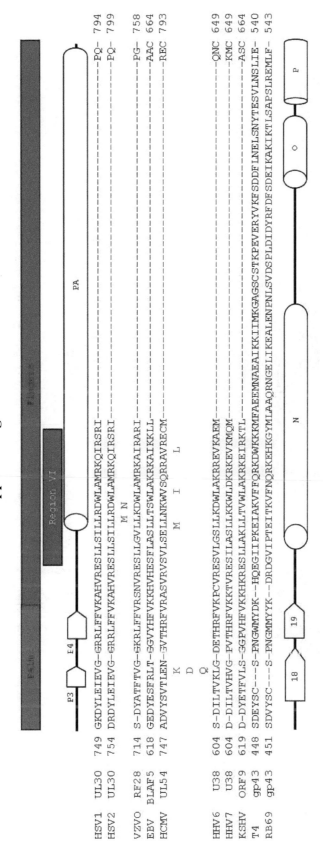

Appendix Figure A1. *Cont.*

Appendix Figure A1. *Cont.*

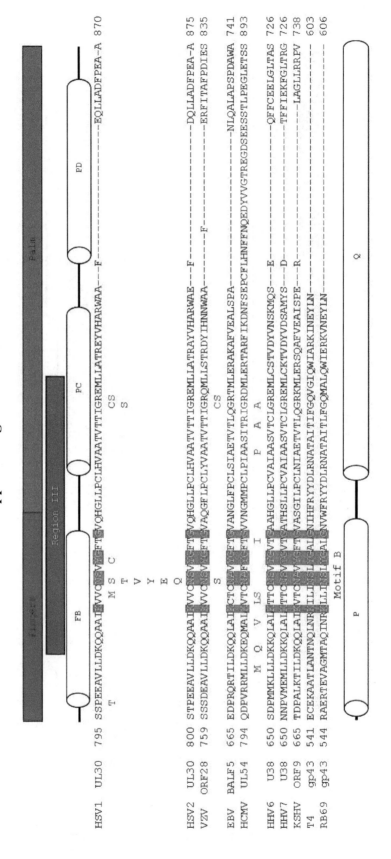

Appendix Figure A1. *Cont.*

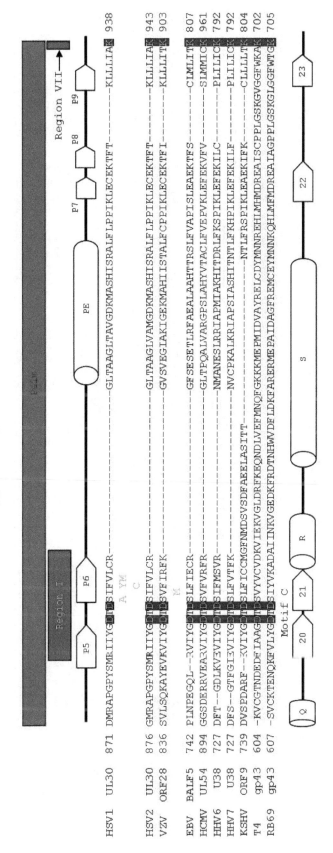

Appendix Figure A1. *Cont.*

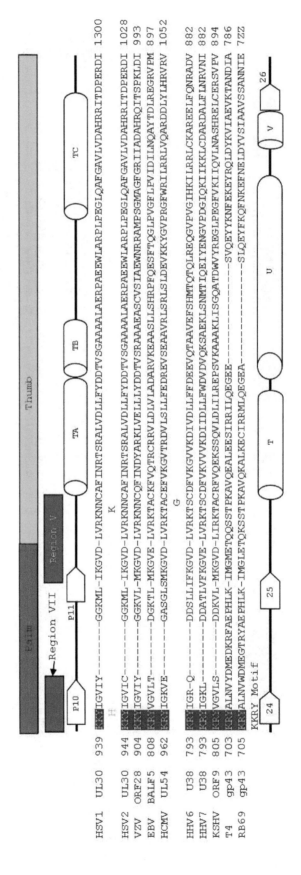

Appendix Figure A1. *Cont.*

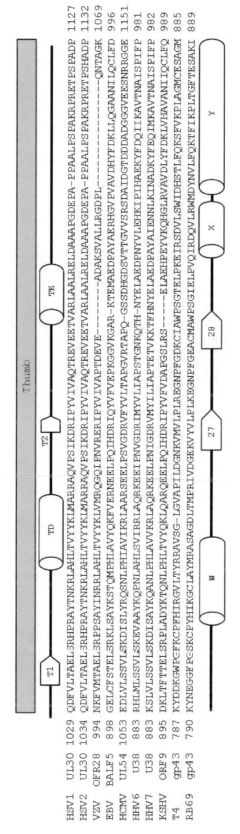

Appendix Figure A1. *Cont.*

Thumb — Non Visible

TF — TG — TH

```
HSV1  UL30  1128  PGGASKPRKLLV----SELAEDPAYAIAHGVALNTDYFSHLLGAACVTFKALFGNNAKITESLLKRFIPEVWHPPDDVAARLRTAGFGAVGAGATAEETR  1224
HSV2  UL30  1133  PGGASKPRKLLV----SELAEDPGYAIARGVPLNTDYFSHLLGAACVTFKALFGNNAKITESLLKRFIPETWHPPDDVAARLRAAGFGPAGAGATAEETR  1229
VZV   ORF28 1070  RCGEAKRKLIIS-----DLAEDPIHVTSHGLSLNIDYFSHLIGTASVTFKALFGNDTKLTERLLKRFIPETRVVNVKMLNRLQAAGFVCTHAPCWDNKMN  1165
EBV   BALF5  997  NNSGAALSVLQNFTARPPF*                                                                                  1016
HCMV  UL54  1152  PAKKRARKPPSAVCNYEVAEDPSYVREHGVPIHADKYFEQVLKAVTNVLSPVFPGGETARKDKFLHMVLPRRLHLEPAFLPYSVKAHECC*            1242
HHV6  U38    982  KTDIKKEKLLLYLLPMKVYLDETFSAIAEVM*                                                                      1013
HHV7  U38    983  KTGIKKEKFLLSILPLKVYVDQSFCDLTDVL*                                                                      1014
KSHV  ORF9   990  NNTSATVAILYNFLDIPVTFPTP*                                                                              1013
RB69  gp43   890  DYEKKATLFDMFDF*                                                                                        903
T4    gp43   886  DYEEKASL--DFLF*                                                                                        897
```

Non Visible

```
HSV1  UL30  1225  RMLHRAFDTLA*------------------  1236
HSV2  UL30  1230  RMLHRAFDTLA*------------------  1241
VZV   ORF28 1166  TEAEITEEQSHQIMRRVFCIPKAILHQS*-  1195
EBV   BALF5       ------------------------------
HCMV  UL54        ------------------------------
HHV6  U38         ------------------------------
HHV7  U38         ------------------------------
KSHV  ORF9        ------------------------------
RB69  gp43        ------------------------------
T4    gp43        ------------------------------
```

Permissions

List of Contributors

Philip Serwer
Department of Biochemistry, The University of Texas Health Science Center, San Antonio, Texas 78229-3900, USA

Ana R. Rama
Department of Health Science, University of Jaén, Jaén 23071, Spain
Institute of Biopathology and Regenerative Medicine (IBIMER), Granada 18100, Spain

Rosa Hernandez, Consolación Melguizo, Celia Vélez and Jose Prados
Institute of Biopathology and Regenerative Medicine (IBIMER), Granada 18100, Spain
Biosanitary Institute of Granada (ibs.GRANADA), SAS-University of Granada, Granada 18071, Spain
Department of Human Anatomy and Embryology, School of Medicine, University of Granada, Granada 18012, Spain

Gloria Perazzoli
Institute of Biopathology and Regenerative Medicine (IBIMER), Granada 18100, Spain
Biosanitary Institute of Granada (ibs.GRANADA), SAS-University of Granada, Granada 18071, Spain

Miguel Burgos
Institute of Biotechnology and Department of Genetics, University of Granada, Granada 18071, Spain

Bryan R. Lenneman
Committee on Genetics, Genomics, and Systems Biology, The University of Chicago, 920 East 58th Street, Chicago, IL 60637, USA

Lucia B. Rothman-Denes
Committee on Genetics, Genomics, and Systems Biology, The University of Chicago, 920 East 58th Street, Chicago, IL 60637, USA
Department of Molecular Genetics and Cell Biology, The University of Chicago, 920 East 58th Street, Chicago, IL 60637, USA

David R. Harper, Helena M. R. T. Parracho, Susan Lehman and Sandra Morales
AmpliPhi Biosciences, Glen Allen, VA 23060, USA

James Walker and Richard Sharp
Public Health England, Porton Down, Salisbury SP4 0JG, UK

Gavin Hughes
Gavin Hughes—The Surgical Materials Testing Laboratory, Bridgend, South Wales CF31 1RQ, UK

Maria Werthén
Maria Werthén, Mölnlycke Health Care AB, SE-402 52 Gothenburg, Sweden
Department of Biomaterial Science, University of Gothenburg, SE-405 30 Gothenburg, Sweden

Paul Shkilnyj, Michael P. Colon and Gerald B. Koudelka
Department of Biological Sciences, University at Buffalo, Buffalo, NY 14260, USA

Andrew Chibeu and Erika J. Lingohr
Laboratory for Foodborne Zoonoses, Public Health Agency of Canada, Guelph, ON, N1G 3W4, Canada

Luke Masson
Biotechnology Research Institute, National Research Council of Canada
Département de microbiologie et immunologie, Université de Montréal, 2900, boul. Édouard-Montpetit, Montréal, QC H3T 1J4, Canada

Amee Manges
Department of Epidemiology, Biostatistics and Occupational Health, McGill University, 1020 avenue des Pins Ouest, Montréal, QC H3A 1A2, Canada

Josée Harel
Groupe de Recherche sur les Maladies Infectieuses du Porc (GREMIP) and Centre de Recherche en infectiologie porcine (CRIP), Université de Montréal, Faculté de médecine vétérinaire, Saint-Hyacinthe, QC J2S 7C6, Canada

Hans-W. Ackermann
Felix d'Herelle Reference Center for Bacterial Viruses, Department of Microbiology, Immunology and Infectionlogy, Faculty of Medicine, Laval University, QC G1K 4C6, Canada

Andrew M. Kropinski
Department of Molecular and Cellular Biology, University of Guelph, ON N1G 2W1, Canada

Patrick Boerlin
Department of Pathobiology, Ontario Veterinary College, University of Guelph, ON N1G 2W1, Canada

Victor Krylov, Olga Shaburova, Sergey Krylov and Elena Pleteneva
Laboratory for Bacteriophages Genetics. Mechnikov Research Institute of Vaccines and Sera, RAMS, 5a, Maliy Kazenniy per, Moscow 105064, Russia

Sidney Hayes, Craig Erker, Monique A. Horbay, Kristen Marciniuk, Wen Wang and Connie Hayes
Department of Microbiology and Immunology, College of Medicine, University of Saskatchewan, Saskatoon, S7N 5E5 Canada

Stephen T. Abedon
Department of Microbiology, The Ohio State University, 1680 University Dr., Mansfield, OH 44906, USA

Nicholas Bennett
Department of Microbiology and Immunology, McGill University, 3775 University Street, Montreal, Quebec H3A 2B4, Canada

Matthias Götte
Department of Microbiology and Immunology, McGill University, 3775 University Street, Montreal, Quebec H3A 2B4, Canada
Department of Biochemistry, McGill University, 3655 Sir William Osler Promenade, Montreal, Quebec H3G1Y6, Canada
Department of Medicine, Division of Experimental Medicine, McGill University, 1110 Pine Avenue West, Montreal, Quebec H3A 1A3, Canada

Index

A

Adherence, 138, 200

Adsorption, 57, 114-115, 122-123, 127, 130, 132, 138-139, 141, 150, 186, 188-199, 201-203, 208

Alcaligenes Eutrophus, 125, 128, 147

Alginate, 64-66, 118, 120, 133-134, 136, 150

Allele, 84, 156, 160, 163, 165, 173, 180

Ampicillin, 84, 87, 92, 94-95, 102, 160, 176-177

Antibiotic Resistance, 60-61, 64, 70, 92, 105-107, 111, 125, 136, 142

Autocleavage, 74-76, 78, 81-84

Autoproteolysis, 75-76, 81, 83, 87

B

Bacterial Arrangements, 186, 188, 191-192, 194, 196-199, 201-204, 206

Bacterial Cell, 62, 116, 119, 140, 209

Bacterial Density, 195, 197, 202-203, 205

Bacterial Fitness, 202-204

Bacterial Growth, 59, 63-64, 188, 203

Bacterial Host, 116

Bacterial Lysis, 187, 195, 197

Bacterial Population, 140, 200, 202-203

Bacterial Size, 189, 193

Bacterial Species, 65, 111-112, 114, 121-123, 125, 128-129, 131, 136-137, 140, 194

Bacteriophage, 1-7, 10-11, 14-25, 38-39, 43, 50-51, 54-76, 81-84, 88-90, 92, 97-99, 102-104, 106-107, 109, 111, 113, 116, 122-123, 130-131, 133, 135, 142-147, 149-153, 174, 178-181, 183-187, 201, 216, 223, 227, 229-231

Biofilms, 59-72, 91-95, 97, 99-100, 104-109, 186-188, 191, 200-202, 205-210

C

Cellular Filamentation, 152-153, 156, 170-171, 177-178

Cystic Fibrosis, 71, 111, 114, 136, 139, 142-144, 150

D

Diplococcus, 189, 192, 194, 204

Disinfectants, 187, 201

E

Electron, 1, 19, 50, 96, 102-103, 113

Escherichia Coli, 5, 21, 39, 54-57, 62, 65-66, 71, 88-92, 105-107, 124-125, 128-129, 142-143, 147, 179-184

Established Biofilms, 99, 104

F

Free Bacteria, 188, 190, 192-196, 198-200, 202-205

Free Cells, 187, 196

G

Genome, 2, 4, 14-15, 20, 23, 38-42, 48, 50-53, 57-58, 61-63, 66, 72, 89, 92, 97-99, 107, 113-119, 122-131, 135-136, 138-142, 144-149, 151, 153, 156, 159, 163, 166, 188, 214, 221

Genome Sequence, 89, 135, 144-146, 149, 163

Gram-negative Bacteria, 111, 114, 125, 129

Growth Rate, 204, 210

H

Herpesviridae, 212-214, 217, 219, 222-223, 226

Herpesvirus, 212-214, 216, 221, 223, 225, 227, 231

Homology, 39, 43, 48-49, 51-52, 56, 98-99, 122, 127-131, 136, 145, 148, 214, 223

Host Cells, 62, 66, 152, 156, 159, 165

I

Immune System, 67, 92, 151, 201, 210, 213

Incubation, 63, 77, 85, 87, 92, 99, 104, 120-121, 133, 157, 161, 167, 176

Individual Bacteria, 186, 190-196, 198-199, 203-204, 206

L

Lambdoid Phage, 75, 184

Lysogenic Bacteria, 121, 123, 126, 128

Lysogenization, 116, 118, 121, 123-126, 139-140, 146

Lysogeny, 74-75, 77, 83-84, 117, 125, 128, 144-145, 147

Lytic Growth, 75, 83, 125, 127

M

Microcolonies, 186-188, 190-191, 199-202, 206

Microcolony Formation, 200, 203

Morphology, 66, 96, 130-131, 135, 209

N

Non-canonical Interaction, 115

Nucleotide Inhibitor, 224-226

P

Pathogen, 34, 114, 124, 132-133, 206

Pathogenic Strain, 115

Personalized Medicine, 110-111, 115

Phage Adsorption, 130, 150, 186, 188-190, 192-198, 202-203

Phage Densities, 186, 190-191, 194, 198-199, 204

Phage Diffusion, 188-190

Phage Infection, 139, 187, 190, 194, 196-197, 199, 203, 206

Phage Interaction, 70, 187, 191

Phage Particle, 115

Phage Propagation, 191, 194-197, 199, 202

Phage Target, 189

Phage Therapy, 60, 70, 97, 105, 110-119, 121, 127, 131-133, 135-138, 140, 143, 147, 149, 186-187, 202, 206-209, 211

Phage-resistant, 114-115, 132, 137, 201, 204-205

Phenotype, 48, 88, 121, 124, 128-129, 145, 147, 158-159, 163, 165, 168, 177, 179-180, 183, 185, 221, 225-226

Plasmid, 29, 36, 43, 79-81, 83, 86, 111, 116-118, 123, 125-128, 130-131, 136, 139, 145-147, 150-161, 163-178, 180-182, 184

Pseudolysogeny, 111, 116-119, 144-145

Pseudomonas Aeruginosa, 39, 60, 63-65, 70-72, 106-107, 110-111, 127, 133, 141-151, 210

Pseudomonas Putida, 71, 124-125, 128, 141-142, 147

Pseudovirulence, 111, 116, 121-122

R

Replication, 39-40, 50-51, 55, 57, 59, 62, 66, 69, 116, 119, 139, 148, 152-156, 158-160, 163-170, 172, 177-185, 203, 205, 209, 213-214, 217, 221, 225, 227-230

Repressor, 74-76, 78, 81-84, 87-89, 116-117, 119-121, 124-126, 128, 131, 153-154, 156-159, 167, 172-173, 180, 185

S

Spatial Vulnerability, 186-187, 203

T

Therapeutic Phage, 115, 132

Transposable Phage, 125, 129, 133, 138, 147-148

V

Vegetative Growth, 128, 156, 158-160, 169

Virulent Mutant, 120

Virulent Phage, 72, 110, 115, 119, 123-124, 132, 137

Printed in the USA
CPSIA information can be obtained
at www.ICGtesting.com
JSHW051623061123
51533JS00005B/78